Introduction to Veterinary Science

Join us on the web at

Agriscience.delmar.com

Introduction to Veterinary Science

James B. Lawhead, V.M.D.
and
MeeCee Baker, Ph.D.

THOMSON
DELMAR LEARNING

Australia Canada Mexico Singapore Spain United Kingdom United States

Introduction to Veterinary Science
James B. Lawhead, V.M.D. and MeeCee Baker, Ph.D.

Vice President, Career Education Strategic Business Unit:
Dawn Gerrain

Director of Editorial:
Sherry Gomoll

Acquisitions Editor:
Zina M. Lawrence

Developmental Editor:
Andrea Edwards

Editorial Assistant:
Rebecca Switts

Director of Production:
Wendy A. Troeger

Production Manager:
Carolyn Miller

Production Editor:
Kathryn B. Kucharek

Technology Project Manager:
Joseph Saba

Director of Marketing:
Donna J. Lewis

Channel Manager:
Nigar Hale

Cover Images:
Getty One

Cover Design:
TDB Publishing Services

Library of Congress Cataloging-in-Publication Data

Lawhead, James B.
   Introduction to veterinary science / James B. Lawhead and MeeCee Baker.
      p. cm.
   Includes bibliographical references (p.).
   ISBN 0-7668-3302-X
   1. Veterinary medicine. I. Title.
SF745.L45 2003
636.089—dc21

2003041465

NOTICE TO THE READER

Dr. Baker dedicates her efforts in producing this text to her daughter Elizabeth "Libby" Baker-Mikesell, who wants to someday be a veterinarian.

Contents

UNIT IV   SURGERY   271

Introduction to Veterinary Science

Agriscience programs nationwide have undergone extensive curricular changes. Many now include advanced placement-type coursework, such as veterinary science. Prior to the 2000-2001 school year, Dr. Baker searched for materials to be used in a new veterinary science course she planned to teach. After a futile hunt, and hearing similar concerns from other instructors, Dr. Baker teamed with Dr. Lawhead, a practicing veterinarian, in an effort to author this veterinary science text along with a complete set of companion materials.

Although the ancillary materials help make the text user-friendly, the authors believe that two of the most useful features in this book are the "Day in the Life" of a veterinarian coupled with the "Clinical Significance" chapter features. These two elements tie the real-life work of a veterinarian with the gravity of the technical and sometimes dry and difficult text material. Therefore, the next time a student says, "I want to be a veterinarian," a venture into *Introduction to Veterinary Science* will provide the learner with a realistic preview of both veterinary work and the academic rigor needed to achieve success in the profession.

Simply put, the goals of this text are to afford learners a base knowledge of veterinary science by moving through topics ranging from the cell to surgery, and to provide a view of the practice of veterinary medicine through the eyes of an experienced practitioner. Chapters 1 and 2 begin the text with a comprehensive investigation of cells and tissues. Following chapters examine the musculoskeletal, circulatory, respiratory, renal, digestive, reproductive, nervous, and immune systems. The basic physiology learned in the beginning of the text is then applied in concluding chapters covering nutrition, species differentiation in nutrition, infectious disease, disease classification, zoonotic diseases, disease diagnosis, and surgery.

Extension Teaching/Learning Materials

Instructor's Guide to Text
The Instructor's Guide provides answers to the end-of-chapter questions and additional material to assist the instructor in the preparation of lesson plans.

Lab Manual (Order #0-7668-3303-8)
This comprehensive lab manual reinforces the text content. It is recommended that students complete each lab to confirm understanding of essential science content. Great care has been taken to provide instruction with low-cost, strongly science-focused labs.

Lab Manual Instructor's Guide
The Instructor's Guide for the lab manual provides answers to lab manual exercises and additional guidance for the instructor.

ClassMaster CD-ROM (Order #0-7668-3306-2)
The ClassMaster technology supplement provides the instructor with valuable resources to simplify the planning and implementation of the instructional program. It includes the Teacher's Resource Guide, transparency masters, motivational questions and activities, answers to questions in the text, and lesson plans to provide the instructor with a cohesive plan for presenting each topic. Also included is a computerized test bank with more than 1,000 new questions, giving the instructor an expanded capability to create tests.

On-Line Resource
The Web-based supplement allows the instructor the flexibility of receiving much of the supplement package on-line. It includes the text Instructor's Guide; the Lab Manual Instructor's Guide; the Teacher's Resource Guide; transparency masters; and an additional list of professional and educational URLs.

Each chapter in the textbook begins with clear educational **objectives** to be learned by the student in the reading, a list of important **key terms,** and an **introduction** overview of the chapter content.

3

Musculoskeletal System

OBJECTIVES

Upon completion of this chapter, you should be able to:

- Describe the functions of the musculoskeletal system.
- Detail the structure of bone.
- Name joint types and their accompanying role in movement.
- List the two major sections of the skeleton, name the corresponding bones, and compare species differentiation.
- Explain how bone grows and remodels.
- Relate bone and muscle groups to movement.
- Connect the text materials pertaining to the musculoskeletal system to clinical practice.

KEY TERMS

herd check
radiograph
orthopedic surgeon
axial skeleton
appendicular skeleton
intervertebral disk
disease

high-rise syndrome
cranial drawer sign
ossification
subluxate
x-ray
radiology

simple fracture
comminuted fracture
compound (open)
fracture
intramedullary pin

hip dysplasia
degenerative joint
disease
joint ill

INTRODUCTION

The skeleton gives mammals shape and support. Combining bones and muscles allows movement. Bones are active tissues that adapt to changes within the animal. The skeleton, although very hard, allows the animal to adapt and grow.

28

A DAY IN THE LIFE

Reproduction…

Preparing this chapter made me think about how much of my career revolves around the reproductive system. I just finished my morning surgeries and had a break until my afternoon office appointments. My morning's surgeries included a cat **spay** and front declaw, a dog spay, and two dog **castrations**.

In a spay, technically called an **ovariohysterectomy,** the ovaries and uterus are removed. In castration the testes are removed. Both spaying and castration prevent any unwanted pregnancies. In addition, the neutering procedure provides other health benefits.

Earlier in the text I mentioned that herd checks make up a large portion of my job. Dairy cattle begin producing milk after their first calves are born. Subsequently producers attempt to have the cattle calve every 12 to 13 months. At herd checks I help facilitate and check for pregnancy. I perform more herd checks than any other dairy-related task.

Yesterday afternoon I received a call from one of our farm clients. He had a down cow. The cow had begun to calve and did not have the strength to rise. She had developed hypocalcemia, or milk fever. (Review the role of calcium in the function of muscles in Chapter 2.) I gave her two bottles of calcium intravenously. Afterward the farmer and I examined another cow that was not eating well. When we returned to the down cow, the calcium had entered the muscles and the cow was able to rise.

I then thoroughly cleaned her vulva and reached into the uterus to find that the calf was ready to be born. Many cows with milk fever do not deliver normally because they lack muscle strength. Having been treated, this cow may have calved normally, but I could not be positive. I placed calving chains around the legs of the calf and, along with the farmer (with the help of the cow pushing), pulled the calf. Fortunately, the calf was born alive and the mother was standing (Figure 8–1). To this day, cases such as this continue to be the most rewarding in my profession. I still enjoy helping to bring a newborn into this world.

I received an emergency call as I was writing this chapter. The receptionist at the office called to tell me that one of our Amish clients had a cow that had thrown

FIGURE 8–1 Newborn calf and dam.

her calf bed, or in other words, had a **prolapsed uterus.** This heifer had just delivered her first calf 2 hours earlier. The calf was quite large, and following the delivery, the cow continued to strain. Consequently, the uterus turned inside out and was pushed through the vulva.

When I arrived, the cow was down and her uterus protruded from out behind her. Because our Amish clients have no electricity on their farm, the farmer lit a lantern for me to see. I felt like an old timer, working in the barn by lantern light.

I gave the cow an **epidural,** in which I inject a local anesthetic **(lidocaine)** into the fluid around the region spinal cord. The epidural relieves pain in the region around the vulva and rectum. Because the cow felt no pain, she resisted my efforts less. I then pushed her large, swollen uterus back through the vulva. Fortunately, this case went very well. The farmer and I hope for a good outcome.

Each chapter features **"A Day in the Life"** of a veterinarian vignette that relays James Herriot–type stories with relevance to clinical practice and the real-life work of a veterinarian.

Each chapter contains a combination of **charts, illustrations, photographs,** and **radiographs** that help to illustrate and enhance the concepts presented.

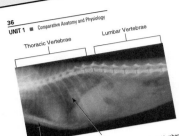

Thoracic Vertebrae

Lumbar Vertebrae

Sternum

Rib

FIGURE 3–12 Radiograph of a cat showing the thoracic and lumbar spine. Ribs and sternum are also visible.

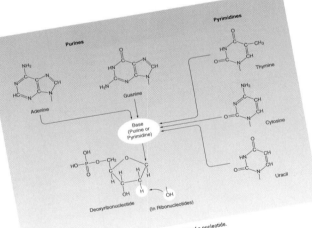

Spine

FIGURE 3–14 A scapula.

the scapula to move over the rib cage. This rotation ranges as high as 25 degrees in animals such as cats while running. This flexibility is also useful in cats as they land after a jump. As the cat falls, it extends its front legs fully, both at the scapula and the elbow. As the front feet hit the ground, the elbow flexes and the scapula rotates. The cat makes this very coordinated act look quite graceful. Clinically this is of significance when cats fall from extreme distances. In large cities this happens often as cats tumble from balconies or windows of tall buildings. In **high-rise syndrome** the falling cat rarely breaks a leg; however, it will often break its lower jaw. The high speed of the falling cat forces the jaw to contact the ground.

The humerus is the upper bone of the forelimb. The scapula joins the humerus through a shallow ball and socket joint that allows for a wide range of motion (Figure 3–15). The humerus then joins at the elbow with the radius and ulna. The ulna runs to the point of the elbow, where a groove accepts the end of the humerus. The radius closely attaches to the ulna and forms the remainder of the elbow joint (hinge joint that permits motion in only one plane). The forearm can be rotated, but this occurs between the radius and ulna, not at the elbow joint.

The radius and ulna run to the level of the carpus (Figure 3–16). The carpus in animals corresponds to the wrist in humans. The carpus, a group of bones, arranges in two rows. Table 3–1 lists the number of carpal bones found in several species. The carpal bones join to the long metacarpal bones. There special differences become very dramatic. Dogs and cats have four long metacarpal bones and one much smaller. The smaller bone associates with the first digit, called the dewclaw (Figure 3–17). As previously stated, horses have only one major metacarpal bone, which

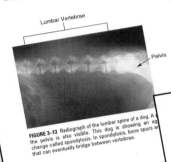

Lumbar Vertebrae

Pelvis

FIGURE 3–13 Radiograph of the lumbar spine of a dog. A the pelvis is also visible. This dog is showing an ag change called spondylosis. In spondylosis, bone spurs a that can eventually bridge between vertebrae.

Purines

Pyrimidines

NH₂

Adenine

Guanine

Thymine

Cytosine

Uracil

Base
(Purine or
Pyrimidine)

Deoxyribonucleotide

(In Ribonucleotides)

FIGURE 1–5 Chemical structure of a nucleotide.

the blood, thus allowing veterinarians to diagnose what specific organism is causing the sickness.

Nucleic acids provide the plans for the differing construction of proteins. Nucleic acids are fabricated with a series of nucleotides. The nucleotides are made up of a five-carbon sugar, a phosphate group, and a nitrogen-containing base (Figure 1–5). Ribonucleic acid (RNA) claims ribose as its sugar, whereas deoxyribonucleic acid (DNA) has deoxyribose as its sugar. There are four different bases for RNA and for DNA (Table 1–1). Notice that the bases are the same

except for thymine and uracil. The order of base combination determines what amino acids are used to make proteins. This information is stored in the cell's genetic material.

CELL STRUCTURE

OBJECTIVE

■ *Identify the Basic Structures of the Cell and Their Corresponding Functions*

Large varieties of types of cells exist. These cells not only look different but function differently as well. Nevertheless, many features remain common among cells. Specialized structures within the cells are called organelles. These organelles are present in most but not all cells. Red blood cells, for example, lack a nucleus.

The cell membrane (or plasma membrane) is a feature common to all cells. It serves as the boundary that

TABLE 1–1 RNA and DNA Bases

DNA Bases	RNA Bases
1. Adenine	1. Adenine
2. Cytosine	2. Cytosine
3. Guanine	3. Guanine
4. Thymine	4. Uracil

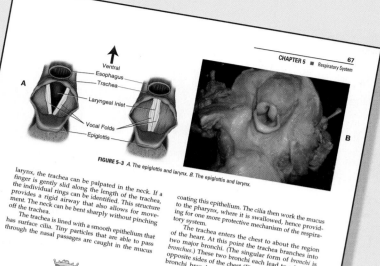

Ventral
Esophagus
Trachea
Laryngeal Inlet
Vocal Folds
Epiglottis

A

B

FIGURE 5–3 *A.* The epiglottis and larynx. *B.* The epiglottis and larynx.

larynx, the trachea can be palpated in the neck. If a finger is gently slid along the length of the trachea, the individual rings can be identified. This structure provides a rigid airway that also allows for movement. The neck can be bent sharply without pinching off the trachea.

The trachea is lined with a smooth epithelium that has surface cilia. Tiny particles that are able to pass through the nasal passages are caught in the mucus

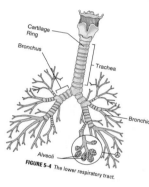

Cartilage
Ring

Bronchus

Trachea

Bronchiole

Alveoli

FIGURE 5–4 The lower respiratory tract.

coating this epithelium. The cilia then work the mucus to the pharynx, where it is swallowed, hence providing for one more protective mechanism of the respiratory system.

The trachea enters the chest to about the region of the heart. At this point the trachea branches into two major bronchi. (The singular form of *bronchi* is *bronchus*.) These two bronchi each lead to a lung on opposite sides of the chest (Figure 5–4). These major bronchi branch into smaller bronchi, dividing and entering different areas of the lungs. The two lungs surround the centrally located heart.

The bronchi continue to divide into smaller and smaller airways, forming the bronchioles (Figure 5–5). The bronchioles have smooth muscle in their walls.

FIGURE 5–5 Trachea opening into the bronchi. The trachea and lung tissue have been incised to expose the openings. Note the tracheal rings and the openings into smaller bronchi.

Each chapter is further enhanced by the addition of repeat objectives to aid in student comprehension.

MUSCULOSKELETAL SYSTEM FUNCTIONS

OBJECTIVE
■ *Describe the Functions of the Musculoskeletal System*

Bones furnish four basic functions: structure, protection, minerals reserve, and blood cell production. The most visible function of bones in the animal forms the collection of bones in the animal forms the skeleton. This provides the framework that defines an animal's shape and size. Differences in both size and shape are very obvious in veterinary medicine. The skull provides a clear example of this variation. When seeing only the bones of the skull, it is easy to distinguish the skull of a cat from that of a horse.

The strength of bones also protects more fragile tissues. The rib cage provides protection for the heart and lungs, whereas the skull protects the delicate brain. Bone acts as a reservoir for calcium and phosphorus. In times of need, the minerals are moved from the bone and sent into the bloodstream. Excess minerals can be stored in the bone. Calcium plays an essential role in muscle contraction and enzyme activity. Phosphorus is necessary for energy metabolism within the cell. Bone, in response to several hormones, maintains a tight regulation on the blood level of these minerals.

The long bones are present in the legs (and arms in humans). The femur and humerus are classified as long bones. They have a dense outer shell and a hollow shaft. Bone marrow is made in this hollow center, the medullary cavity. Bone marrow in turn produces blood cells.

BONE STRUCTURE

OBJECTIVE
■ *Detail the Structure of Bone*

Splitting a long bone along its length shows the typical structure of bone (Figure 3–3). The outer shell is composed of dense or compact bone. The more forces placed on a bone, the thicker this layer will be. In the femur this compact bone is thickest in the middle of the shaft, where greatest strain occurs.

Within compact bone lies a more loosely arranged bone, called spongy or cancellous bone. Spongy bone is found within the long bones but not inside the flat bones of the skull or pelvis. It only fills the ends of these bones. Spongy bone is made up of tiny spicules and plates of bone. The spicules look random but are actually arranged to maximize strength. The medullary cavity is located in the hollow center of

the shaft. The bone marrow lies within the medullary cavity and the spaces of the spongy bone. As mentioned earlier, bone marrow produces blood cells.

Bones are covered with a thin connective tissue called the periosteum. The periosteum blends into tendons and ligaments, binding them to the bone. The portion of bone covered with cartilage is not covered by periosteum. The ends of bones within joints have this cartilage protection. The open spaces within bone are covered with a similar connective tissue, the endosteum. Both the periosteum and endosteum provide cells necessary for the repair of damage.

A dried bone is composed of about 70% inorganic minerals and 30% organic components. The inorganic minerals have a high level of calcium and phosphorus. This is found as crystals of hydroxyapatite $(3Ca_3(PO_4)_2 \cdot Ca(OH)_2)$. The organic portion contains collagen fibers and cells. The fibers provide a framework on which the hydroxyapatite crystals can be deposited. Whereas organic fibers give the bone a small amount of elasticity, minerals give bone its typical hardness and strength. The collagen fibers and the hydroxyapatite crystals make up the matrix that surrounds the cells.

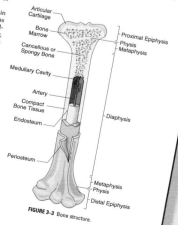

Articular
Cartilage

Bone
Marrow

Cancellous or
Spongy Bone

Medullary Cavity

Artery

Compact
Bone Tissue

Endosteum

Periosteum

Proximal Epiphysis
Physis
Metaphysis

Diaphysis

Metaphysis
Physis
Distal Epiphysis

FIGURE 3–3 Bone structure.

A chapter **summary** highlights the topics that have been presented, and the end of each chapter is also followed by a series of **review questions** and student **activities.**

The addendum on **Career Focus** supplies valuable information on careers in veterinary science including veterinary technicians, veterinary assistants, and private practitioners.

142
UNIT 1 ■ Comparative Anatomy and Physiology

but include head tilt, circling, paralysis of facial muscles, and difficulty chewing and swallowing.

Treatment of horses with EPM has had relatively poor and inconsistent results. A positive diagnosis is possible only with an autopsy. However, blood tests are available to show exposure to the organism but not disease confirmation. The clinical signs and the positive blood test do help support a diagnosis. Recently a new medication has gained approval and offers hope in treating these horses.

SUMMARY

Understanding of the nervous system begins with the ability to describe the neuron, nerve impulse, synapse, and components of a reflex arc. An ability to explain brain structures and their associated functions is also important.

The brain and the spinal column comprise the central nervous system, whereas the nerves in the limbs constitute the peripheral nervous system. Conversely, the nervous system can also be divided into the sensory somatic and autonomic nervous systems and between the two branches of the autonomic system. The nervous system in its entirety controls many body functions.

REVIEW QUESTIONS

1. Define any 10 of the following terms:
   listeriosis (circling disease)
   epilepsy
   cervical disk disease
   equine protozoal myeloencephalitis
   neuron
   volt
   polarization
   myelinated nerves
   coma
   myelogram
   sensory somatic system
   autonomic system
   plexus
   sympathetic system
   parasympathetic system
   dilate
   constrict
   nystagmus
   ataxia
   atrophy

2. True or False: All nerves have a myelin sheath.
3. True or False: Damage to the right side of the brain causes weakness in the left side of the body.
4. The long, thin extension of a neuron is called the _____.
5. Gray matter is housed in the _____ region of the spinal cord.
6. The dorsal root, which exits the spinal cord, carries _____ nerves.
7. Sympathetic stimulation causes the heart rate to _____.
8. Does a reflex occur with conscious thought?
9. Are the rod or cone cells of the eyes receptive to colors?
10. Can an underlying cause of epilepsy seizures be found?
11. How can humans contract listeriosis?
12. Name the two systems of the peripheral nervous system.
13. Name the junctions where nerve impulses are transmitted.
14. List the three regions of the brain.
15. List the four types of taste buds found on the tongue.

ACTIVITIES

Materials needed for completion of activities:
flashlight
reflex hammer
six small paper bags
six small scoops of mini jelly beans in three different flavors (lemon, grape, cherry)
six marking pens
yardstick
graph paper

1. To evaluate the reflexes of the eye, have a subject seated in a dimly lit room. The dim light allows the pupil to dilate, but it should be light enough to observe the size of the pupil.

296
ADDENDUM B

Dr. Lawhead performing a displaced abomasum surgery on a cow.

In addition, Dr. Lawhead spends approximately one third of his time in a small animal clinic consisting primarily of dog and cat patients.

When Dr. Lawhead joined the practice, two partners employed him. These two veterinarians were extremely supportive and patient as he gained experience in private practice. They helped to train Dr. Lawhead. Dr. Lawhead is now a partner as well, and the practice has grown to employ five veterinarians.

Dr. Lawhead experiences a tremendous variety of situations in his career. As a mixed practitioner, he finds value in each aspect of his job. The work with dairy farmers proves very rewarding. While working with the same clients for 15 years, a special relationship between Dr. Lawhead and his clients has developed, because both the veterinarian and farmer strive to maximize productivity and profitability. Dr. Lawhead finds that process challenging and stimulating. Dr. Lawhead has a special interest in nutrition and formulates the rations for several dairy clients.

The small animal portion of the profession also offers challenges and rewards. Although similar goals exist for maintaining the health of all animals, the companion animal portion of his job allows him to work on challenging medical and surgical cases. Further, the group practice environment is rewarding and allows for collaboration on cases.

Private practice does have its drawbacks. Dr. Lawhead is often on call and often is called at night. Even so, Dr. Lawhead must be at work the next day. Fortunately, the on-call responsibility is shared equally in his multiveterinarian practice. The work-day varies and is unpredictable. Dr. Lawhead's workday often extends well beyond 12 hours.

CAREER FOCUS: ACADEMIA

Dr. Abby Maxson Sage is a 1987 graduate of the University of Pennsylvania School of Veterinary Medicine. Following her graduation, Dr. Sage elected to participate in advanced training in an internship and residency, also at the University of Pennsylvania. After the

Dr. Sage, with the assistance of Dr. Ormond, is performing an ultrasound examination on a horse. (Courtesy Dr. Abby Maxson Sage.)

Acknowledgments

Although only two authors are listed for this text, the number of people responsible for the final product is quite large. The authors would like to thank all of those people who supported and contributed to the text, especially the Delmar Thomson Team. Zina Lawrence, acquisitions editor at Delmar Thomson, deserves special recognition for her faith in the authors.

We would like to thank all the veterinarians and staff at Millerstown Veterinary Associates for their assistance and contributions. Their help in obtaining case material and photographs for the text was invaluable. Likewise, we appreciate the support of the clients that encouraged the use of their case material for the text. Special thanks are in order for Leesa Landis, Dr. Robert Mikesell, and Krista Pontius for their long hours of technical help in putting together the text. We appreciate the use of reference material supplied by Mechelle Regester. The veterinary science students at Greenwood High School completed activities, lessons, and accompanying assignments to help fine tune the text and ancillary material. We appreciate their thoughtful consideration.

In addition, we would like to thank Dr. David Sweet, Dr. Cathy Hanlon, Dr. Abby Maxson Sage, and Dr. Lawrence Hutchinson for their contributions of photographs and support to the project.

Having input from experts in various fields helped to strengthen the core material of the text. Our utmost thanks to Dr. William Bacha Jr., Dr. Linda Bacha, and Dr. Arthur Hattel for the photographic material provided. The histology and pathology photographs are a tremendous benefit to the text.

About the Authors

Dr. James Lawhead is a veterinarian in a private mixed animal practice. He works primarily with dairy cattle, dogs, and cats. Dr. Lawhead joined this practice in 1987 following graduation from the University of Pennsylvania School of Veterinary Medicine. He gained acceptance to veterinary school following completion of his bachelor's degree at Juniata College. Since that time, Dr. Lawhead has become a partner in this practice. He has a special interest in dairy cattle nutrition, providing nutritional service to a number of his clients. Dr. Lawhead enjoys teaching as well and actively supports the local school districts with lectures and demonstrations.

Dr. MeeCee Baker teaches agricultural education at Greenwood High School in Millerstown, Pennsylvania. In addition, Dr. Baker serves as an adjunct professor at the North Carolina State University. She earned both her bachelor's and doctorate degrees from Pennsylvania State University in agricultural education and a master's of science degree from the University of Delaware in agricultural economics. Dr. Baker was the first woman to be elected president of the National Vocational Agriculture Teachers' Association (now known as the National Association of Agricultural Educators). Baker lives on her family farm in Port Royal, Pennsylvania, with her husband, Dr. Robert Mikesell, her daughter Elizabeth ("Libby") Baker-Mikesell, and her parents, Bob and Dorothy Baker. She and her family raise Maine-Anjou cattle and are active members of the Port Royal Lutheran Church.

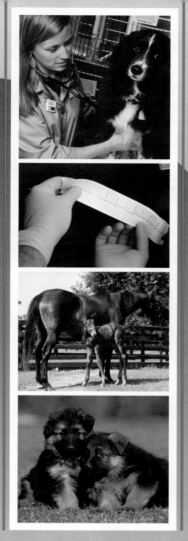

UNIT I

Comparative
Anatomy
and Physiology

1

Basic Cell Biology

OBJECTIVES

Upon completion of this chapter, you should be able to:

- *Explain the molecular makeup of cells.*
- *Identify the basic structures of the cell and their corresponding functions.*
- *Review the basic function of the cell.*
- *Discuss mitosis and its clinical significance in diseases such as cancer.*
- *Detail meiosis in mammalian reproduction.*
- *Connect cellular parts and function to clinical veterinary practice.*

KEY TERMS

anesthetize	glycogen	catabolism	benign
antibiotics	enzymes	homeostasis	malignant
cancer	antibodies	diffusion	pathologists
lipid	exocytosis	osmosis	
glucose	metabolism	active transport	
diabetes	anabolism	endocytosis	

INTRODUCTION

The cell is the basic structure of animal life. However, the cell contains other structures and molecules. Cells conduct many functions and are also able to reproduce. Animals not only have millions of cells that comprise the body, but also many different cell types. The combination of these cell types makes an animal function. This chapter will discuss the structure of cells and how they work.

A DAY IN THE LIFE

There Just Never Seems To Be a Typical Day...

I headed to the office with the thought of doing only cow work this day. However, the job had other plans. Shortly after I got to work, two nervous owners walked through the door with their Labrador retriever. Poor Jake had just been run over by the owner's car! Amazingly Jake was doing very well, although he was a bit excited. Besides a couple cuts on his jaw, he was ready to go home and play.

At my first farm call, the farmer wanted me to look at his dog, Millie. Millie had a grapefruit-size lump under her jaw. The lump felt like it was full of fluid. I asked him to bring Millie to the office so I could work on her there. I finished my farm calls and headed back to the small animal clinic.

Once there, I **anesthetized** Millie and made an incision into the skin. Pus flowed from the lump (Figure 1–1). I flushed the large pocket left behind and started Millie on a course of **antibiotics,** drugs that fight bacterial infections. Although I don't know why it started, I do know Millie was fighting an infection with her body's cells.

Next I had the opportunity to remove a tumor from Penny, a 12-year-old cocker spaniel. Last week I gave Penny a physical and administered blood tests.

FIGURE 1–1 Draining an abscess on the side of the face of an anesthetized cat.

The surgery went well, and I was able to remove the entire lump.

In private practice, cells affect me every day. Today I saw Millie's cells attacking the bacteria in her neck. Penny, on the other hand, had **cancer**-causing cells dividing uncontrollably. To understand how mammals work and how to treat them, I first had to learn how cells function.

CELL MAKEUP

OBJECTIVE

■ *Explain the Molecular Makeup of Cells*

The cells and their structures are composed of molecules. Biochemistry is the study of these molecules in living creatures. One goal of this chapter is to point out the differing types of molecules and their properties.

Lipids, or fats, combine hydrogen, carbon, and oxygen in a form that is poorly dissolved in water. (This is why fat floats to the top of water.) Fat consists of a molecule of glycerol and three fatty acid molecules (Figure 1–2). Fats are stored in the cells of the body as a source of high energy.

Phospholipids are similar but have only two fatty acid groups and a phosphate group (PO_4). This is significant because one end of the molecule is attracted to water and the other end repelled by water (important in the discussion of the cell membrane).

Carbohydrates supply energy and provide structure within the cell. Monosaccharides are the simplest of these molecules. They possess the basic structure of $(CH_2O)_n$ (Figure 1–3). In this formula, n describes the number of carbon atoms in the molecule. The genetic material in the cell has the five-carbon sugars, ribose, and deoxyribose. **Glucose** (blood sugar), a six-carbon sugar, is used for energy in the cells. Glucose is routinely measured with blood tests. If there is too much or too little glucose in the blood, the animal will not function normally. In **diabetes** the blood sugar becomes very high and the animal does not utilize it properly. Diabetes requires treatment to lower the blood sugar.

Polysaccharides are composed of many monosaccharides. One example of a polysaccharide is starch, such as **glycogen,** that is used to store energy within the cell. Glycogen is made when monosaccharides are taken into the cell and then assembled into a long chain. Polysaccharides can be joined with protein

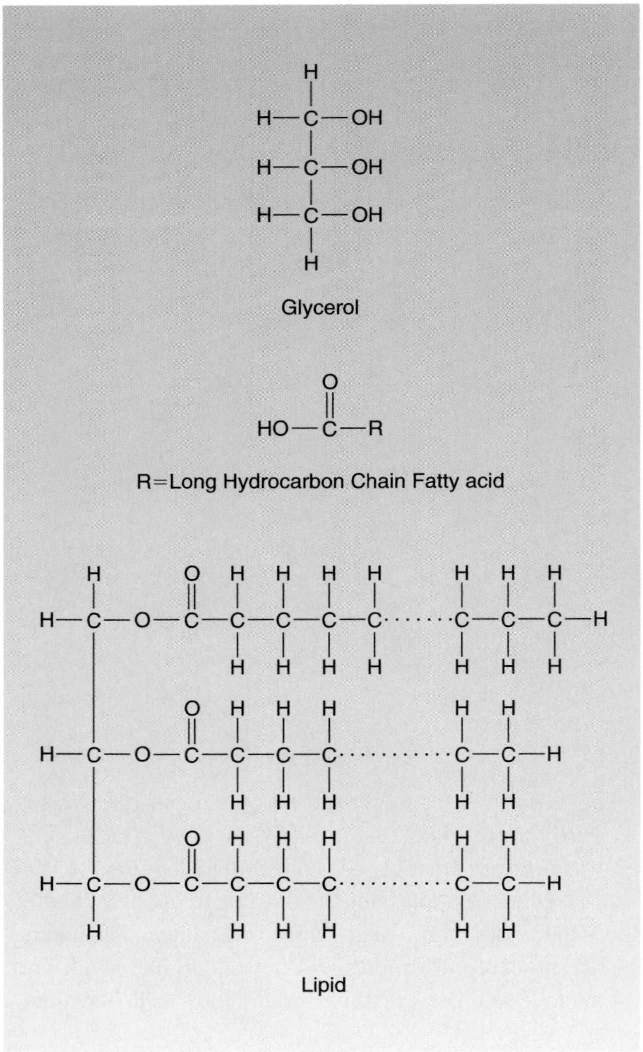

Glycerol

R=Long Hydrocarbon Chain Fatty acid

Lipid

FIGURE 1–2 Chemical structure of glycerol, fatty acid and a typical lipid.

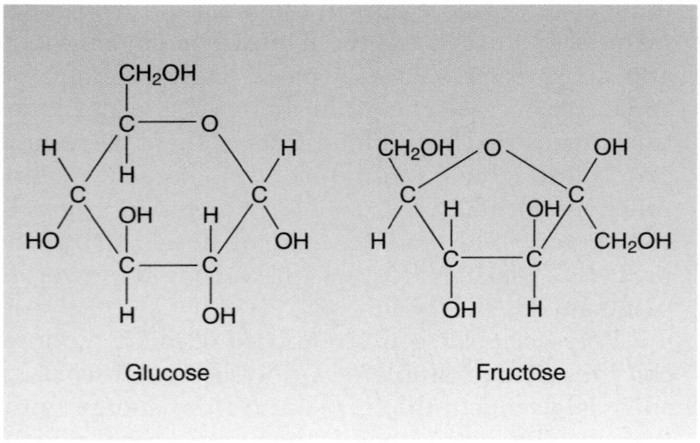

Glucose

Fructose

FIGURE 1–3 Chemical structure of selected sugars.

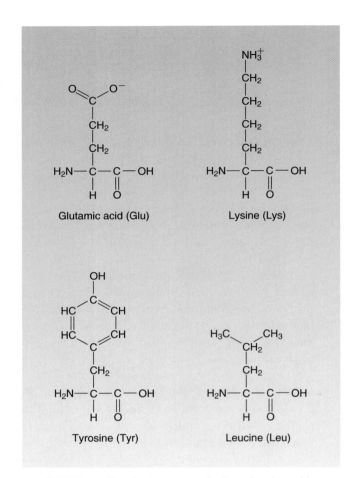

Glutamic acid (Glu)

Lysine (Lys)

Tyrosine (Tyr)

Leucine (Leu)

FIGURE 1–4 Chemical structure of selected amino acids.

molecules to form glycoproteins, which assist in building cell structure.

Proteins play a key role in the structure and function of cells. Proteins make up 50% of the dry weight of animals. Proteins are large molecules made up of many amino acids. (Twenty-two different amino acids are used to make proteins; Figure 1–4.) A single protein can include 200 to 300 of these amino acids. It was mentioned earlier that proteins could be joined to sugars. The may also be joined with lipids and phosphate groups.

Proteins have many functions in cells. Muscle is largely composed of protein that is specially arranged to allow cells to contract and move. Further, **enzymes** are protein molecules that speed the chemical reactions in the body. Proteins also add strength to many of the structures in the body. In addition, proteins found in blood help to carry oxygen, stop bleeding, and fight off infection. These infection-fighting proteins are called **antibodies**. In practice, specific antibodies unique for different diseases are measured in

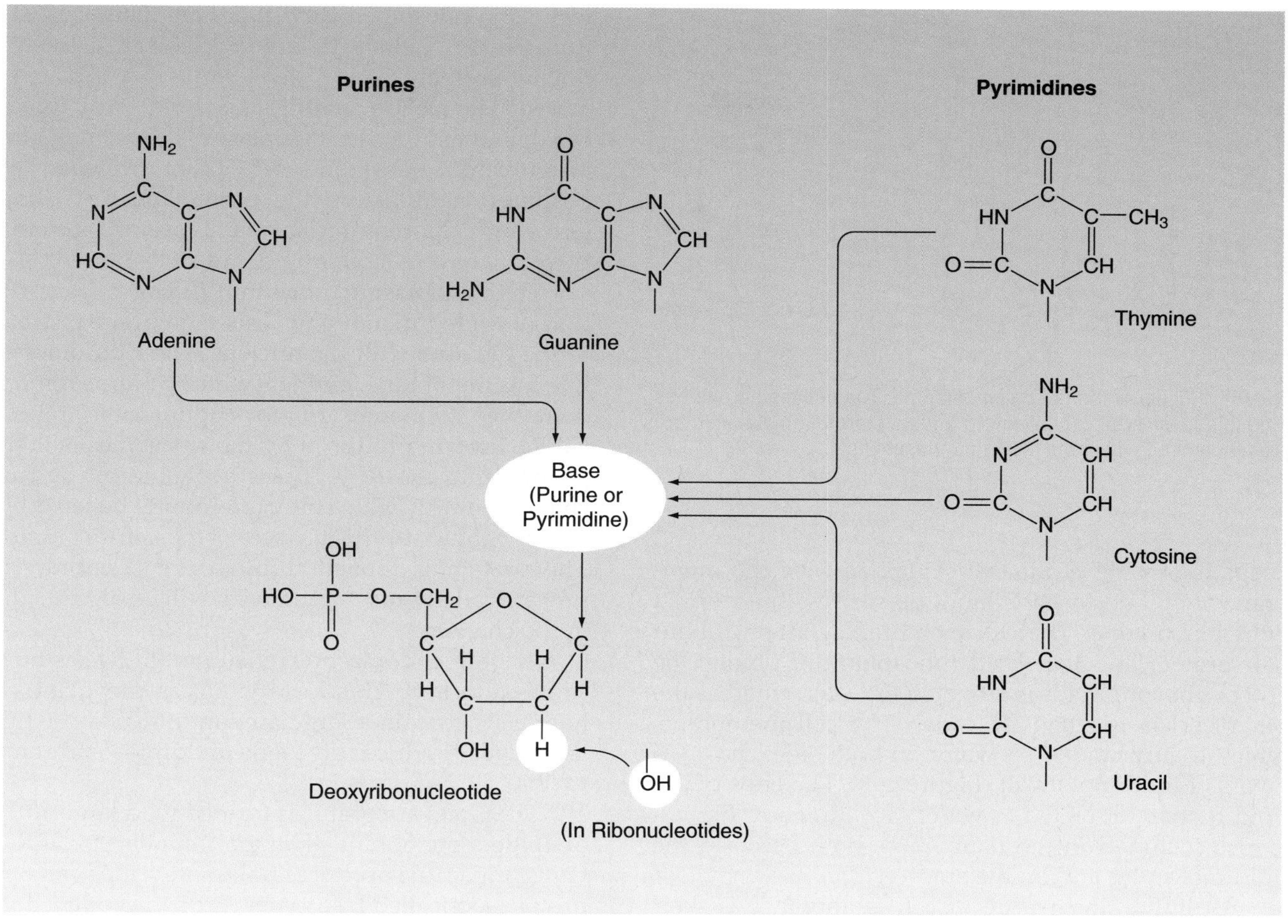

FIGURE 1–5 Chemical structure of a nucleotide.

the blood, thus allowing veterinarians to diagnose what specific organism is causing the sickness.

Nucleic acids provide the plans for the differing construction of proteins. Nucleic acids are fabricated with a series of nucleotides. The nucleotides are made up of a five-carbon sugar, a phosphate group, and a nitrogen-containing base (Figure 1–5). Ribonucleic acid (RNA) claims ribose as its sugar, whereas deoxyribonucleic acid (DNA) has deoxyribose as its sugar. There are four different bases for RNA and for DNA (Table 1–1). Notice that the bases are the same except for thymine and uracil. The order of base combination determines what amino acids are used to make proteins. This information is stored in the cell's genetic material.

CELL STRUCTURE

OBJECTIVE

■ *Identify the Basic Structures of the Cell and Their Corresponding Functions*

Large varieties of types of cells exist. These cells not only look different but function differently as well. Nevertheless, many features remain common among cells. Specialized structures within the cells are called organelles. These organelles are present in most but not all cells. Red blood cells, for example, lack a nucleus.

The cell membrane (or plasma membrane) is a feature common to all cells. It serves as the boundary that

TABLE 1–1 RNA and DNA Bases

DNA Bases	RNA Bases
1. Adenine	1. Adenine
2. Cytosine	2. Cytosine
3. Guanine	3. Guanine
4. Thymine	4. Uracil

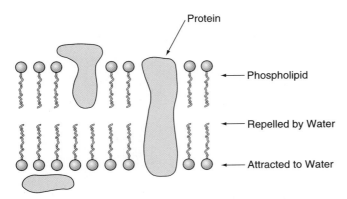

FIGURE 1–6 Illustration of cell membrane. The cell membrane has a double layer of phospholipid. In addition, protein molecules are present on and within the layers of phospholipid.

keeps the inside of the cell contained. The cell membrane is so fine that it cannot be seen with a normal light microscope. The cell membrane contains about half protein and half lipid (phospholipid type). One end of phospholipids is attracted to water, whereas the other end is repelled by water. The cell membrane, which is surrounded by water on both sides, has two layers of lipid in its wall (Figure 1–6). The ends of the lipid that are attracted to water face outward. Protein is also included in the membrane, both between the lipid molecules and on the surface.

An important feature of cell membranes is their ability to allow certain items to pass through and not others. This property is called semipermeability. Certain molecules, such as water, are able to pass through easily. Other molecules, such as proteins, starches, and some ions, are unable to pass.

Many of the organelles within the cell also have a membrane around them. The basic structure is the same for all the membranes. The specifics of the makeup are different depending on their function.

The cell contents are divided into the nucleus and the cytoplasm. *Cytoplasm* is a general term used for the organelles and fluid in the cell. A nucleus comes as a standard part of most cells (with few exceptions, such as the red blood cell; Figure 1–7). The nucleus controls the cellular activities and contains the genetic material (DNA) of the cell. The DNA in the nucleus is called chromatin. As the cell divides, the chromatin clumps into chromosomes. All the cells in the body have the same chromatin. Cells take on different roles by using certain areas of the chromatin more than others.

A membrane surrounds the nucleus. This membrane is often joined to other organelles, such as endoplasmic reticulum and ribosomes. Such a close association aids the nucleus in controlling cell function.

In cells not dividing, there is often a nucleolus seen in the nucleus. The nucleolus produces RNA that forms the ribosomes. Ribosomes are then used to produce protein. Cells with large nucleoli are actively producing protein.

Ribosomes are small granular-like structures that can be found in the cytoplasm. Ribosomes contain roughly 60% RNA and 40% other protein. The ribosomes manufacture the protein used in the cell. Growing cells require large amounts of protein and therefore have a large number of ribosomes.

The endoplasmic reticulum (ER) is a large collection of folded membrane. This membrane is attached to the membrane of the nucleus. The ribosomes often line this membrane, giving it a bumpy appearance and therefore its name, rough endoplasmic reticulum (RER). Protein produced by the ribosomes is then deposited into the RER. These proteins can be further changed in the RER. This protein may be used by the cell or moved to the surface of the cell for secretion. The protein is moved through the membrane in a process called **exocytosis,** which will be discussed later in the chapter.

Smooth endoplasmic reticulum (SER) has no ribosomes attached. This form is not as prevalent. Some liver cells contain a large amount of SER. The SER in these cells produces glycogen and lipids and removes toxins.

The Golgi apparatus is formed with large amounts of folded membrane that look similar to SER. The Golgi apparatus produces polysaccharides and special protein sacs called lysosomes. Protein produced in the RER is moved to the Golgi apparatus. The Golgi apparatus then changes the protein and collects it into the lysosomes. These sacs are pinched from the Golgi apparatus and then moved to the surface of the cell and released.

The proteins contained in the lysosomes are enzymes (remember, enzymes are molecules that help speed chemical reactions in the body). Lysosomes contain enzymes that help to breakdown other molecules. Varying enzymes match with differing molecules. The membrane surrounding the lysosome prevents enzymes from attacking other parts of the cell.

Lysosomes are used to digest food taken in by the cell and to destroy cell structures no longer needed. In Millie, the dog with the abscess, her white blood cells were using lysosomes to destroy bacteria. Cells that die in the body are eliminated by lysosomes to make room for replacement cells.

Mitochondria are small rod-shaped organelles found in varying numbers in cells. The more active the cell, the more mitochondria are present. Mitochondria have a double membrane, which is very similar to the cell membrane. The outer membrane is smooth and forms the shape of the mitochondria. The inner membrane is highly folded. These shelflike infoldings are called cristae.

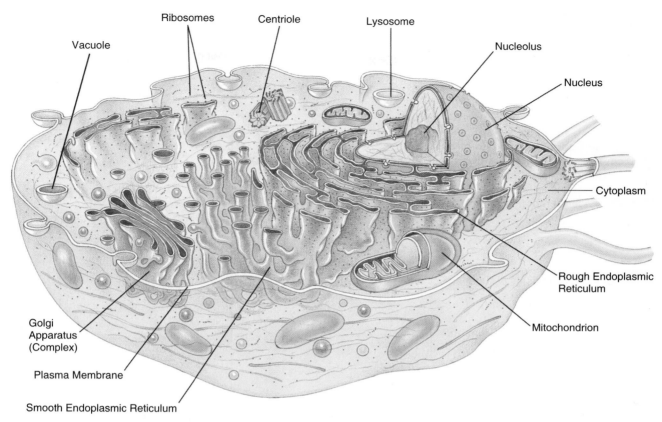

FIGURE 1–7 Illustration of cell structure.

The role of mitochondria is well defined. The mitochondria are involved in converting food substances into a form of energy that can be used by the cell. Mitochondria contain the enzymes necessary for this process. Because of this role, mitochondria are called the powerhouses of the cell. The mitochondria are found within the cells at regions of the highest activity.

CELL FUNCTION

OBJECTIVE

■ *Review the Basic Function of the Cell*

The cell constantly reacts with its environment. **Metabolism** describes all the reactions conducted in cells. Metabolism can be categorized into two main types. **Anabolism** describes reactions in which smaller molecules are combined into larger ones. The joining of amino acids to form proteins serves as an example. **Catabolism,** the opposite, occurs when large molecules are broken down into smaller ones. Glycogen being broken down to release energy is an example of catabolism.

A liquid called extracellular fluid (ECF) surrounds living cells. The ECF supplies the cells with all the products necessary for function. ECF is derived from the blood. The outermost skin cells are not covered in liquid because they are no longer living.

Other cells exposed to the surface, such as those of the eye, need moisture. In the eye, tears produced by glands act as the source of moisture and nutrients. The eyelids help to sweep the tears across the surface of the eye. Certain breeds of dogs, such as the pug, have eyes that bulge from the eye socket. This bulging can be so severe that the eyelids cannot keep the surface of the eye moist with tears. This results in a disease condition on the surface of the eye. Artificial tears are often used to keep the surface moist.

Table 1–2 summarizes the makeup of ECF. Water is the major component of ECF. Oxygen passes to the cells through the ECF, and carbon dioxide passes from the cells through it. There are many inorganic ions in the ECF. Some ions, macrominerals, are present in large amounts. Trace minerals are present in much smaller amounts. Both macrominerals and trace minerals are essential for cellular function. Many of the trace minerals are needed for enzymes to function. Organic compounds, the lipids, proteins, carbohydrates, are also delivered by the ECF. Metabolism produces waste products that must be removed from the cells.

TABLE 1–2 Components of the Extracellular Fluid

1. Water
2. Dissolved gases: oxygen, carbon dioxide
3. Inorganic ions
   Macrominerals: sodium, potassium, chloride, phosphate, calcium, bicarbonate
   Trace minerals: copper, zinc, manganese, cobalt, selenium, fluoride, iron
4. Organic compounds (carbon containing compounds): proteins, amino acids, lipids, carbohydrates, vitamins
5. Hormones: compounds produced by glands to influence metabolism of cells
6. Waste products

TABLE 1–3 Mechanisms of Cellular Exchange

1. Diffusion
2. Osmosis
3. Active transport
4. Endocytosis
5. Exocytosis

These waste products are eliminated by the ECF. Without elimination, the waste products actually become toxic to the cell.

Many of the products in ECF must be maintained at constant normal concentrations. Cells will be unable to function properly if there is too much or too little of certain products. Glucose provides an excellent example. Small puppies can become low in blood sugar if they have too many parasites robbing them of their nutrients. When the sugar in the ECF becomes too low, the cells do not have adequate energy. The puppy can become weak or in severe cases can develop a seizure. **Homeostasis** is the maintenance of the ECF. Homeostasis allows for the maintenance of normal concentrations of molecules in spite of a wide variety of external conditions.

Cells must be able to obtain products from the ECF. It is not adequate for the chemicals to just exist in the ECF, there must be means for exchanging them with the cell. Table 1–3 summarizes the mechanisms by which materials are exchanged across the cell membrane. The first mechanism is a process called **diffusion** (Figure 1–8), in which molecules move from higher to lower concentrations. Because molecules are always moving, there is a greater chance that they will move toward areas of lower concentration. This movement continues until the concentrations are equalized.

The cell membrane does not allow totally free diffusion. Diffusion is influenced by the size of the mole-

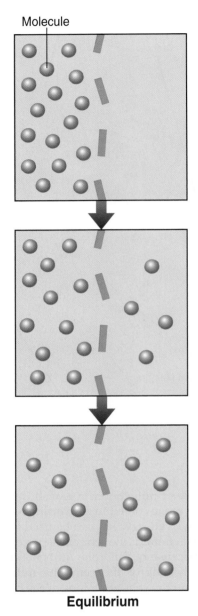

Molecule

Equilibrium

FIGURE 1–8 Diffusion: The random movement of molecules allows for the equalization of concentrations across a membrane.

cule, its charge, and its ability to dissolve in lipid. In general, the smaller the molecule, the more easily diffusion occurs. Some large molecules such as proteins are unable to diffuse through the membrane and must be transported in other ways.

This property of allowing only certain molecules to diffuse through the membrane is called semipermeability. These characteristics set the stage for a special type of diffusion, called **osmosis**. A solvent (in the following case, water) moves across the membrane to equalize the concentration; however, the molecules dissolved in the water (called solutes) cannot pass

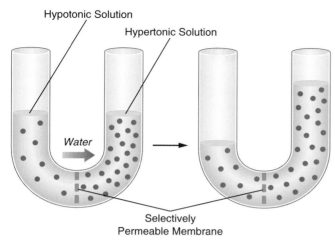

FIGURE 1–9 Osmosis: The semipermeable membrane prevents the passage of large molecules. In this situation, water moves across the membrane to equalize the concentration.

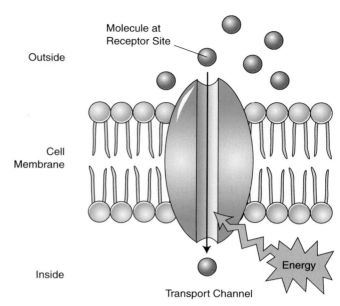

FIGURE 1–10 Active transport: Energy is used to actively pump molecules into a region of higher concentration.

through the membrane (Figure 1–9). This process can be observed in red blood cells when they are placed in a concentrated solution. The water from the cell moves outward into the solution. Microscopically the red blood cells can be seen to shrink.

In certain situations a cell may require a higher concentration of a molecule than found in the ECF. For example, red blood cells have higher levels of potassium than the surrounding fluid. Diffusion constantly attempts to equalize the concentrations (potassium continually diffuses from the cell). In this case the potassium is being pumped back into the cell keeping the higher concentration. This process is referred to as **active transport** (Figure 1–10). Active transport does require the cell to burn energy and use enzymes to aid the process. Many different cell types perform the process. Another example occurs in intestinal cells, which transport glucose into the bloodstream, where it is present at higher levels.

Large molecules, such as proteins, must be moved through the membrane in a process called **endocytosis** (Figure 1–11). During endocytosis, the cell membrane wraps around the particle, pinches off, and moves into the cytoplasm as a vacuole. Lysosomes then join with the vacuole, providing the enzymes necessary to break down the particle. The smaller fragments produced are then released into the cell.

In cells producing protein, the opposite process occurs. In exocytosis a membrane-bound sac containing the protein joins with the cell membrane and releases it into the ECF (Figure 1–12). These sacs are produced within the Golgi apparatus. In intestinal cells, fat droplets can be taken into the cell through endocytosis. The vacuole is transported across the cell and released into the bloodstream by exocytosis.

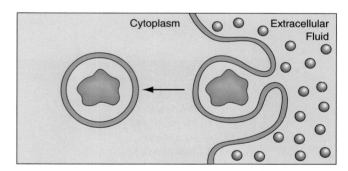

FIGURE 1–11 Endocytosis: A large particle is engulfed by the cell membrane and brought into the cytoplasm within a vacuole.

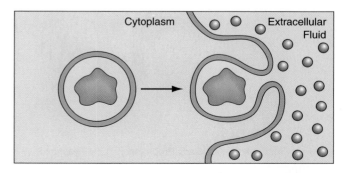

FIGURE 1–12 Exocytosis: A membrane bound sac joins with the cell membrane to release the particle.

MITOSIS AND CANCER

OBJECTIVE

■ *Discuss Mitosis and its Clinical Significance in Diseases Such As Cancer*

Cells must be capable of reproducing themselves. In mitosis the cells divide, producing two identical cells. The genetic material contained in every cell is the same. Mitosis is necessary for the animal to grow and maintain. Cell division is controlled by a number of factors that are present within the cell and the extracellular fluid. The rate of cell division is adapted to the needs of the animal. Some cells, such as those epithelia lining the intestinal tract, divide frequently to maintain the integrity of the layer. Other cells, such as skeletal muscle, do not divide in an adult.

When these normal controls break down, the cells can begin to undergo frequent mitosis. Uncontrolled mitosis results in cancer. New cells are produced more quickly than needed, resulting in an accumulation or mass of cells in a region. This mass of rapidly dividing cells is called a tumor.

In a nondividing cell the genetic material is called chromatin. In this form the chromatin is loosely arranged in the nucleus. The individual chromosomes cannot be seen with a light microscope. These cells are described as being in the interphase. In this stage the cell is in the process of doubling its DNA. The steps of division are divided into four phases (Table 1–4). The phases are identified to help the understanding of the process. However, actual cell division is a continuous process, as seen in Figure 1–13.

Prophase begins as the chromatin thickens into visible chromosomes. This is the first time that the individual chromosomes can be seen with a light microscope. Along with this process, the nucleoli and nuclear membrane begin to disappear. At this point the chromosomes show the doubling that occurred during interphase. The chromosomes have an X shape. The two identical halves are joined at a small point called the centromere. Two small organelles, the centrioles, separate and move to opposite ends of the cell.

In metaphase a spindle is formed between the two centrioles. This is a collection of microtubules that stretch between the two centrioles. The chromosomes move to the center of the cell and align themselves on the spindle.

As anaphase begins, the chromosomes split at the centromere. At this point the chromosomes are still on the spindle. Each half of the chromosome begins to move outward. The centromere portion moves first, giving the chromosome a V shape. The chromosomes move to opposite ends of the cell.

Telophase is basically the reverse of prophase. The chromosomes become loosely organized into chromatin. The nuclear membrane and nucleoli return. A groove then forms down the center of the cell. This groove deepens until two identical cells are produced.

TABLE 1–4 Stages of Mitosis
1. Interphase
2. Prophase
3. Metaphase
4. Anaphase
5. Telophase

Interphase   Early Prophase   Late Prophase   Metaphase   Anaphase   Telophase   Interphase

FIGURE 1–13 Mitosis. *Interphase:* Cell in its normal state, as the chromosomes begin to replicate. *Prophase:* The chromatin thickens and becomes visible, taking on an X shape. The nucleoli and nuclear membrane begin to disappear. *Metaphase:* The spindle forms between two centrioles. The chromosomes align on the spindle. *Anaphase:* Chromosomes split at the centromeres, with each half moving to opposite ends. *Telophase:* The nucleus reforms and a groove divides the two new cells.

MAMMALIAN REPRODUCTION

OBJECTIVE

■ *Detail Meiosis in Mammalian Reproduction*

Mammals rely on sexual reproduction for species survival. In sexual reproduction a sperm cell and egg cell join to form the new embryo. In this process, half the genetic material is provided by each of the cells. Meiosis is the division in which the resulting cells contain only half the genetic material.

There are two cell divisions during meiosis, with only one doubling of the chromatin. The final result is four cells, each with half the number of chromosomes. Just as in mitosis, meiosis divides into phases (Figure 1–14).

Prophase I: Mammals should have an even number of chromosomes. The chromosomes come in pairs, and each member of a pair is called a homologue. Before prophase I, the homologue of each pair replicates and is formed by two strands (or chromatids). The chromatids are joined by a centromere. Prophase I is very complex. The basic process allows the homologues pair up near the center of the cell. In this arrangement the homologues are joined at several points. At these points, an exchange of DNA fragments occurs.

Metaphase I: This step is very similar to mitosis. The nuclear membrane and nucleoli begin to disappear. The paired chromosomes move into alignment on the spindle. The important distinction in meiosis is that the homologues align themselves where they will be divided into opposite cells.

Anaphase I and Telophase I: In anaphase the chromosomes begin to move to opposite ends of the cell. In this step the centromere does not split. Rather, the pairs of chromosomes are divided.

The length of interphase between the two divisions is variable and may not occur. The two cells produced enter into the second division. The second division of meiosis is basically the same as that of mitosis. In this division the chromosomes align on the spindle, separate at the centromere, and send one strand to each new cell. The stages are named just as they are in mitosis. The final result of meiosis is four cells, each with half the number of chromosomes of the original cell.

Meiosis allows genetic material to be provided from each parent. The exchange of genetic material between homologues in Prophase I produces variability in each cell. Offspring acquire traits from each parent. With the variation, no two sperm or egg cells will provide the same genetic material.

Twinning can result in two animals having the same genetic makeup. Identical twins occur when an embryo splits. Each half then develops into a new

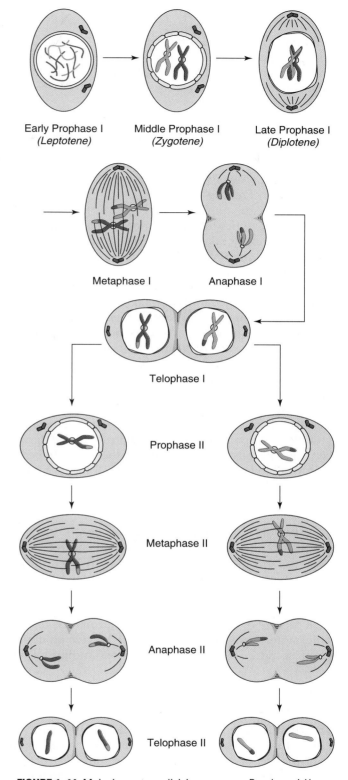

FIGURE 1–14 Meiosis—a two–division process. *Prophase I:* Homologous chromosomes align in the center of the cell. The homologues exchange segments of genetic material. *Metaphase I:* Similar to the step in mitosis, except the homologues will separate into opposite cells. *Anaphase I:* The chromosomes move to opposite ends of the cell. *Telophase I:* The cell is divided into two daughter cells, each with half of the chromosome numbers of the original cell. The second division of meiosis is similar to mitosis. The net effect of meiosis is four cells, each with half of the original number of chromosomes.

embryo. The resulting offspring begin life with identical chromosomes. Even identical twins, however, do not appear completely identical. There is variation in the way the genes are expressed.

CLINICAL PRACTICE

OBJECTIVE

■ *Connect Cellular Parts and Function to Clinical Veterinary Practice*

In clinical practice the appearance of cells is often evaluated. A biopsy takes tissues or cells from an animal for microscopic review. This procedure allows cancer diagnosis. In cancer the cells divide without normal control. This leads to a mass in tissue or in an organ. Tumors are divided into two major groups, **benign** and **malignant**. Benign tumors are localized to one area and do not spread to other parts of the body. Malignant tumors are more likely to invade surrounding tissues and spread to other parts of the body. For example, cells from a tumor may break off into the bloodstream and then settle into a new location. With biopsy, cells are evaluated to determine the type of tumor that is present. Penny's tumor was submitted for biopsy. Fortunately the **pathologists** who interpret and diagnose changes in cells and tissues found the tumor to be benign and did not find any tumor cells at the margins of the sample. This is great news for Penny, and we are optimistic that her tumor will not cause her any more problems.

Tumors develop because the cells are growing rapidly and dividing without normal control. Many features of cancer can be predicted with this information. In cancerous cells there is often a large nucleus with many nucleoli. The chromatin is often clumped and visible. There are many more cells involved in the process of cell division than in normal tissue. Furthermore, many of the dividing cells have an abnormal spindle. Only with an understanding of the

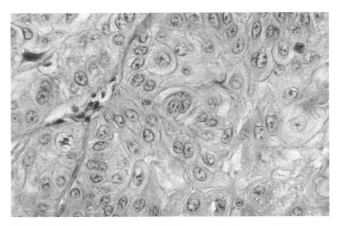

FIGURE 1–15 Photomicrograph (high power) of cancer (squamous cell carcinoma) in the skin of a horse. There is variation in the shape and size of the nucleus and cells. Many mitotic figures can also be seen. *(Photograph courtesy of Dr. Arthur Hattel, Pennsylvania State University.)*

normal cell cycle can we interpret these abnormal findings (Figure 1–15).

Understanding cellular function is essential for sound veterinary practice. Although there is great similarity between all mammals, there are species differences. This chapter has discussed enzymes and their importance in cellular function. Cells contain enzymes that break down medication (that is why medicines must be given more than once). Furthermore, differences between animals can lead to reactions to medications. Cats possess much less of certain enzymes that break down many medications. Acetaminophen (Tylenol), a common over-the-counter pain reliever, has been proven quite safe for use in humans. Because of the differences in the enzymes between species, this product is very dangerous for cats. As little as half a tablet designed for adult humans can make a cat sick. Cellular detail seems far detached from a complete animal. However, the details of cells guide treatment of animals in clinical practice.

SUMMARY

Successful students of veterinary science begin study by mastering the understanding of cells. A thorough knowledge of cellular makeup, including structures and functions, will allow veterinary students to build a foundation of information on which to start a more comprehensive investigation of body processes. Moreover, knowing how cellular activities such as meiosis take place gives a more complete understanding of the reproductive system, which in turn can help veterinarians assist producers in making financial decisions concerning their livestock. Along that same line, another cellular activity, mitosis, can result in cancer if uncontrolled. Therefore veterinary practitioners again rely on their cellular understanding to assist in diagnosis and treatment of that disease. Examination of the cell sets the groundwork for the study of veterinary science.

REVIEW QUESTIONS

1. Define any 10 of the following terms:
   anesthetize
   antibiotics
   cancer
   lipid
   glucose
   diabetes
   glycogen
   enzymes
   antibodies
   exocytosis
   metabolism
   anabolism
   catabolism
   homeostasis
   diffusion
   osmosis
   active transport
   endocytosis
   benign
   malignant
   pathologists

2. True or False: Fats easily dissolve in water.

3. True or False: Larger molecules diffuse more readily than smaller molecules.

4. True or False: Smooth endoplasmic reticulum (SER) contains ribosomes.

5. All cellular reactions are collectively called _____.

6. Extracellular fluid (ECF) surrounds all living cells and is derived from the _____.

7. Too much blood sugar indicates which disease?

8. How many different types of amino acids are used to make proteins?

9. Give another name for the cell membrane.

10. Where is chromatin found within a cell?

11. Do mammals have an even or odd number of chromosomes?

12. Do enzymes function identically in all species?

13. How might cancer cells differ from that of a normally dividing cell?

14. List five mechanisms of cellular exchange.

15. List the four stages of mitosis.

ACTIVITIES

Materials needed for completion of activities:
   several beakers
   assorted fats, such as cooking and motor oil
   water
   food coloring
   heat source
   eggs
   dissecting pins

1. Does fat really float as explained in the cellular molecular component section? Assemble several beakers of water. Add drops of assorted fats, such as cooking and motor oil. Observe the results.

2. Pretend the classroom is a cell, with the walls being the cell membrane. Your instructor will assign you a cellular part. Review your function within the cell and your relationship to other cell parts (other students). Present your information to the group.

3. Add one drop of food coloring to a beaker of water. Observe for evidence of diffusion. Does heat influence diffusion? Investigate by trying the activity with water samples of varying temperature.

4. Using a small dissecting pin, pick away the shell at the air sac end of an egg. Leave the inner shell membrane intact. Place the egg under water in a beaker. Over time watch osmosis cause the inner shell membrane to rupture.

5. Find the following Web site: <http://www.biology.arizona.edu>. Key search word is cell bio. Try to identify the stages of mitosis by taking the interactive quiz.

2

Tissue Types and Functions

OBJECTIVES

Upon completion of this chapter, you should be able to:

- *Describe the properties, locations, functions, and varieties of epithelial tissues.*
- *Describe the properties, locations, functions, and varieties of connective tissues.*
- *Describe the properties, locations, functions, and varieties of muscle tissues.*
- *Describe the properties, locations, functions, and varieties of nerve tissues.*
- *Link knowledge of tissues to clinical practice.*

KEY TERMS

tissue
organs
displaced abomasum
foot and mouth
 disease
epithelial tissues
integument

keratin
tendons
ligaments
adipose tissue
myofiber
porcine stress
 syndrome

rigor mortis
hypocalcemia
sweeny
central nervous
 system
peripheral nervous
 system

neurons
tying-up or Monday
 morning disease
Horner's syndrome

INTRODUCTION

In the previous chapter, learners examined the cell, the basic unit of life. Cells develop specialized structure and function. A collection of cells, organized for a particular function, is called a tissue. Collections of tissues are then arranged into organs. Mammals have four basic tissue types: epithelial, connective, muscle, and nerve.

A DAY IN THE LIFE

Some Days Seem Relatively Easy...

Today was a beautiful day in early June. The farmers were busy making hay and planting the last of their crops. On hectic days like this, the last people farmers want to call are veterinarians, but sometimes they just do not have a choice. The welfare of their cattle still takes priority despite the pressures of spring planting. I received a call from a farmer who had a cow that was not doing well.

A **displaced abomasum** (cows have four stomachs, the fourth being the abomasum), a common circumstance in dairy cows, often occurs shortly after a calf has been born. This condition, often abbreviated DA, is commonly called twisted stomach. In this disease the stomach fills with gas and is pulled upward. Instead of lying in the normal position at the bottom of the abdomen, gas pulls the abomasum up toward the side. Cows with this problem do not eat well and therefore are not very productive.

Several procedures exist to correct displaced abomasums. The simplest procedure is to roll the cow. In this process the gas that pulls the stomach upward is used to pull it back into place by rolling the cow from the right side to the left. This procedure is often effective in correcting the problem but does not keep the stomach in place. Recurrence is common after this technique, which works best in a cow that has other problems to be corrected at the same time, such as an infected uterus.

Another technique involves rolling the cow onto its back. A blind stitch is then used to fasten the stomach to the body wall. The gas in the stomach is used to hold the stomach against the body wall. Either a large needle or a trocar (a metal tube over a sharp metal rod) is used to place suture material into the abomasum. The success rate is higher than the simple rolling technique, and it is relatively quick to perform. The disadvantage of this procedure is that the exact location of the attachment to the stomach cannot be determined.

Several surgical procedures are available to correct a DA as well. The method that I use most commonly utilizes a surgical approach into the lower part of the abdomen (Figure 2–1). I make an incision into the abdomen to identify all the areas of the abomasum. This is the most time-consuming procedure (and therefore the most ex-

FIGURE 2–1 A Holstein cow, shown following surgery to correct a displaced abomasum.

pensive), but in my experience, surgery has the highest success rate. The farmer elected surgery for this cow.

With the farmer's help, I rolled the cow onto her back and tied her to a gate. I then used a local anesthetic before making an incision into the cow's abdomen. The abomasum was then sutured to the body wall and the remaining layers closed. Once this procedure is completed, the cow in question will not get another displaced abomasum.

Several weeks ago, friends of mine called because their dog had just tangled with a groundhog. Remmy, a small 12-pound Jack Russell terrier with a big attitude, attacked the groundhog that was invading his yard. I do not know how badly the groundhog was injured before it escaped, but Remmy had several bite wounds on his neck and face. The bite of the groundhog had torn the skin of his cheek away from the underlying tissues and now a pocket of blood was now forming. I also commonly see bite wounds when a larger dog attacks a smaller dog. In these cases the tissue structure guides my treatment.

The recent news of the **foot and mouth disease** outbreak in the United Kingdom has been quite the source of conversation lately. Many farmers have become concerned over the risk to their farms in Pennsylvania. The disease seemed so far away until one of my clients

continued

became concerned about European visitors. She had friends coming to visit from Germany, and she wondered about the risks of inviting them to stay at the farm. Fortunately, because foot and mouth had not spread to Germany, the risk was extremely low. However, foot and mouth disease (FMD) is a highly contagious viral disease that attacks **epithelial tissues.** Veterinarians have to be aware of such reportable diseases and understand how the disease affects tissues.

Knowledge of tissues affects my everyday life as a veterinarian. In the surgery just described, I had to know what tissues I would encounter, along with their properties and functions. For example, I needed to know about nerve tissue in order to use anesthetic to eliminate pain for the cow. Moreover, my surgical incision passed through epithelial, connective, and muscle tissue. Each of these tissue layers required special handling procedures.

EPITHELIAL TISSUES

OBJECTIVE

■ *Describe the Properties, Locations, Functions, and Varieties of Epithelial Tissues*

Epithelial tissues are collections of cells packed together in sheets. These sheets line the body's surface and openings. These tissues also cover all the openings of the intestinal, reproductive, and urinary tracts. In addition, they line tubes in the body, such as blood vessels and the heart.

Epithelial tissues perform multiple functions. Skin, an important epithelium, offers defense in many forms. Skin protects the body from trauma, the sun's ultraviolet light, extremes of temperature, drying, and bacterial invasion. The cells lining the respiratory, intestinal, urinary, and reproductive tracts also provide protection. Specialized cells in the respiratory tract have cilia on their surface. These cilia, tiny motile filaments on the surface of the cell, are able to move particles from the large airways. For example, dust or bacteria is moved upward so it can be coughed from the airway.

Epithelial tissues produce a variety of secretions. Tears and saliva help to moisten and protect the epithelium. In the airways, mucus secretions help to trap the particles mentioned in the previous paragraph. Also, urine and sweat are forms of epithelial excretion. In addition, the mammary system is lined with epithelial cells that secrete milk. Dairy cows have very well developed mammary systems that produce large volumes of milk.

Epithelial cells can also absorb materials in a highly selective manner. Cells lining the intestines, lungs, and kidneys all absorb materials from the surrounding fluids. Cells lining the blood vessels provide points of exchange for materials from the blood and extracellular fluid (ECF).

Along with keeping substances from entering the body, the epithelial cells conserve materials within the body. These cells help to prevent excessive loss of fluid and nutrients from the ECF.

Every epithelial lining has an underlying connective tissue layer. The epithelium itself has no direct blood supply. Therefore the connective tissue not only provides support but also supplies nutrients and removes wastes for the epithelium. The blood supply within the connective tissue allows for these functions.

Epithelial tissues are classified based on the shape of the cells and the number of layers (Figure 2–2). The epithelial tissue may be described as follows: (1) simple, with one cell layer; (2) stratified, with multiple layers; (3) transitional, with multiple layers (the shape of the cells can change). Descriptive terms also identify the shape of the cells: (1) squamous (very flat); (2) cuboidal (cube shaped); (3) columnar (more tall than wide).

The two most appropriate terms are then combined to describe the epithelium. Simple squamous epithelium contains a single layer of flat cells. These cells can be so thin that the nucleus forms a bulge in the surface of the cell. Simple squamous epithelium is found where there is need for exchange across the border. Blood vessels are lined with simple squamous epithelium, allowing for transfer of fluids, nutrients, gases, and wastes. Likewise, the small air spaces of the respiratory system have a similar lining, which provides for the exchange of oxygen and carbon dioxide.

Simple cuboidal epithelium consists of a single layer of cells that are almost square when observed from the side. Simple cuboidal epithelium is found in many glands and in the tubules of the kidney. This type of epithelium is often associated with secretion or absorption. It may, however, be found in tubules that are only responsible for transport.

Simple columnar epithelium contains a single layer of cells that is taller than it is wide. This type of epithelium (often associated with secretion or absorption) can be found in many glands, the stomach, and the intestines. Stratified cuboidal and columnar also exist in certain glands and ducts.

Skin provides the classic example of stratified squamous epithelium, with many layers of very flat cells (Figure 2–3). All these cells originate from a basal layer. As the cells divide, they move outward and un-

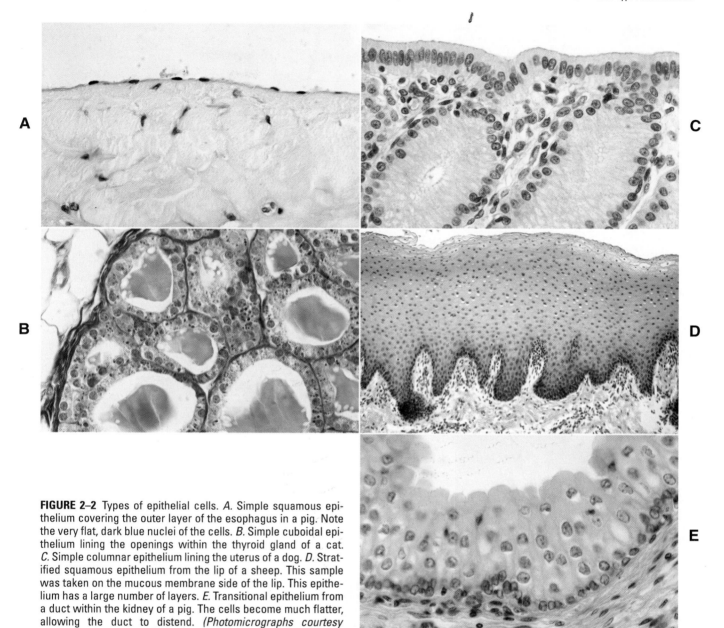

FIGURE 2–2 Types of epithelial cells. *A.* Simple squamous epithelium covering the outer layer of the esophagus in a pig. Note the very flat, dark blue nuclei of the cells. *B.* Simple cuboidal epithelium lining the openings within the thyroid gland of a cat. *C.* Simple columnar epithelium lining the uterus of a dog. *D.* Stratified squamous epithelium from the lip of a sheep. This sample was taken on the mucous membrane side of the lip. This epithelium has a large number of layers. *E.* Transitional epithelium from a duct within the kidney of a pig. The cells become much flatter, allowing the duct to distend. *(Photomicrographs courtesy William J. Bacha, PhD, and Linda M. Bacha, MS, VMD.)*

dergo changes. The most outer layer is covered in dead squamous cells that are constantly being shed. The thickness of the layer depends on the stresses in the area. In human terms, the sole of the foot has a thick layer, adding protection, relative to an area such as the back.

Transitional epithelium remains specific to the urinary tract. This epithelium possesses the ability to stretch. In its relaxed state the epithelium appears to have at least six or seven layers of cells. As the urinary bladder fills, the epithelium stretches, allowing for storage of urine. With the bladder full, the epithelium seems to be only a few cells thick. In addition to its ability to stretch, transitional epithelium prevents exchange of fluid between the urine and the underlying tissue.

The **integument,** or skin, performs many important functions. As generally discussed with epithelium, skin offers a two-way barrier over the body. Skin keeps damaging agents out of the body and fluids and nutrients in the body. Specialization of the skin (hair, fur, and sweat glands) helps animals to maintain a stable body temperature. Moreover, pigment in the skin helps to protect the body from damaging ultraviolet radiation. Clinically this is important in animals with little skin pigment. Gray horses have a much higher

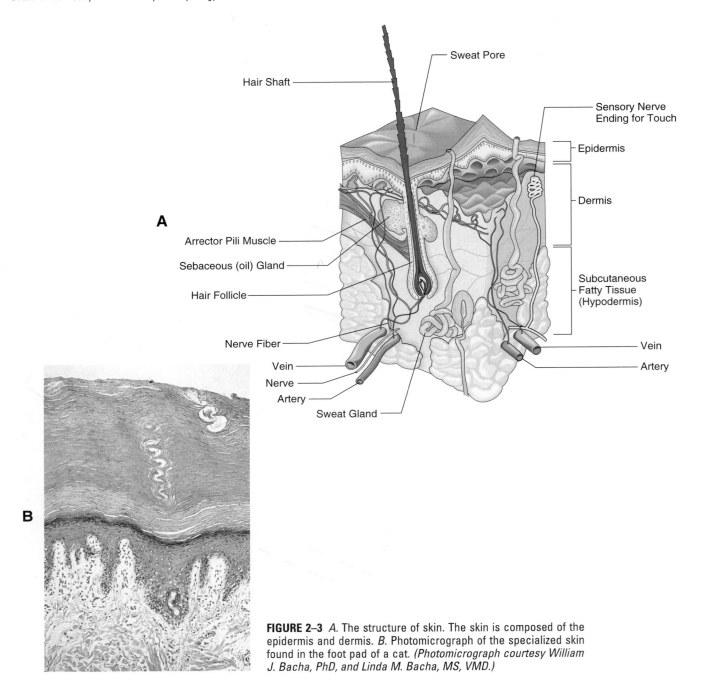

FIGURE 2–3 *A.* The structure of skin. The skin is composed of the epidermis and dermis. *B.* Photomicrograph of the specialized skin found in the foot pad of a cat. *(Photomicrograph courtesy William J. Bacha, PhD, and Linda M. Bacha, MS, VMD.)*

incidence of malignant melanoma than darkly pigmented horses. As these horses age, accumulated sun exposure causes highly malignant melanoma tumors to develop in the skin. Skin, a sensory organ, detects pain, pressure, and temperature, thus adding further means of protection. Furthermore, skin's flexibility allows for movement.

Clinically the skin provides the first visible impression of an animal's health. Many diseases give evidence in the appearance of the skin. (For example, fleas commonly cause hair loss and skin sores.) Many

other parasitic, nutritional, and endocrine diseases affect the appearance of the skin.

The skin consists of the epidermis and the underlying dermis. The epidermis is a stratified squamous epithelium. The cells originate from the basal layer and move outward. The outermost cells of the epidermis are dead and continually shed from the surface. The dermis is a connective tissue layer that contains blood vessels, nerves, and glands. Another layer of connective tissue, called the hypodermis, then supports the skin, epidermis with dermis. (Note the relationship to the term *hy-*

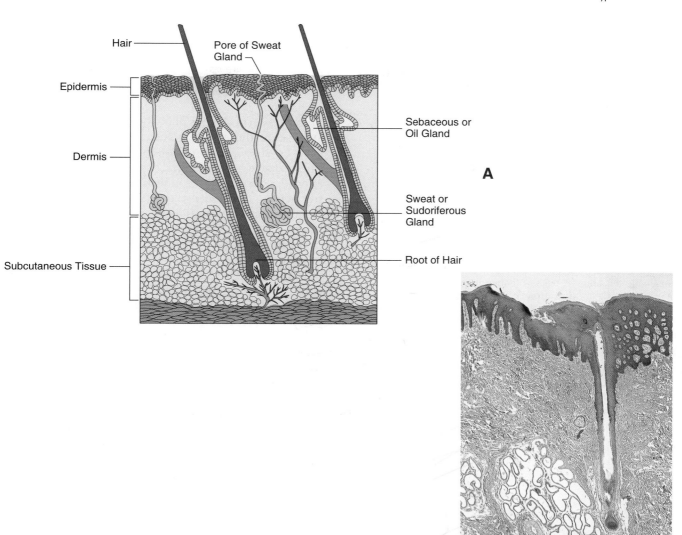

FIGURE 2–4 *A.* Hair follicles and surrounding structures. *B.* Photomicrograph of a region of thin skin in a pig, showing a section through a hair follicle and a nearby sweat gland. *(Photomicrograph courtesy William J. Bacha, PhD, and Linda M. Bacha, MS, VMD.)*

podermic needle, which allows for injection to layers beneath the skin.) Connective tissues are discussed more thoroughly later in this chapter. Connective tissues are mentioned at this point to aid in the understanding of the close connection with the epithelium.

The footpads or digital pads of dogs and cats are extremely thickened and hardened specialized areas of skin. This thickened skin resists physical trauma. A pad of fat found under this skin acts as a cushion for the foot.

Hair serves as another modification of the epidermis. Hair provides insulation, protection, and sensation. The hair originates from a follicle in the dermis (Figure 2–4). The shaft of the hair is made of epithelial cells, much like the outer layers of the epidermis. Growth occurs as cells are attached to the base of the hair. The arrector pili muscle is attached to the connective tissue around the hair follicle. When contracted, this muscle makes the hair stand upright. In cold weather this is used to improve the insulation effect of the hair. Dogs also use this as a signal of aggression and fear. Many dogs make the hair stand on their back when they are being aggressive. It should be recognized as a sign that the aggressive dog should be approached with caution!

Claws and hooves are regions of modified epidermis. **Keratin,** a specialized protein, is deposited in the cells, giving the typical hardness and durability. The claws of dogs and cats surround the last bone of the toe. A rich blood supply surrounds the bone and subsequently nourishes the claw. Hooves have a similar anatomy. A section of the equine foot shows how the last bone of the foot is actually suspended within the hoof (Figure 2–5). The laminar corium has a rich blood supply and connective tissue that supports and nourishes the hoof wall. Growth of the hoof rises from the coronary band, where additional cells are deposited.

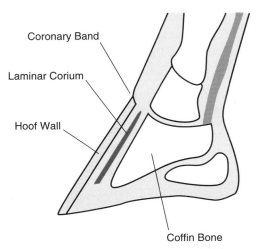

FIGURE 2–5 The structure of the equine foot. The laminar corium suspends the bone of the toe (coffin bone) within the hoof and nourishes the hoof wall. Growth of the hoof originates from the coronary band.

Horns have a similar structure to claws and hooves. The center of the horn, a bone, extends from the skull. Blood in the surrounding tissue feeds the horn. The horn material, called keratinized epithelium, is similar to a hoof. The keratin makes the epithelium hard and durable.

CONNECTIVE TISSUES

OBJECTIVE

■ *Describe the Properties, Locations, Functions, and Varieties of Connective Tissues*

There are several types of connective tissues. These tissues share the common features of specialized cells embedded in large amounts of extracellular material. This extracellular material, the matrix, is produced and deposited by the connective tissue cells. The matrix may be fibrous or smooth in appearance.

Connective tissues have a number of functions (Figure 2–6). As the name implies, connective tissues connect one organ or tissue to another. **Tendons** serve to connect muscles to bones. The matrix in tendons is mainly composed of the protein called collagen. The collagen is arranged in bundles of fibers, which provide great strength. **Ligaments,** another fibrous connective tissue, connect bones to bones. In addition to collagen, ligaments have another protein called elastin. The collagen provides the strength, whereas the elastin provides the ability to stretch and return.

Connective tissues provide both support and protection. Bone and cartilage are two of the supporting connective tissues. (These topics are covered in detail in Chapter 3.) Just as in other connective tissues, bone and cartilage are cellular tissues with a large amount

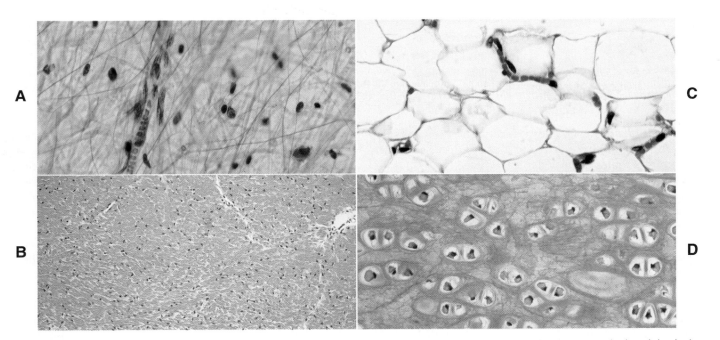

FIGURE 2–6 Types of connective tissues. *A.* Mesentery of a cat. The mesentery is a loosely arranged connective tissue attached to abdominal organs. This shows collagen fibers (pink) and elastic fibers (thin dark fibers). *B.* Densely packed elastic fibers found within a ligament of a sheep. *C.* Adipose tissue. The actual lipid content of the cells is no longer present. Also visible are capillaries with red blood cells. *D.* Cartilage found in a pig, showing the large cartilage cells, some elastic fibers, and the lightly stained matrix. *(Photomicrographs courtesy William J. Bacha, PhD, and Linda M. Bacha, MS, VMD.)*

of matrix. Cartilage lacks blood vessels and is nourished by the surrounding fluid. Cartilage provides a durable contact surface between the two bones of a joint. It also supports the shape of other organs, such as the larynx or voice box.

Bone has similar functions, but the matrix is mineralized. This mineral gives bone its characteristic hardness. Bones give the body shape, allow for movement, and protect internal organs. In contrast to cartilage, bone does have blood vessels. The vessels provide the nourishment for the bone cells.

Connective tissues support organs and hold tissues together. For example, connective tissue holds muscles together and attaches the skin to underlying tissues. This type of tissue quite obviously appeared in the DA surgery discussed at the beginning of the chapter. After opening the skin, the connective tissue layers were encountered next. **Adipose tissue** (fat) was also found deposited in this layer. Adipose tissue, another form of connective tissue, consists of cells filled with lipid.

Blood is considered a special connective tissue. The cells are suspended in a large volume of liquid matrix. The liquid portion of the blood is called plasma. Three formed elements are found in blood. Red blood cells transport oxygen and carbon dioxide. White blood cells fight infection. Platelets aid the blood in clotting. Further details on blood can be found in Chapter 4.

MUSCLE TISSUES

OBJECTIVE

■ *Describe the Properties, Locations, Functions, and Varieties of Muscle Tissues*

Muscles allow mammals to move. Three muscle types exist in mammals: skeletal, smooth, and cardiac. Skeletal muscle attaches to the skeleton and allows for motion. Skeletal muscle is under voluntary control. The animal controls the movement of skeletal muscle with nerve signals from the nervous system. The animal can control which muscles it will move.

Smooth muscle (involuntary muscle) is located in many of the hollow organs of the body, including the gastrointestinal tract, urinary bladder, and blood vessels. The third type, cardiac muscle, is found in the heart. This type is also an involuntary muscle. Involuntary muscle functions without the conscious thought of the animal. These muscles continue to function at all times, including while the animal sleeps.

Skeletal muscle is a striated voluntary muscle (Figure 2–7A). The description of striation comes from its appearance under the microscope. A muscle consists of thousands of muscle fibers, or muscle cells. An entire muscle cell is called a **myofiber**. Myofibers have several nuclei and a large number of mitochondria. Also, myofibers organize in parallel rows. They are separated by connective tissue that includes blood vessels and nerves.

The ability of a muscle fiber to contract comes from a very complicated system. Within the fiber, a highly organized system of myofilaments exists (Figure 2–8). Two proteins, actin and myosin, make up these filaments. These units are organized along the entire length of the cell. During contraction, the actin and myosin filaments slide along each other. The filaments have small bridges between them that bind and release as they slide.

The contraction of a muscle fiber begins with stimulation from a nerve cell. The impulse stimulates the release of calcium that is stored in the endoplasmic reticulum. This flow of calcium ions causes the filaments to slide across each other. Energy is required for this entire process. A large number of mitochondria are present to supply the needed energy. During relaxation, the cell actively transports calcium back into the endoplasmic reticulum. This process also requires energy.

Porcine stress syndrome (PSS) is a genetically transmitted disease in pigs in which calcium is not transported back into the endoplasmic reticulum. Therefore the muscles don't relax normally. Bouts of PSS typically occur when pigs are stressed from heat or transportation. Extreme muscling and leanness can generally identify pigs with a predisposition for this condition.

Rigor mortis (muscle stiffness) occurs after death because there is no supply of energy to pump calcium back into the endoplasmic reticulum. Without the energy, muscles cannot relax and remain stiff.

Obviously calcium plays an essential role in muscle activity. Dairy cows can develop a lack of calcium around calving time. At this point the cow dramatically increases the use of calcium for milk production. So much calcium can be excreted in the milk that calcium levels in blood and muscles become too low. This is technically called **hypocalcemia** but is commonly referred to as milk fever. A cow with milk fever becomes weak and unable to rise. Veterinarians treat milk fever with a calcium solution placed directly into the bloodstream. As the calcium transfers to the muscles, the cow responds. Within a few minutes after treatment, the affected cow often stands.

A nerve cell stimulates more than one muscle fiber. A motor unit is the collection of the nerve cell and all the muscle fibers it stimulates. When fine control of movement is necessary, the number of muscle fibers involved per nerve cell remains small. This is seen in the muscles of the eye and larynx. Muscles engaged for gross movement, such as the upper leg, have a large

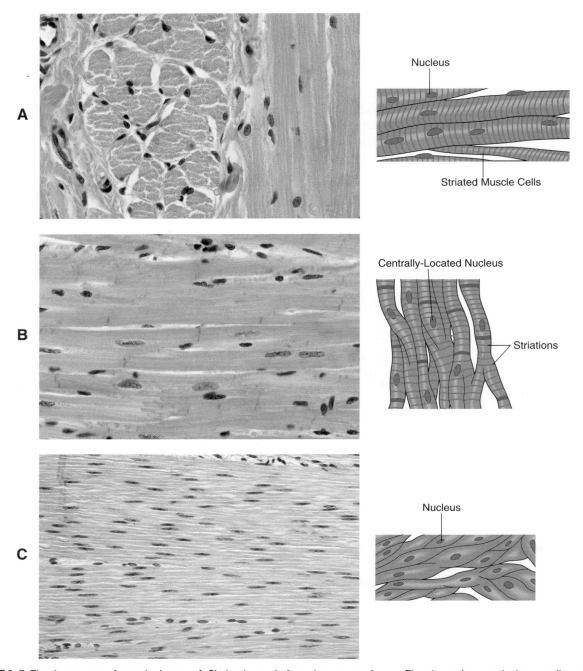

FIGURE 2–7 The three types of muscle tissues. *A.* Skeletal muscle from the tongue of a cat. The photomicrograph shows cells cut along their length and in cross section. *B.* Cardiac muscle from the heart of a goat. Note the intercalated disks that connect the cells. These structures allow the cells to act together with an organized contraction. *C.* Smooth muscle from the colon of a horse. Note that smooth muscle lacks the striations found in cardiac and skeletal muscle. *(Photomicrographs courtesy William J. Bacha, PhD, and Linda M. Bacha, MS, VMD.)*

number of fibers for each nerve cell. When a muscle fiber is stimulated, the contraction is complete. Muscles can partially contract. The more motor units used, the more completely the muscle contracts; conversely, the fewer motor units used, the less the muscle contracts.

The number of muscle fibers in a muscle stays basically constant. Muscles do get larger in response to their usage. This occurs because individual fibers add more myofilaments to become larger. When a muscle is not used, it decreases in size. This can occur from in-

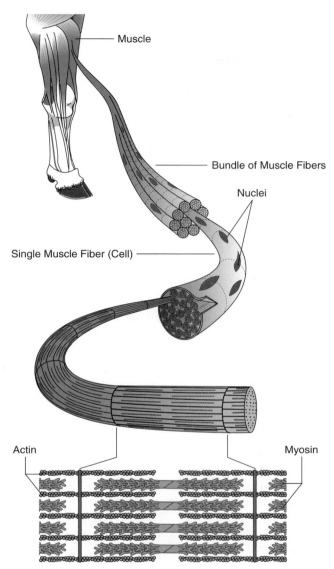

Muscle

Bundle of Muscle Fibers

Nuclei

Single Muscle Fiber (Cell)

Actin

Myosin

FIGURE 2–8 Muscle structure. A muscle is a collection of muscle fibers or cells. Each muscle fiber is a multinucleated cell. Actin and myosin filaments are arranged in an organized manner to allow for contraction.

activity. A house pet that gets little exercise will experience a decrease in muscle size. Having a limb immobilized in a cast also causes the limb muscles to shrink. After a cast is removed, these same muscles increase in size as the limb is exercised. If the nerve supply to a muscle is damaged, the muscle also shrinks. An example of this can occur in draft horses used for pulling. A nerve in front of the shoulder can be damaged from the pulling harness. When this nerve is damaged, the muscles on top of the shoulder blade shrink. This condition is called **sweeny**.

Cardiac muscle is also striated in appearance but is involuntary in action (Figure 2–7B). The appearance

of the cardiac myofilaments is very similar to that of skeletal muscle. Moreover, the mechanism that allows for cardiac muscle contraction is identical to skeletal muscle. The myofibers (muscle cells) are branched in cardiac muscle and contain even more mitochondria than skeletal muscle.

Cardiac muscle cells have the unique ability to initiate their own contraction. No nerve cell stimulation is necessary for a contraction to begin. Specialized pacemaker cells are responsible for establishing the rate of contraction. However, nerve cells are present to influence the rate of contraction. The autonomic nervous system can increase or decrease the rate. As long as oxygen and glucose are provided to the cardiac muscle cells, the heart continues to beat.

For the heart to effectively serve as a pump, the cells must contract in a very organized manner. The cardiac muscle cells have a specialized connection between them. When one cell contracts, the electrical signal immediately passes to the next cell through this junction. This allows for a chamber of the heart to function as one unit.

Smooth muscle gets its name because it lacks the striated appearance of skeletal muscle (Figure 2–7c). Smooth muscle contains actin and myosin filaments, but not in the same arrangement as skeletal muscle. Each myofiber is a spindle-shaped cell, tapered at each end. In addition, each cell has one nucleus.

Smooth muscle is arranged in sheets around hollow openings such as those in the gastrointestinal tract. Contraction of this sheet of muscle may make the opening smaller. In blood vessels this occurrence is called constriction. In an organ such as the esophagus, the contraction actually aids in propelling food toward the stomach. This organized contraction that propels the food is called peristalsis.

Smooth muscle contracts much more slowly than skeletal muscle. Moreover, the cells are able to maintain contraction for prolonged periods without tiring. By keeping this muscle tone, the opening within an organ can be kept at the same diameter. Blood vessels may also be maintained at the same diameter for long periods. The autonomic nervous system controls the action of smooth muscle.

NERVE TISSUES

OBJECTIVE

■ *Describe the Properties, Locations, Functions, and Varieties of Nerve Tissues*

Nerves allow communication among areas of the body. Nerve tissue is found in the brain and spinal cord. Together the brain and spinal cord are called the

central nervous system. In addition, nerves extend from these areas to other locations. The **peripheral nervous system** includes all the nerves outside the brain and spinal cord.

Nerve tissue contains cells called **neurons** (Figure 2–9). Neurons can be very large cells. The body of the neuron houses the nucleus and many other organelles. The axon, a hairlike extension from the cell body, carries the nerve impulse. The axon may end on other neurons or other tissues, such as muscles. The axons from many neurons bundle together to form a nerve. The axon can be very long. In a horse, for example, some axons may be more than 2 meters long. The neuron has other extensions called dendrites. An axon often ends on the dendrite of another neuron. This site of connection is called a synapse. When stimulated, the dendrite begins the nerve impulse.

The nerve impulse occurs as a flow of ions passes through the cell membrane. In a resting nerve cell, sodium ions are actively transported into the extracellular fluid. At the same time, potassium is pumped into the cytoplasm. Once stimulated, the ions flow rapidly across the membrane. Stimulation at one point then moves down the axon in a rapid progression. Afterward the neuron prepares for the next impulse. Using microelectrodes, the nerve impulse can be measured as an electrical event.

There are three basic types of neurons.

1. Sensory neurons: These neurons have receptors that stimulate in response to a change in the animals' environment. Once stimulated, the nerve signal transmits through the neuron back to the central nervous system. There are many types of receptors in the body. Table 2–1 lists the common receptor types. The signal from these neurons is then transmitted to a motor neuron or an interneuron. These neurons give feedback on changes occurring outside and within the animal.

2. Interneurons: These neurons are found within the central nervous system. A sensory neuron or another interneuron stimulates them. There are a tremendous number of connections within the brain and spinal cord. The connections between all these neurons provide the pathways that allow the central nervous system to control the animal's activities.

3. Motor neurons: These neurons begin in the central nervous system and extend to a muscle or gland. When stimulated by the motor neuron, an action occurs. For example, the muscle contracts or the gland releases its secretion. A sensory neuron or an interneuron stimulates the motor neuron.

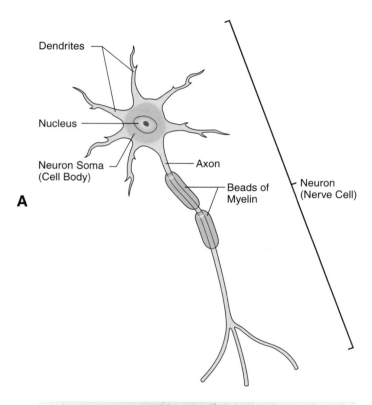

Dendrites

Nucleus

Neuron Soma (Cell Body)

Axon

Beads of Myelin

Neuron (Nerve Cell)

A

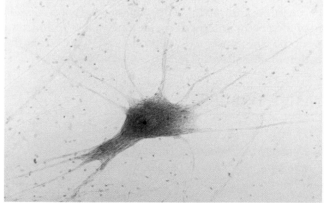

B

FIGURE 2–9 *A.* The structures of a neuron. *B.* Photomicrograph of a motor neuron from the spinal cord of a cow. *(Photomicrograph courtesy William J. Bacha, PhD, and Linda M. Bacha, MS, VMD.)*

TABLE 2–1 Receptor Types
Light
Heat
Touch/pressure
Hearing
Balance
Taste
Smell
Internal chemical receptor (results in feeling thirst)

CLINICAL PRACTICE

OBJECTIVE

■ *Link Knowledge of Tissues to Clinical Practice*

Certain infectious diseases may infect specific tissue types. One such disease, foot and mouth disease, selectively attacks epithelial tissue. This disease attracted international media attention when an outbreak occurred in the United Kingdom. This highly infectious disease spreads very rapidly. It commonly infects cattle, sheep, goats, and swine. FMD was last diagnosed in the United States in 1929 but has been present in other countries ever since. The recent outbreak in Europe raised the concern that FMD might reoccur in the United States.

Although it is usually not fatal, FMD, a viral infection, causes very serious signs in affected animals. The epithelium in the mouth and tongue develops blisters. These blisters cause great pain, making the animal drool and quit eating. The epithelium around the hooves is also commonly involved, making movement painful as well. Other epithelial linings, such as those deeper in the intestinal tract, can also be infected.

This disease spreads rapidly by contact with infected animals. Humans can also transport the virus (for instance, on clothing and shoes or in the respiratory tract), which makes its spread more difficult to control. Fortunately for my client, her friends from Germany were not likely to be a source of introducing the disease into her farm. This tendency for rapid spread is the major concern of potential FMD reintroduction into the United States.

In the United Kingdom, infected and exposed animals had to be destroyed in an attempt to stop the disease. No treatment is available for the disease, although most animals eventually recover. Vaccines are available to control the spread of the disease but are not currently being used in the United States. Because the United States has been free of FMD for such a long time, no animals would show positive on a blood test for this disease. If vaccination were started, it would be more difficult to use screening tests to monitor the incidence of the disease because these tests may show positive in vaccinated animals. Distinguishing between the natural infection and a vaccinated animal can be difficult. If FMD were to occur in the United States, vaccination might be used to stop the spread. Note that FMD is not related to hand, foot, and mouth disease, which commonly affects small children and is known to spread rapidly in preschools and day care centers.

A common problem encountered in companion animal medicine occurs when a larger dog attacks a small dog. In Remmy's case the groundhog happened to be the culprit. In these situations the skin receives lacerations, which need repair. In addition to the lacerations, the skin may pull loose from the underlying connective tissue. The preceding discussion of tissues emphasized the significance of the connective tissue holding the skin to the underlying muscle. In this type of trauma, this bond has been destroyed.

If the torn skin is pulled back together without further repair of the connective tissue, dead space is created. Dead space, a pocket under the skin, allows tissue fluid to accumulate. This buildup of fluid often prevents complete healing. To avoid this problem, the dead space may be closed with an additional layer of sutures. Placing a drain into the dead space can also aid in repair. This drain provides an opening that allows any fluid to drain. Both these methods keep the skin in contact with the connective tissue. Over time the two tissues heal together.

Horses may be affected by a disease called **tying-up** or **Monday morning disease**. This condition often occurs on a Monday after a weekend of rest, when the horse is consuming a full diet. As the horse begins working or exercising, it develops severe cramping. The cramps are often so severe that damage occurs to the muscle tissue. Products from the muscle leak from the cells and eventually are cleared from the bloodstream by the kidneys. These breakdown products may damage the kidneys.

The cause of this disease process is very complex. The muscle tissue requires a great deal of energy and oxygen for the work involved. High metabolism also produces waste products that must be cleared from the cells. With the right combination of rest, diet, and exercise, the normal balance is altered. The buildup of waste products results in damage to the muscle and results in tying-up syndrome.

Nerves often provide signals to several muscles in one area. Damage to a nerve can be detected based on the group of observable signs. **Horner's syndrome** results from damage to a nerve in the autonomic nervous system. This nerve comes through the neck and the base of the skull to control several eye functions. When damage occurs, several signs are observed, including the following: (1) the pupil is constricted; (2) the upper eyelid droops; (3) the third eyelid protrudes; (4) the eye is sunken in the socket. When these signs are present, the animal has Horner's syndrome. These signs help to identify what nerve is damaged. Knowing this, the veterinarian can direct attention to diagnosing the underlying cause.

SUMMARY

In Chapter 2, four tissue types were examined: epithelial, connective, muscle, and nerve. Epithelial tissues line the body's surfaces; openings, including the intestinal, reproductive, and urinary tracts; and tubes, such as blood vessels and the heart. Connective tissues vary in type but share the common feature of specialized cells embedded in vast amounts of extracellular material. Moreover, three types of muscle tissue exist: skeletal, cardiac, and smooth. Lastly, nerve tissues provide for communication within the body. These four tissue types allow the body to function efficiently and effectively.

REVIEW QUESTIONS

1. Define any 10 of the following terms:
   tissue
   organs
   displaced abomasum
   foot and mouth disease
   epithelial tissues
   integument
   keratin
   tendons
   ligaments
   adipose tissue
   myofiber
   porcine stress syndrome
   rigor mortis
   hypocalcemia
   sweeny
   central nervous system
   peripheral nervous system
   neurons
   foot and mouth disease
   tying-up or Monday morning disease
   Horner's syndrome

2. True or False: Hair is epidermal tissue.

3. True or False: Kidney damage may occur in Monday morning disease.

4. Which of the four stomachs of a cow becomes displaced when a twisted stomach occurs?

5. What type of tissue lines the body's surface and openings?

6. What type of epithelial tissue lines the urinary tract?

7. What type of tissue is under attack in foot and mouth disease?

8. Name the hairlike extension from the nerve cell body that carries the nerve impulse.

9. After death, the body lacks energy to pump calcium back into the endoplasmic reticulum. Consequently, the body stiffens. Name this condition.

10. Why do light-colored horses have a higher incidence of melanoma than dark-colored horses?

11. Describe the shape of squamous cells.

12. Differentiate between tendons and ligaments.

13. List the three muscle types.

14. List two involuntary muscle types.

15. List three types of neurons.

ACTIVITIES

Materials needed for completion of activities:
   light microscope
   slides and cover slips
   toothpicks
   iodine
   cardiac, smooth, and skeletal muscle samples
   dissecting kits
   rulers
   chicken legs
   paper plates

1. Perform a simple cheek cell scrape. After securing a light microscope, toothpicks, slides, cover slips, and iodine (or another suitable stain), prepare slides in the following manner: Scrape toothpick across the inside of the cheek. Place the scraped cells onto a slide. Place a drop of iodine on the cells, and cover with a cover slip. View under low and high power. Try to identify the epithelial cells.

2. Dissect a chicken drumstick and attempt to identify and classify connective tissue as tendons or ligaments.

3. From a local butcher shop, secure cardiac muscle (portion of the heart), smooth muscle (such as from the wall of the gastrointestinal tract or the urinary bladder), and skeletal muscle samples. Cut each sample. Compare and contrast muscle tissue types.

4. Receptors are not evenly spaced throughout the body. The pressure receptors in the skin are more common in sensitive areas, such as a finger, than on the back. Tape a toothpick to the end of a ruler, with the end sticking below the ruler. Hold a second toothpick close to the first one (Figure 2–10). Begin with the two toothpicks next to each other. Touch the toothpicks to the skin of another student. The toothpicks should contact the skin at the same time. The student being touched should not be able to see the contact. Move the second toothpick farther apart until the student being touched can tell that there are two contact points. Repeat this experiment at different sites on the body, such as the finger and back. Is there a difference in the distance detected between the two areas? If so, why?

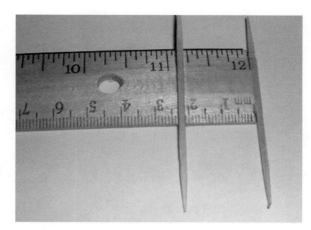

FIGURE 2–10

5. Research the current status of foot and mouth disease. Possible sources for information include the U.S. Department of Agriculture, state departments of agriculture, land grant university departments of animal or veterinary science, or the local cooperative extension office.

3

Musculoskeletal System

OBJECTIVES

Upon completion of this chapter, you should be able to:

- *Describe the functions of the musculoskeletal system.*
- *Detail the structure of bone.*
- *Name joint types and their accompanying role in movement.*
- *List the two major sections of the skeleton, name the corresponding bones, and compare species differentiation.*
- *Explain how bone grows and remodels.*
- *Relate bone and muscle groups to movement.*
- *Connect the text materials pertaining to the musculoskeletal system to clinical practice.*

KEY TERMS

herd check
radiograph
orthopedic surgeon
axial skeleton
appendicular skeleton
intervertebral disk
 disease

high-rise syndrome
cranial drawer sign
ossification
subluxate
x-ray
radiology

simple fracture
comminuted fracture
compound (open)
 fracture
intramedullary pin

hip dysplasia
degenerative joint
 disease
joint ill

INTRODUCTION

The skeleton gives mammals shape and support. Combining bones and muscles allows movement. Bones are active tissues that adapt to changes within the animal. The skeleton, although very hard, allows the animal to adapt and grow.

A DAY IN THE LIFE

Routine Days Aren't Always Routine...

Today seemed predictable as far as my days go. **Herd checks** (reproductive exams and routine health maintenance work) dominated the day. Many of our dairy clients ask a veterinarian in our practice to visit once or twice a month to perform a herd check. When called for a herd check, I examine cows for pregnancy (veterinarians can detect pregnancy in cattle bred for 30 days or more) and also make sure they have recovered from calving. Keep in mind that dairy cows experience a lactation peak shortly after calving. Therefore farmers attempt to have their cows calve every 12 to 13 months to maximize milk production. I help them accomplish this goal.

I heard a familiar bark when I pulled into my second call of the day (another herd check). Missy, a 120-pound Saint Bernard, let everyone know that I had arrived (Figure 3–1). I first met Missy in October 2000 after she was injured while working on the farm. Her owner had driven his pickup to cut firewood. Dutiful Missy rode along in the bed of the truck. When the truck stopped, Missy jumped off, just like she had done many times before. Unfortunately, this time Missy's right hind leg became caught and she fell.

Missy was crying in pain and not able to bear weight on her injured leg when the farmer brought her to our office. My associate, Dr. Deppen, examined her. It was obvious that the leg was very swollen and likely broken. Dr. Deppen took a **radiograph** of Missy's tibia, a bone in the lower leg.

It was apparent from the radiograph that the tibia had been broken into several pieces (Figure 3–2). We offered to refer Missy to an **orthopedic surgeon**, a veterinarian that specializes in surgery of the bones. Knowing a referral of this type can be quite expensive, the owners wondered if we could repair the bone.

Dr. Deppen and I discussed the options. We both felt that, considering the severity of the fracture, a cast or splint was not likely to be successful. Conversely, we could attempt to perform the needed surgery. I called the farmer to offer the choices at hand. I first told him that I am not an orthopedic specialist and that the fracture was quite severe. Then I explained that I could attempt the surgery. He agreed to allow me to perform the surgery, knowing that the operation might not be successful. I obviously needed a thorough knowledge of bones before I could repair them.

FIGURE 3–1 Missy enjoying life at home. *(Photograph courtesy Fred Benner.)*

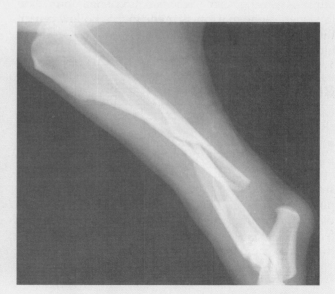

FIGURE 3–2 Radiograph of Missy's fractured leg.

MUSCULOSKELETAL SYSTEM FUNCTIONS

OBJECTIVE

■ *Describe the Functions of the Musculoskeletal System*

Bones furnish four basic functions: structure, protection, minerals reserve, and blood cell production. The most visible function of bones is structure. The collection of bones in the animal forms the skeleton. This provides the framework that defines an animal's shape and size. Differences in both size and shape are very obvious in veterinary medicine. The skull provides a clear example of this variation. When seeing only the bones of the skull, it is easy to distinguish the skull of a cat from that of a horse.

The strength of bones also protects more fragile tissues. The rib cage provides protection for the heart and lungs, whereas the skull protects the delicate brain. Bone acts as a reservoir for calcium and phosphorus. In times of need, the minerals are moved from the bone and sent into the bloodstream. Excess minerals can be stored in the bone. Calcium plays an essential role in muscle contraction and enzyme activity. Phosphorus is necessary for energy metabolism within the cell. Bone, in response to several hormones, maintains a tight regulation on the blood level of these minerals.

The long bones are present in the legs (and arms in humans). The femur and humerus are classified as long bones. They have a dense outer shell and a hollow shaft. Bone marrow is made in this hollow center, the medullary cavity. Bone marrow in turn produces blood cells.

BONE STRUCTURE

OBJECTIVE

■ *Detail the Structure of Bone*

Splitting a long bone along its length shows the typical structure of bone (Figure 3–3). The outer shell is composed of dense or compact bone. The more forces placed on a bone, the thicker this layer will be. In the femur this compact bone is thickest in the middle of the shaft, where greatest strain occurs.

Within compact bone lies a more loosely arranged bone, called spongy or cancellous bone. Spongy bone is found within the long bones but not inside the flat bones of the skull or pelvis. It only fills the ends of these bones. Spongy bone is made up of tiny spicules and plates of bone. The spicules look random but are actually arranged to maximize strength. The medullary cavity is located in the hollow center of the shaft. The bone marrow lies within the medullary cavity and the spaces of the spongy bone. As mentioned earlier, bone marrow produces blood cells.

Bones are covered with a thin connective tissue called the periosteum. The periosteum blends into tendons and ligaments, binding them to the bone. The portion of bone covered with cartilage is not covered by periosteum. The ends of bones within joints have this cartilage protection. The open spaces within bone are covered with a similar connective tissue, the endosteum. Both the periosteum and endosteum provide cells necessary for the repair of damage.

A dried bone is composed of about 70% inorganic minerals and 30% organic components. The inorganic minerals have a high level of calcium and phosphorus. This is found as crystals of hydroxyapatite $(3Ca_3(PO_4)_2 \cdot Ca(OH)_2)$. The organic portion contains collagen fibers and cells. The fibers provide a framework on which the hydroxyapatite crystals can be deposited. Whereas organic fibers give the bone a small amount of elasticity, minerals give bone its typical hardness and strength. The collagen fibers and the hydroxyapatite crystals make up the matrix that surrounds the cells.

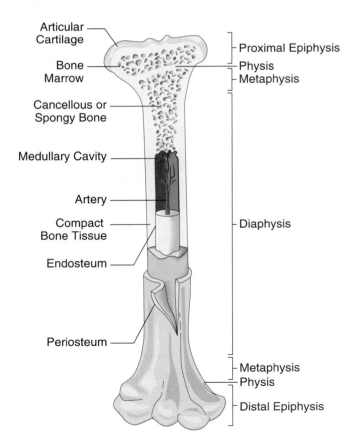

FIGURE 3–3 Bone structure.

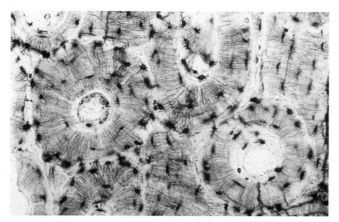

FIGURE 3–4 Microscopic structure of bone, showing osteons with a central canal. The bone matrix is deposited in a circular arrangement. Darkly stained osteocytes are visible within the matrix. *(Photomicrograph courtesy William J. Bacha, PhD, and Linda M. Bacha, MS, VMD.)*

There are three types of bone cells. Osteoblasts lay down the collagen matrix. These osteoblasts become encased in matrix, developing into osteocytes. The osteocytes are responsible for maintaining the bone matrix. Osteoclasts are large multinucleated cells that release the minerals from bone. Bone is a living tissue that is always being remodeled in response to physical forces on the body and the body's need for calcium.

Osteoporosis is a condition in which the bones lose their normal density. Several disease conditions can result in this decrease of bone mass. The problem can also occur in animals when a limb is not used for long periods (for example, extended time in a cast). Be-

cause the bone is not subjected to physical forces, new mineral is not deposited. Osteoclasts continue to release the minerals into the bloodstream. The bones can become so thin that they can break under normal usage. Osteoporosis that occurs from disuse is reversible once the animal begins to use the leg.

Bone is composed of a collection of microscopic units called osteons. In the center of the osteon there is a canal (Figure 3–4). Blood vessels, nerves, and lymphatics run through this canal. Bone is laid down in circles around this channel. Within these layers are the osteocytes that maintain the bone matrix. Many osteons are joined to form the layers of bone.

JOINT TYPES AND MOVEMENTS

OBJECTIVE

■ *Name Joint Types and their Accompanying Role in Movement*

Joints form where other tissues join two bones. Generally joints are classified by the amount or type of movement allowed (Figure 3–5).

Fibrous joints: The bones in a fibrous joint are joined together with a dense connective tissue. These are also called fixed joints because little movement is possible. This joint can be found in the skull, where it is called a suture.

Cartilage joints: As the name describes, the bones in this joint type are connected with cartilage. The growth plate of young animals serves as an example of this joint. The growth plate exists within a

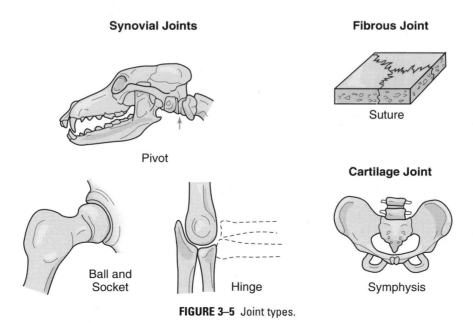

Synovial Joints

Pivot

Ball and Socket

Hinge

Fibrous Joint

Suture

Cartilage Joint

Symphysis

FIGURE 3–5 Joint types.

bone and allows for rapid growth. The cartilage layer within the growth plate is eventually replaced by bone as the animal reaches adulthood. A symphysis is another type of cartilage joint. Examples of symphysis are found between the halves of the pelvis or the lower jaw.

Synovial joints: Synovial joints are the true moveable joints. A dense layer of bone at the joint is covered with a layer of cartilage. This articular cartilage covers the contact surfaces of the two bones. The joint is enclosed with a joint capsule. The outer layer of the joint capsule is strong connective tissue. This is lined with a synovial membrane, which produces synovial fluid. Synovial fluid provides lubrication to the joint and carries nutrients to the cartilage. Ligaments are also present to provide strength to the joint. Some joints may have a meniscus, which is a hard cartilage pad. The

meniscus acts as a cushion between the bone ends. The knee joint has ligaments both inside and outside the joint capsule. Menisci are also present as cushions for the wide range of motion in this joint.

Several terms are used to describe the motion within a joint. The same description can be used for the muscle group that causes that motion. *Flexion* occurs when the angle between the two bones gets smaller. The opposite motion, *extension,* occurs as the angle between the bones increases. *Abduction* occurs when a part is moved away from the body and *adduction* as the part is moved closer.

The type of motion allowed also describes specific types of joints. The simplest, the hinge joint, allows movement in one axis. The classic example of a hinge joint is the elbow. The arm can be flexed and extended at the elbow in only one plane. A pivot joint allows

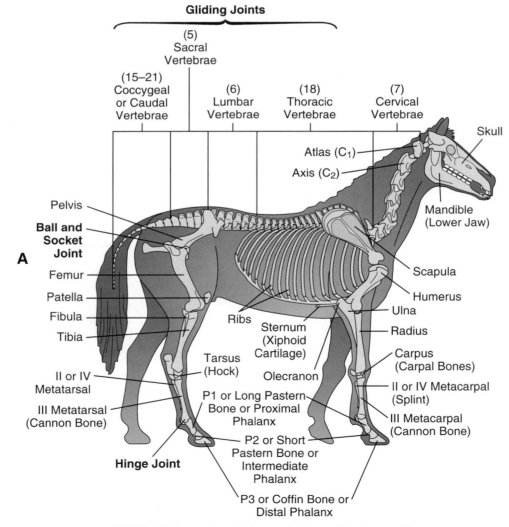

Gliding Joints

(5)
Sacral
Vertebrae

(15–21)
Coccygeal
or Caudal
Vertebrae

(6)
Lumbar
Vertebrae

(18)
Thoracic
Vertebrae

(7)
Cervical
Vertebrae

Skull

Atlas (C₁)

Axis (C₂)

Mandible
(Lower Jaw)

Pelvis

**Ball and
Socket
Joint**

A

Femur

Patella

Fibula

Tibia

Ribs

Sternum
(Xiphoid
Cartilage)

Scapula

Humerus

Ulna

Radius

Carpus
(Carpal Bones)

II or IV
Metatarsal

III Metatarsal
(Cannon Bone)

Tarsus
(Hock)

Olecranon

P1 or Long Pastern
Bone or Proximal
Phalanx

II or IV Metacarpal
(Splint)

III Metacarpal
(Cannon Bone)

Hinge Joint

P2 or Short
Pastern Bone or
Intermediate
Phalanx

P3 or Coffin Bone or
Distal Phalanx

FIGURE 3–6 Comparison skeletons of a horse *(A)* and a cat *(B).*

rotation around a point. The pivot joint between the first and second vertebrae allows the head to rotate. The wrist is an example of an ellipsoid joint, which allows motion not only in hinge fashion but also in rotation. The ball and socket joint, such as the hip or shoulder, allows motion in any direction.

AXIAL AND APPENDICULAR SKELETONS

OBJECTIVE

■ *List the Two Major Sections of the Skeleton, Name the Corresponding Bones, and Compare Species Differentiation*

The skeleton can be divided into two major sections (Figure 3–6). The **axial skeleton** contains the skull, vertebrae, ribs, and sternum. The **appendicular skeleton** consists of the bones of the limbs. The total number of bones varies between species and even within a species. For example, dogs may or may not be born without the first digit (dewclaw) on their legs. Moreover, some dogs have very short tails with small numbers of vertebrae, whereas others have long tails and more vertebrae. A "typical" dog has about 320 bones (134 in the axial skeleton, 186 in the ap-

pendicular skeleton). On the other hand, horses have fewer bones in the distal limbs than dogs and total only 205 bones.

Each bone is given a distinct name. In addition, the parts of the bone are also named. These sites often serve as points of muscle attachments, and naming allows for the communication of details about bones. For example, before operating on Missy's leg, a review of the surgical procedure was necessary. This protocol mentioned several points of the tibia that would act as landmarks during the surgery.

Found in the axial skeleton, the skull (cranium) combines numerous flat bones (approximately 50 in the dog, Figure 3–7). The skull performs many functions, most notably protection for the brain and other organs of special senses (sight, hearing, taste, and smell). A moveable mandible (lower jaw) allows animals to obtain and chew food. The bones of the jaws hold the teeth, which vary tremendously among species due to diet adaptations (Figure 3–8). The shape of the skull also differs among species and within species. For instance, consider the long narrow nose of the collie and the short broad nose of the pug. Regardless of appearance, the basic skull anatomy remains quite similar. The size of the bones in the skull accounts for differentiation.

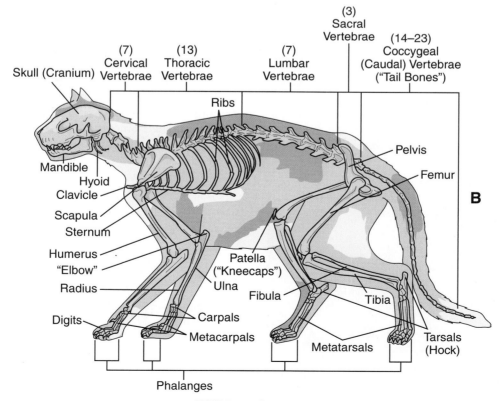

FIGURE 3–6 *Continued*

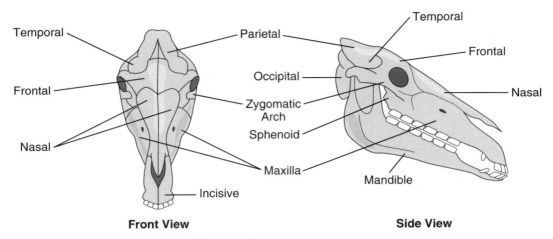

FIGURE 3–7 Skull structure of a horse.

The vertebral column extends the length of the body from the skull. The vertebral column protects the spinal cord and allows for movement. Force from the hind limbs transfers through the spinal column to propel an animal.

The body of the vertebra is covered with a bony arch (Figure 3–9). The series of arches, with all the vertebrae, forms a tube that houses the spinal cord. The spinous process and transverse processes are sites of attachment for tendons and ligaments. Between the bodies of the vertebrae (except C1 to C2 and within the sacrum) lie intervertebral disks. The disks have strong, fibrous outer rings and soft, spongy centers. Disks provide cushioning between the vertebrae bones.

The intervertebral disk has clinical significance in many dogs. Although **intervertebral disk disease** can occur in any breed, certain breeds, such as the dachshund, Pekingese, and cocker spaniel, have higher incidences of the disease. In this condition the center of the disk becomes less spongy. Pressure between the vertebrae causes the center to rupture through the fibrous

outer layer. The bulging disk can press on nerves or the spinal cord. Dogs with this condition may experience severe pain and can be paralyzed. Depending on the severity, the afflicted dogs may be treated medically or surgically.

The vertebral column is broken down into anatomic divisions. At the head end, the cervical vertebrae make up the neck (Figure 3–10). Mammals, from cats to horses to giraffes, have seven cervical vertebrae. The first cervical vertebra is also called the atlas. This vertebra possesses a unique shape and allows for up and down motion of the head, as in nodding to signify yes. The axis, the second vertebra, permits the head to rotate back and forth, as seen in shaking the head to indicate no. The remaining five cervical vertebrae are all similar in size and shape.

The thoracic vertebrae are found in the next division of the spinal column. These vertebrae have attached ribs (Figure 3–11). The bones of the thorax protect the heart and lungs. This framework also allows for the expansion of the lungs that occurs during breathing. Dogs have

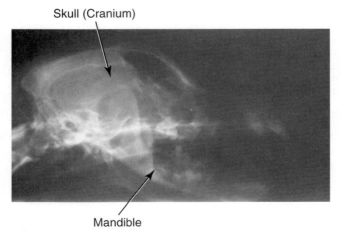

FIGURE 3–8 Radiograph of the skull of a dog.

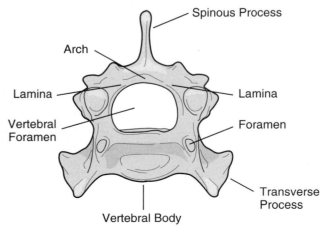

FIGURE 3–9 The structure of a vertebra.

Cervical Vertebrae

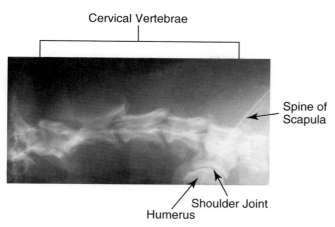

Spine of Scapula

Shoulder Joint

Humerus

FIGURE 3–10 Radiograph of the cervical spine of a dog.

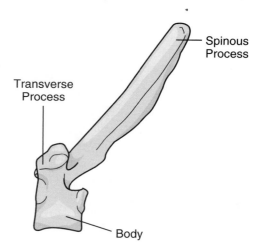

Spinous Process

Transverse Process

Body

Lumbar Vertebra

13 ribs. The first nine join to the sternum with cartilage. The sternum, a group of bones, forms the floor of the thorax. In older animals the cartilage joining the ribs to the sternum can be replaced with bone.

The lumbar vertebrae, located in the lower back, lie between the thoracic vertebrae and the pelvis (Figure 3–12). This area flexes and extends as animals walk and run. In addition, these vertebrae support the organs in the abdomen. The muscles of the abdomen attach to these vertebrae, forming a sling that supports internal organs.

The sacrum, a group of three sacral vertebrae, fuses to support the pelvis (Figure 3–13). The sacrum then joins with the pelvis, allowing the hind limbs to support the weight of the body. This connection can be damaged. The pelvis may split away from the sacrum when dogs and cats are hit by cars (HBC). During this type of accident, fracture of the pelvis itself is also common. Very painful lameness often results from a split pelvis or pelvic fracture. Many of these fractures heal if the animal's activities are restricted. In severe cases, surgeries may be required.

The final group of vertebrae is called caudal or coccygeal. These small vertebrae comprise the tail. The numbers of vertebrae vary among species and within a species. The typical dog has 20 caudal vertebrae, but this can range from 6 to 23.

The appendicular skeleton includes the bones of the forelimbs and hind limbs. Study of this part of the skeleton provides a clear examination of comparative anatomy. Although the same anatomic terms are used for all mammals, great differences exist in the numbers and sizes of bones in the mammalian appendicular skeleton. For instance, a dog has four or five toes, whereas a horse has only one.

The forelimb, or thoracic limb, does not have a bony connection to the axial skeleton. The scapula, or shoulder blade, lays flat against the rib cage (Figure 3–14). The scapula connects to the axial skeleton with a group of muscles. This attachment allows for

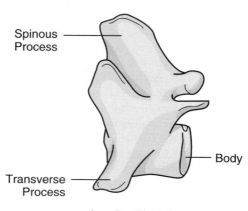

Spinous Process

Transverse Process

Body

Lumbar Vertebra

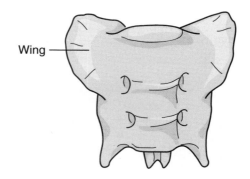

Wing

Sacrum, Ventral View

FIGURE 3–11 Comparison of the structure of thoracic, lumbar, and sacral vertebrae.

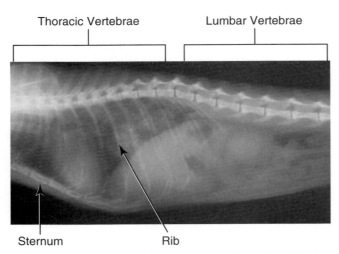

FIGURE 3–12 Radiograph of a cat showing the thoracic and lumbar spine. Ribs and sternum are also visible.

the scapula to move over the rib cage. This rotation ranges as high as 25 degrees in animals such as cats while running. This flexibility is also useful in cats as they land after a jump. As the cat falls, it extends its front legs fully, both at the scapula and the elbow. As the front feet hit the ground, the elbow flexes and the scapula rotates. The cat makes this very coordinated act look quite graceful. Clinically this is of significance when cats fall from extreme distances. In large cities this happens often as cats tumble from balconies or windows of tall buildings. In **high-rise syndrome** the falling cat rarely breaks a leg; however, it will often break its lower jaw. The high speed of the falling cat forces the jaw to contact the ground.

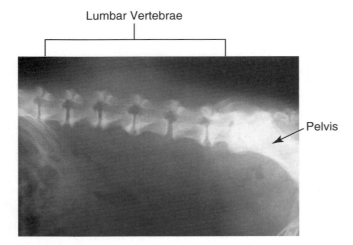

FIGURE 3–13 Radiograph of the lumbar spine of a dog. A portion of the pelvis is also visible. This dog is showing an age-related change called spondylosis. In spondylosis, bone spurs are formed that can eventually bridge between vertebrae.

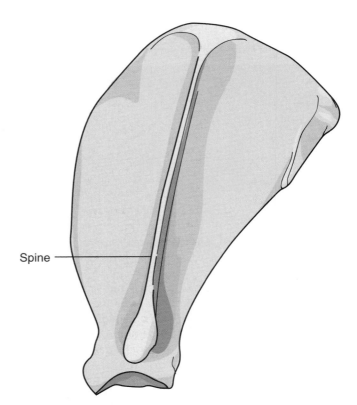

FIGURE 3–14 A scapula.

The humerus is the upper bone of the forelimb. The scapula joins the humerus through a shallow ball and socket joint that allows for a wide range of motion (Figure 3–15). The humerus then joins at the elbow with the radius and ulna. The ulna runs to the point of the elbow, where a groove accepts the end of the humerus. The radius closely attaches to the ulna and forms the remainder of the elbow joint (hinge joint that permits motion in only one plane). The forearm can be rotated, but this occurs between the radius and ulna, not at the elbow joint.

The radius and ulna run to the level of the carpus (Figure 3–16). The carpus in animals corresponds to the wrist in humans. The carpus, a group of bones, arranges in two rows. Table 3–1 lists the number of carpal bones found in several species. The carpal bones join to the long metacarpal bones. There special differences become very dramatic. Dogs and cats have four long metacarpal bones and one much smaller. The smaller bone associates with the first digit, called the dewclaw (Figure 3–17). As previously stated, horses have only one major metacarpal bone, which corresponds to the third one in other species. The horse has two smaller metacarpal bones, also called the splint bones. Ruminants such as cattle and sheep also have one very large metacarpal bone. As the ruminant embryo develops, the third and fourth metacarpal bones fuse.

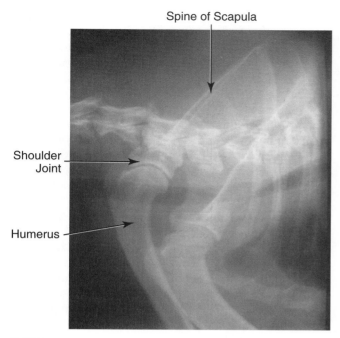

FIGURE 3–15 Radiograph of the scapula and shoulder joint of a dog.

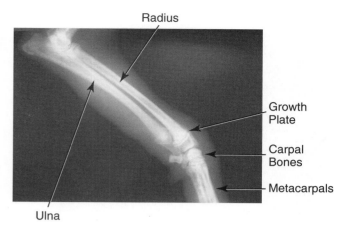

FIGURE 3–16 Radiograph of the radius and ulna of a young dog. Note the growth plates visible in the radius and ulna.

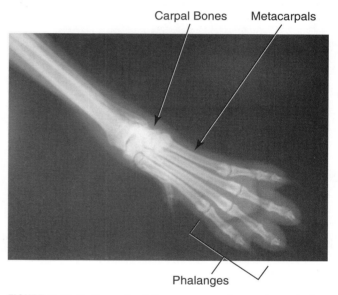

FIGURE 3–17 Radiograph of the carpus, metacarpals, and phalanges of a dog.

The toes contain three bones, the phalanges. The number of toes corresponds to the number of metacarpal bones. The singular form of the word *phalanges* is *phalanx.* The last phalanx is covered by the nail or hoof (Figure 3–18).

In contrast to the thoracic limb, the pelvic limb does form a bony connection with the spine. As discussed earlier, the fused vertebrae of the sacrum join with the pelvis (Figure 3–19). The pelvis is made of two halves. Each half divides into regions named the ilium, ischium, and pubis. The ilium joins with the sacrum at the iliosacral joint. The halves of the pelvis join in the middle, forming the pubic symphysis, a basically immovable cartilage joint. During parturition the cartilage softens in response to hormonal changes, thus permitting the newborn to move through the pelvic canal.

The acetabulum, or socket portion of the hip joint, lies on either side of the pelvis. This socket accepts the ball portion of the femur. The femoral head holds in place by a strong ligament within the acetabulum and the surrounding muscles. The hip joint serves as a classic example of a ball and socket joint.

The femur extends down the leg to the level of the knee or stile. At this point the femur joins with the

TABLE 3–1 Number of Carpal Bones

Dog—Seven
Ruminants—Six
Horse—Seven or Eight

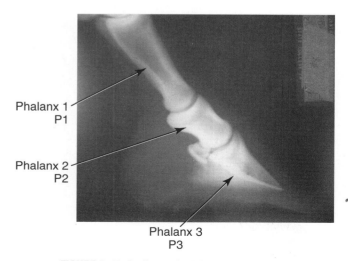

FIGURE 3–18 Radiograph of the foot of a horse.

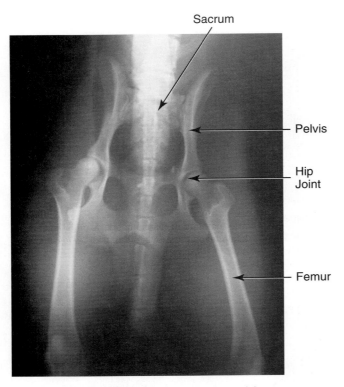

FIGURE 3–19 Radiograph of the pelvis.

tibia (Figure 3–20). Also present in the lower leg is a fibula. This bone is much smaller than the tibia and plays a less significant role in our domestic species. The complicated knee joint is supported by several external ligaments and also by the tendon of the patella, or kneecap.

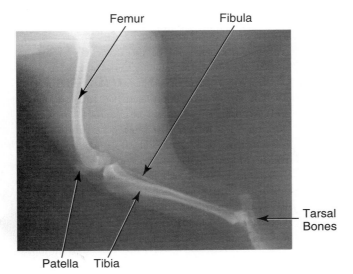

FIGURE 3–20 Radiograph of the hind limb, showing the femur and tibia.

Internally two ligaments located within the knee cross in an X. These cruciate ligaments provide great stability to the joint but can be damaged and torn. A common cause of lameness in dogs is a torn cranial cruciate ligament. Veterinarians diagnose this problem by pressing back on the end of the femur and forward on the tibia. If the ligament is torn, the tibia will slide forward, much more than in a normal leg. This **cranial drawer sign** indicates a diagnosis of torn cruciate ligament. In small-breed dogs this condition often shows good improvement with restricted activity. In larger dogs, surgery is often required to correct the condition.

The tibia, the heavy bone of the lower leg, closely attaches to the much smaller fibula. The tibia is the main weight-bearing bone of this region. The fibula supports little weight but does provide for muscle attachments in that region. The tibia extends down the leg to the hock joint (which corresponds to the ankle in humans). There a hinge joint between the tibia and tarsal bones exists. The many tarsal bones are arranged in much the same fashion as the carpal bones of the front leg. (It is a common mistake for students to confuse the legs associated with carpal or tarsal bones. It is helpful to remember *t*, as in toes and tarsal, which are both associated with the hind limb.)

In the hind foot the metatarsal bones and phalanges order identically to the front foot. The number of bones in the hind foot generally matches that of the thoracic limb.

BONE GROWTH AND REMODELING

OBJECTIVE

■ *Explain How Bone Grows and Remodels*

Obviously there must be a mechanism for an immature animal's bones to grow longer. A long bone, such as the femur, divides into the shaft, or diaphysis, and the ends, or epiphysis. In a young animal a cartilage plate separates the diaphysis from the epiphysis. These growth plates are the sites of elongation. Bones also increase in diameter, with bone being deposited under the periosteum.

In young animals the cartilage growth plate continues to grow. At the same time, at the edge, osteoblasts move into the cartilage. In this process, called ossification, the cartilage is replaced by bone tissue. **Ossification** involves more than just mineral being deposited. The cartilage is replaced by true bone, with all the appropriate cells, structure, and blood supply. As maturity approaches, the cartilage grows more slowly and the growth plate narrows. Eventually bone completely replaces the growth plate. After the growth plates close, bones do not increase in length.

For proper development to occur, the rate of limb growth must match. Injury to a young animal may result in premature closure of a growth plate (the weakest site in the bone). With sufficient trauma, the bone fractures at the growth plate. As this injury heals, the growth plate may be replaced with bone, causing the wounded site to cease growth in length. If this happens, a shorter limb on the injured side results. However, animals with mismatched limbs compensate very well by adjusting the positioning of the shorter leg. For example, if one pelvic limb is shorter, the hock and stifle are kept in greater extension.

A significant problem can result when damage occurs to either the radius or ulna. If only one bone in the forelimb stops growing, the increasing length of the other forces the leg to curve. This problem can be a result of trauma but also "naturally" occurs in certain breeds of dogs. Basset hounds typically have premature closure of the distal growth plate of the ulna. The radius is bound between the humerus and the carpus. Because it grows faster than the ulna, it develops a curve to fit between the other bones (Figure 3–21). In most of these dogs, the only effect is the typical short curved leg. In certain animals the pressure of the radius partially dislocates (**subluxates**) the humerus from the elbow joint. This can cause such severe lameness that surgical intervention may be necessary.

Radiographs are produced when a stream of **x-rays** are passed through a body part, exposing a piece of photographic film. X-rays are a form of electromagnetic radiation that can pass through living tissue. (The term *x-ray* is commonly used to describe the resulting picture. However, it is technically correct to refer to the photograph as the radiograph. Invisible x-rays expose the film.)

Dense structures allow the smallest amount of x-rays to pass and are termed radiopaque. Tissues, such as teeth and bone, fit into this category and appear light on the radiograph. (Refer to Missy's radiograph, Figure 3–2, at the beginning of the chapter to see how white the bone appears.) Radiolucent tissues allow much more of the x-ray energy to pass through and show up dark on the radiograph. Variation among tissues permits the radiograph to be interpreted. The fracture in Missy's tibia shows up as a dark line through the center of the white bone.

Many radiographs are shown in this text. In these radiographs there are basically five stages of density. Ranging from the most radiolucent to the most radiopaque are: (1) air (such as in the lung); (2) fat; (3) soft tissue or muscle; (4) bone; and (5) mineral (such as in teeth). Remember, air is darkest and mineral whitest on a radiograph.

X-rays are capable of damaging living tissues when used at high dosages and with repeated usage. **Radiology,** the study of radiographs, is an essential part of veterinary and human medicine. However, the application of radiology must be done with judgment to minimize the exposure of both the animal and the human doing the procedure (Figure 3–22).

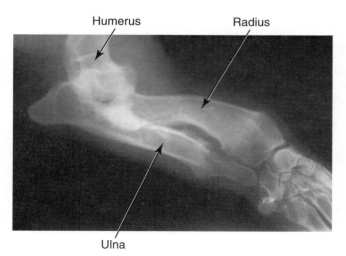

Humerus Radius

Ulna

FIGURE 3–21 Radiograph of the radius and ulna of a basset hound. The growth plate of the ulna closed prematurely, forcing the radius to curve as it continued to grow.

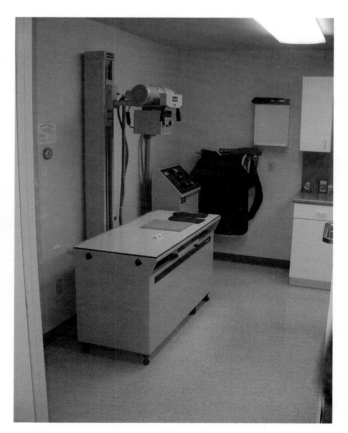

FIGURE 3–22 Radiograph machine.

Film and the cassettes that carry films are designed for minimum x-ray use. Also, specially designed lead aprons and gloves are used to minimize exposure to humans. Lead effectively prevents x-ray penetration. To ensure safe working conditions, technicians and veterinarians are monitored for their x-ray exposure level.

Radiographs throughout this text show anatomy in both healthy and diseased animals. Keep in mind that radiographs show a two-dimensional picture of a three-dimensional object. Although not always shown in the text, standard procedure mandates that two views (at 90 degrees to each other) of a single body part are radiographed. Two views give the veterinarian a better understanding of the structure in question.

RELATION OF BONES, MUSCLES, AND MOVEMENT

OBJECTIVE

■ *Relate Bone and Muscle Groups as Associated with Movement*

Muscles are included in this chapter because of their close association with bone. Together, bones and muscles provide the ability to move. Skeletal muscles attach to bone or cartilage by connective tissue. In most locations this tendon appears as a narrow cord. Some muscles end so close to the bone that no obvious tendon can be seen. Muscles are described by the location of their attachments. The more fixed point is called the origin. The more moveable point is called the insertion. Muscles of the limbs always have the most distal point as the insertion. For example, the biceps (the muscle on the front of the upper arm) originate on the scapula and insert on the radius and ulna.

Muscles that cross a joint are also described based on the type of motion that they cause. Extensors cause the bones to move into straighter alignment or open the joint. Flexors on the opposing side bend the joint or decrease the angle between the bones.

CLINICAL PRACTICE

OBJECTIVE

■ *Connect the Text Materials Pertaining to the Musculoskeletal System to Clinical Practice*

As in Missy the Saint Bernard, fractured bones are a common abnormality. If the bone is broken into two pieces, a clean break, it is called a **simple fracture**. A **comminuted fracture** results in several fragments of

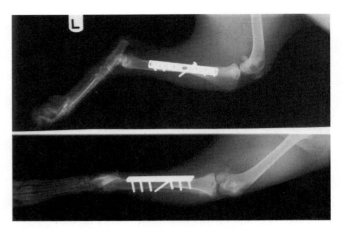

FIGURE 3–23 Radiograph showing a repair of tibial fracture using a bone plate and screws.

bone. Missy's radiograph showed that she had a comminuted fracture. A **compound (open) fracture** results when one of the bone ends punctures through the skin. Higher risk of bone infection exists with compound fractures.

In order for a fracture to heal, the bone ends must be put back in alignment and held without movement. Just as in humans, many fractures can be repaired with a cast or splint. In veterinary medicine the support must be made to hold the weight of the animal. After correction, the animal must be able to use the leg. Very active pets have a hard time keeping a cast in place, clean, and dry.

Surgical correction is often used to repair fractures. With surgery, some form of surgical stainless steel, such as a bone plate, is used to support the bone. A bone plate is applied to the outside edge of the bone and attached with screws (Figure 3–23). We offered to refer Missy for this type of surgery. Bone plates offer a very stable form of correction. Proper size plates must be correctly shaped to fit the bone.

Another method of repair is termed an **intramedullary pin**. I used this type of correction on Missy. During the surgery, I drove a stainless steel pin into the center of Missy's broken tibia (Figure 3–24). The pin entered the bone on the top of the tibia and then entered the medullary cavity. The pin was driven to extend in the distal piece of tibia, very close to the hock joint. Because the bone was shattered into many pieces, the pin alone would not have been sufficient to support the bone. To stabilize the pieces I also added several wires wrapped around the bone fragments. These are called cerclage wires. Radiographs are taken to ensure proper placement of the pin before completing the surgery. One possible mistake would have been to drive the pin too far, causing irritation of the hock joint. Satisfied with the

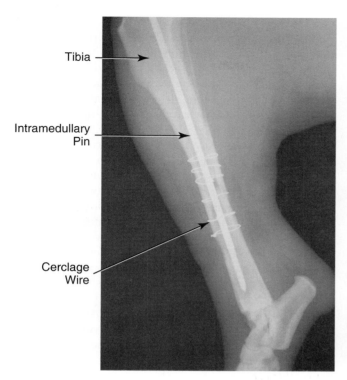

FIGURE 3–24 Radiograph showing the repair of Missy's tibia. The bone was repaired with an intramedullary pin and six cerclage wires.

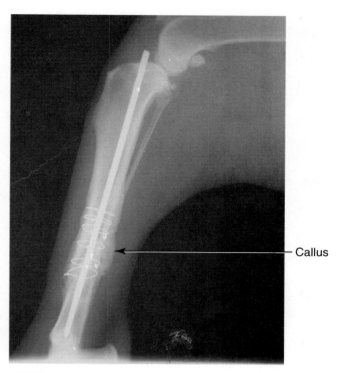

FIGURE 3–25 Radiograph of healing bone 2 months following surgery. Note the callus formation surrounding the site of the fracture.

placement and security of the repair, I finished the surgery. For the first 2 weeks, I also kept a padded splint on Missy's leg. This gave a little more support and helped to minimize swelling.

After the surgery, Missy's bone began to repair itself. The process of bone healing is very similar to that of bone growth in an immature animal. Because of the trauma to the bone, blood fills the gaps between the fracture ends. Cells move into this clot, laying down cartilage. This initial structure, which helps to stabilize the fracture site, is the beginning of a firm structure called a callus.

After approximately two weeks, the cells develop into osteoblasts. This cartilage callus gradually is replaced by bone. This bony callus not only fills in between the fragments but also extends into the marrow cavity and outside the edge of the bone. Two months after the surgery the bony callus in Missy's leg has filled the gaps between the fracture ends (Figure 3–25). It also extends outside the original bone margin and has actually covered the cerclage wires. Over the following months and years the callus will continue to be remodeled.

Missy's case was quite successful. Not all bone repairs end as favorably. Had my repair allowed for even a small amount of movement after the surgery, the fracture would not have healed correctly. Having the edges in alignment and completely immobilized is essential for healing. Bone infection, especially in an open fracture, is another reason that fractures do not heal.

Hip dysplasia commonly occurs in dogs. In this condition the ball and socket joint of the hip becomes diseased. A normal animal should have a deep socket that holds the head of the femur. In hip dysplasia the socket is very shallow (Figure 3–26). Due to the poor structure, the joint subluxates as the dog moves. Over time the cartilage lining the acetabulum and the head of the femur become worn. This is called **degenerative joint disease**. Dogs with this condition develop pain and lameness and often have trouble rising.

Hip dysplasia is generally thought of in large- and giant-breed dogs, but any dog can have it. Genetics play a role in this disease. If two dogs with hip dysplasia are bred, the resulting puppies are likely to have it as well. Nutrition has also been shown to influence the severity of the disease. Larger-breed dogs that are fed to maximize growth seem to have an increased risk for the disease.

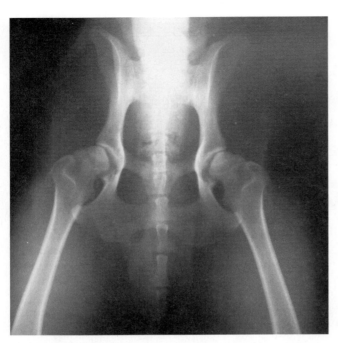

FIGURE 3–26 Radiograph of a dog with hip dysplasia. Note the shallow acetabulum and the subluxation (partial dislocation) of the femur.

The clinical signs of hip dysplasia can vary from a very mild lameness to such severe pain that the dog will not stand. The most common treatment for mild cases is antiinflammatory medications. An over-the-counter medicine, aspirin, can be used in mild cases. More potent drugs are available for severe pain. In the worst cases, total hip replacements are performed.

Foals, calves, pigs, and lambs all can develop a lameness called **joint ill**. After birth the umbilical cord or navel has an open end, because these umbilical vessels are torn to separate the newborn from its mother. Farm animals are often born into conditions that allow for bacteria to enter these vessels. An infection can result. Occasionally the bacteria enter the bloodstream and spread throughout the body. The bacteria can land in the joints. The joint then becomes the new site of infection. On examination, these animals have an elevated body temperature and swollen painful joints. They are often reluctant to rise or walk. Although only one joint may be affected, the condition often involves multiple joints. If caught early, antibiotics will correct the problem. Permanent damage to the joint cartilage is possible.

SUMMARY

Bones provide form for animals. Moreover, bones are composed of a compact dense outer layer and a more loosely arranged center. Three basic types of joints (fibrous, cartilage, and synovial) serve to join bones together. The movement they allow further classifies synovial joints. The bones of the skeleton divide into two main sections, the axial (centerline) bones and the appendicular (limb) bones. A thorough understanding of bones allow practitioners to treat disease conditions and injuries of the musculoskeletal system.

REVIEW QUESTIONS

1. Define any 10 of the following terms:
   herd check
   radiograph
   orthopedic surgeon
   axial skeleton
   appendicular skeleton
   intervertebral disk disease
   high-rise syndrome
   cranial drawer sign
   ossification
   subluxate
   x-rays
   radiology
   simple fracture
   comminuted fracture
   compound (open) fracture
   intramedullary pin
   canine hip dysplasia
   degenerative joint disease
   joint ill

2. True or False: Aspirin can be used to treat hip dysplasia.

3. True or False: The hinge joint allows rotation.

4. Dried bone consists of _____% inorganic and _____% organic material.

5. The appendicular skeleton contains the bones of the _____.

6. The smallest metacarpal bone in a dog is called the _____.

7. The cranial drawer sign diagnosis indicates a torn _____ ligament.

8. In immature animals the growth plate consists of
_____.

9. Canine hip dysplasia is a degenerative _____
disease.

10. Joint ill occurs when bacteria enter the _____
of newborns.

11. What type of joint is the pelvis?

12. Where are the cervical vertebrae located?

13. Does abduction occur when a body part is moved
closer to or further away from the body?

14. Do teeth and bones appear light or dark on a
radiograph?

15. What is the scapula?

16. Name the fractured bone(s) in Figure 3–27.

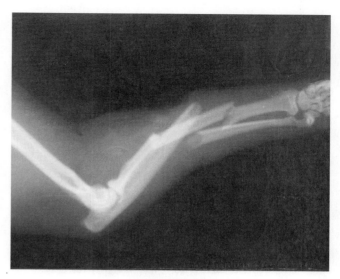

FIGURE 3–27

ACTIVITIES

Materials needed for completion of activities:
long bone from butcher shop
two same-type chicken bones
vinegar
beaker
clay
pipe cleaners
Popsicle sticks
toothpicks

1. The following demonstration is adapted from
several found at <http://www.eecs.umich.edu>.
Clean two similar-type chicken bones. Allow one
to dry on a windowsill and submerge the other in
vinegar for 4 days. At the end of the 4 days, check
the hardness of each. The bone that sat in vinegar
should be pliable. Which mineral did the vinegar
deplete?

2. Obtain a long bone from a butcher shop. Ask the
butcher to section it lengthwise. Examine the bone
for the various structures (compact bone, spongy
bone, medullary cavity, bone marrow).

3. Using clay and Popsicle sticks, model the various
types of joints.

4. Using clay, pipe cleaners, Popsicle sticks, and
toothpicks, fashion a portion of the skeleton.

4

Circulatory System

OBJECTIVES

Upon completion of this chapter, you should be able to:

- *List blood components and explain the functions of blood.*
- *Identify the basic structures of the mammalian heart.*
- *Trace the flow of blood through the heart and body while detailing the parts of blood vessels and their structural significance.*
- *Use knowledge of heart function and control to explain the clinical significance of the electrocardiogram; heart sounds, including heart murmurs; and blood pressure.*
- *Discuss the clinical significance of the academic material learned in this chapter.*

KEY TERMS

hardware disease	veins	electrocardiogram	hypo-
centrifuge	pacemaker system	arrhythmia	hyper-
erythropoiesis	cardiac cycle	tachycardia	-emia
cranial	systole	cardiopulmonary	autoimmune disease
caudal	diastole	resuscitation (CPR)	shock
arteries	electrocardiograph	heart murmur	heart failure

INTRODUCTION

The circulatory system is essential to support the life of each of the millions of cells that make up an animal. The blood itself has a wide range of functions that help to maintain the animal. The heart and blood vessels provide the means to deliver the blood throughout the body.

A DAY IN THE LIFE

ADR—Ain't Doin' Right...

I remember the day in veterinary school when our stethoscopes arrived. The air filled with excitement as we all listened to our own heartbeats. This instrument became a necessary tool in everyday life as I began to examine animals. I must admit I felt cool walking around the hospital in a white lab coat with a stethoscope draped around my neck! It seems like yesterday, even though more than a few years have passed.

Several months ago I examined a cow that was ADR—ain't doin' right. As I walked into the pen, she obviously wasn't feeling well at all. She appeared quite droopy, had lost a lot of weight, and had developed a swelling under her jaw. During the physical, I listened to her heart. It sounded like the noise from a washing machine in mid-cycle. The heart made a sloshing sound with every beat. Using the stethoscope, I diagnosed **hardware disease**. The cow had eaten a piece of metal that migrated from the stomach and lodged close to the heart. The location and structure of the heart provided me with the information necessary to interpret the symptoms of this disease. Hardware disease is often found during my appointed rounds. The next diagnosis is not.

Last year our office received a call from a local school. The sixth grade class mascot, Sonic the hedgehog, had a sore foot. The famous author James Herriot portrayed veterinary work in the view of *All Creatures Great and Small*. Times have changed considerably since Herriot practiced. Much more information and sophisticated medicines and techniques are now readily available. Still, I cannot possibly be an expert on all animals. In this case my vast experience with hedgehogs was limited to reading just one obscure journal article that featured them. I never even met one in real life. Therefore I advised the teacher of my lack of experience but agreed to examine Sonic.

Sonic arrived at the office in a cage (Figure 4–1). He looked just like a miniature porcupine. Because hedgehogs are nocturnal animals, Sonic was apparently taking

FIGURE 4–1 A hedgehog.

his afternoon nap when he arrived at the office. I disturbed him as I tried to examine his leg. Sonic jumped and snorted in an attempt to scare me. To be honest, it worked! His prickly quills were quite sharp. My assistant and I then put on thick leather gloves and proceeded with the examination. Sonic countered with another protective measure. He rolled himself into a tight ball, so tight his legs were completely hidden. I referred to the journal article for help.

Following the recommendations, I anesthetized Sonic with an inhalant anesthetic. We placed him in the large clear mask. The anesthetic was slowly delivered with every breath. Finally Sonic relaxed enough so I was able to have a more thorough look. Once Sonic's leg was exposed, the problem was quite obvious. The rags that Sonic used as a nest had tattered edges with loose strings. One of these strings had wrapped tightly around his foot and stopped the circulation. The foot had turned dark and was oozing. All mammals rely on circulation to maintain their bodies. What happened to Sonic's foot when the blood supply was stopped?

BLOOD COMPONENTS AND FUNCTIONS

OBJECTIVE

■ *List Blood Components and Explain the Functions of Blood*

Blood separates into a fluid portion and a formed element portion. If blood is placed in a tube and spun in a **centrifuge,** the formed elements settle to the bottom of the tube (Figure 4–2). The cellular portion of the blood makes up about 30% to 45% of the total contents. This percentage varies dramatically among species and animal ages. The remaining fluid portion is called plasma.

Plasma consists mainly of water. For most animals, plasma contains 91% to 94% water. Protein makes up 5% to 8% of the plasma. The remainder of the plasma is a combination of electrolytes, gases, nutrients, wastes, and hormones.

The protein portion of the blood can be divided into three types. Albumin, the major protein in blood, maintains the water in the bloodstream. Albumin draws water into the blood vessels through osmosis.

The large albumin molecules do not diffuse from the blood vessels; therefore water is kept in the bloodstream. The liver produces albumin and secretes it into the bloodstream.

Globulins, another type of protein in the plasma, are the antibodies produced to fight disease. The immune system produces these antibodies to fight specific disease-causing organisms. The globulins can be used to make a diagnosis in infectious diseases. A high level of globulin for a particular organism gives evidence that the animal has been exposed to it. Other globulins are used for transporting certain molecules, such as hormones.

Fibrinogen, the third type of protein found in the plasma, aids in clotting blood. When a blood vessel is damaged, fibrinogen is converted to fibrin. The fibrin forms fibrous threads that attach to the damaged vessel. If the clotting proteins are removed from plasma, the resulting fluid is called serum.

The formed elements of the blood are divided into three fractions: red blood cells, white blood cells, and platelets. Red blood cells are also called erythrocytes. White blood cells are also called leukocytes.

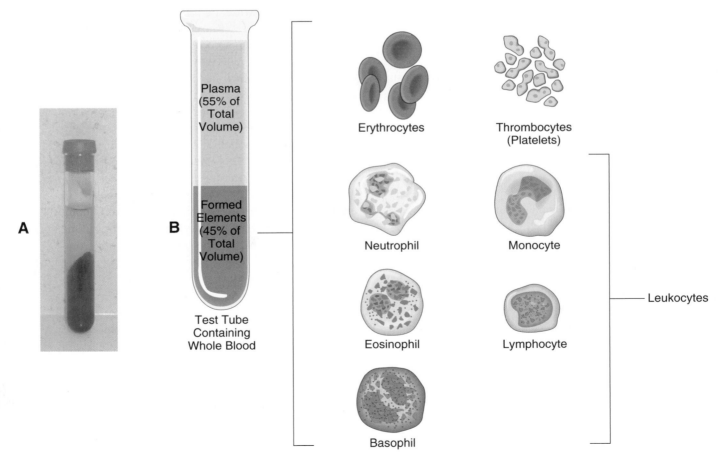

FIGURE 4–2 *A.* Centrifuged blood separating serum from the formed elements. *B.* Blood and its components.

Red blood cells (RBCs) carry oxygen. Mature RBCs do not have a nucleus (Figure 4–3). The shape of an RBC is described as a biconcave disk. This shape provides a large surface area, which allows the cell to exchange oxygen and carbon dioxide. An adult dog has approximately 6 to 8 million RBCs per microliter (μl).

The RBCs are produced in the bone marrow. As the cells are produced, the immature RBCs contain a nucleus. These immature RBCs are occasionally found in the bloodstream. This usually occurs when the body is producing a large amount of RBCs, such as an animal that recently lost a large amount of blood.

TABLE 4–1 Life Span of Red Blood Cells	
Dog	100–110 days
Cat	66–78 days

The production of RBCs by the bone marrow is called **erythropoiesis**. Red blood cells have a limited life span and are constantly being replaced (Table 4–1). In addition to the normal turnover of RBCs, conditions such as blood loss increase the need for more cells. When the number of RBCs decreases, the amount of

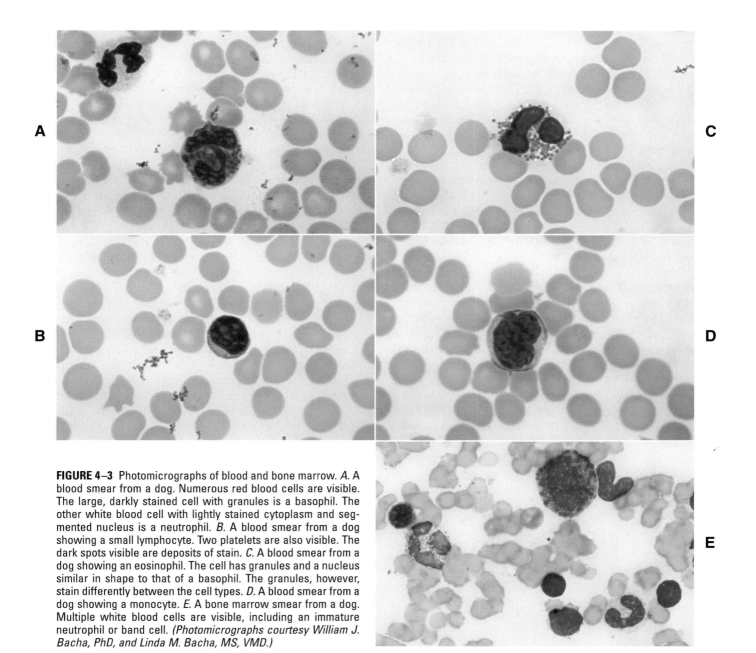

FIGURE 4–3 Photomicrographs of blood and bone marrow. *A.* A blood smear from a dog. Numerous red blood cells are visible. The large, darkly stained cell with granules is a basophil. The other white blood cell with lightly stained cytoplasm and segmented nucleus is a neutrophil. *B.* A blood smear from a dog showing a small lymphocyte. Two platelets are also visible. The dark spots visible are deposits of stain. *C.* A blood smear from a dog showing an eosinophil. The cell has granules and a nucleus similar in shape to that of a basophil. The granules, however, stain differently between the cell types. *D.* A blood smear from a dog showing a monocyte. *E.* A bone marrow smear from a dog. Multiple white blood cells are visible, including an immature neutrophil or band cell. *(Photomicrographs courtesy William J. Bacha, PhD, and Linda M. Bacha, MS, VMD.)*

oxygen delivered to the tissues also declines. Special receptors in the kidney detect such a change and release a hormone called erythropoietin. Erythropoietin stimulates the bone marrow to produce more RBCs.

The RBCs contain the protein hemoglobin. Hemoglobin is the protein responsible for transporting oxygen. Hemoglobin is an iron-containing molecule. The iron gives the red appearance to hemoglobin. Hemoglobin is efficient at binding oxygen, which is extremely important because oxygen is poorly soluble in plasma. Only about 3% of the oxygen in blood is dissolved in the plasma. The hemoglobin carries the remaining 97% of the oxygen.

Hemoglobin has the highest oxygen concentration in the capillaries of the lungs. As the blood pumps to tissues where the oxygen concentration is lower, the oxygen releases from the hemoglobin. The higher the level of carbon dioxide at the tissue level, the more oxygen is released. Some of the carbon dioxide is carried by hemoglobin, which helps to release the oxygen. A lower pH (more acidic) also increases the release of oxygen. Active muscles produce lactic acid, which lowers the pH. As a result, more oxygen is released at the site of these active muscles.

The hemoglobin carries only about 20% of carbon dioxide. Most of the carbon dioxide is converted to bicarbonate and transported in the plasma (Figure 4–4). The hemoglobin, helping to keep the pH normal, absorbs much of the free hydrogen ions. The blood then carries the bicarbonate to the lungs, where the carbon dioxide is released.

Leukocytes, or white blood cells, are present to help fight infection. There are five major types of white blood cells (WBCs): neutrophils, lymphocytes, eosinophils, monocytes, and basophils. These cells are also produced in the bone marrow. WBCs spend only a portion of their time in the bloodstream. The remainder of the time the cells move into the tissues to fight infection.

The total count of WBCs in the blood of a normal dog ranges from 6,000 to 17,000 per microliter. The total count and the types of WBCs present can be used to help diagnose infectious conditions in animals. The different WBCs have different functions, and changes in their number are useful in understanding the disease process.

The main function of neutrophils is to phagocytize (ingest in a form of endocytosis) and destroy microorganisms. The organism is taken into the cell in a membrane-bound sac that joins with granules within the cytoplasm. These granules contain enzymes that can destroy organisms. Neutrophils generally carry out this function in the tissues, not in the blood.

The neutrophil has a nucleus that appears segmented or divided (Figure 4–3A). This is the typical appearance of a mature neutrophil. If the body is attacked by an infection, the neutrophils move to the tissue infected (such as the lungs in pneumonia). Within hours the bone marrow releases a large number of neutrophils that have been held in reserve. The bone marrow then begins to increase production of the neutrophils. This higher production level takes three to four days to be fully transferred to the bloodstream. To speed up production, the bone marrow releases less mature neutrophils into the blood. These immature neutrophils, also called band cells, have a nucleus that is shaped like a U (Figure 4–3E). A high percentage of band cells in blood tells the veterinarian that the animal is actively fighting an infectious agent.

Monocytes are another WBC that actively phagocytize microbes (Figure 4–3D). Monocytes are produced in the marrow and move into the bloodstream and then into the tissues. In the tissues the monocytes mature into macrophages. Some of these macrophages are established in places such as the lung or liver. They remove microorganisms, dead cells, and foreign particles (such as inhaled dust in the lungs).

Eosinophils look similar in appearance to neutrophils with the segmented nucleus. The eosinophils also have a large number of visible granules in the cytoplasm (Figure 4–3C). Eosinophils play roles in fighting parasites and also in allergic reactions. The granules within the eosinophils help to control inflammation.

Basophils are darkly staining cells with many granules and a segmented nucleus (Figure 4–3A). Basophils, like eosinophils, are involved in allergic reactions. Some of the granules in basophils contain histamine. Many students are familiar with taking antihistamines for hay fever. Antihistamines block the effects of histamine. Histamine causes inflammation in the linings of the nasal passages and respiratory tract. This inflammation produces the signs of sneezing and runny nose common to hay fever sufferers. Histamine causes similar effects in animals.

$$CO_2 + H_2O \longrightarrow H_2CO_3 \longrightarrow H^+ + HCO_3^-$$

$$\left(\text{Carbon Dioxide} + \text{Water} \longrightarrow \text{Carbonic Acid} \longrightarrow \text{Hydrogen Ion} + \text{Bicarbonate Ion} \right)$$

FIGURE 4–4 The chemical form of carbon dioxide in blood.

Lymphocytes, which have a single nucleus, are essential in immune function (Figure 4–3B). Lymphocytes produce the antibodies that help to fight disease. These antibodies make up a portion of the globulin that is found in plasma. Lymphocytes are found in all the tissues and organs used in fighting infection. They are present in the tonsils, lymph nodes, spleen, and thymus. A more detailed discussion of the immune system is found in Chapter 11.

Neutrophils and lymphocytes make up the largest number of WBCs found in the circulating blood. The blood smears stained and examined in the laboratory exercise of Chapter 1 can be reexamined. Using the illustrations in this text and the laboratory manual, the individual types of cells can be identified. It will be more difficult to find the basophils, eosinophils, and monocytes than the common neutrophils and lymphocytes. Table 4–2 shows the normal ranges found in our domestic species.

Platelets are the third type of formed element. Platelets, produced in the bone marrow, aid in the normal clotting of blood. Blood clotting is a very complicated process involving the platelets and numerous proteins and factors in the blood.

Immediately after a blood vessel is cut, there is a constriction of the vessel. In this simple reflex the size of the leak is automatically decreased. Platelets then begin to attach to the edges of the damaged vessel, plugging the hole. In addition, a number of clotting factors help to convert the fibrinogen protein found in the plasma into fibrin, which completes the plug. Over time the fibrin clot is replaced with repaired blood vessel.

The ability of an animal to clot vessels is quite impressive. Farm cats, with legs amputated by farm equipment, have been known to stop bleeding without intervention. One cow with a hole in its mammary vein (which can be more than an inch in diameter) large enough for a thumb to fit stopped bleeding before the veterinarian arrived at the farm. However, the laceration was sutured for extra insurance against bleeding. The ability of animals to clot unaided is amazing.

Only the damaged blood vessel wall stimulates the formation of a clot. The normal smooth epithelial lining of the vessels does not stimulate clotting. Abnormal clots potentially cause damage to tissues by stopping blood flow.

The first major function of blood is to transport substances throughout the body. The list of transported components includes oxygen, carbon dioxide, nutrients, wastes, electrolytes, and hormones. Secondly, the blood helps to protect the body from infectious diseases. The white blood cells stand as the first line of defense against organisms attacking the body.

MAMMALIAN HEART STRUCTURES

OBJECTIVE

■ *Identify the Basic Structures of the Mammalian Heart*

Mammals have a four-chambered heart. This provides for two separate circulatory paths. The first is the pulmonary side, where blood pumps to the lungs to exchange oxygen and carbon dioxide. The second pathway delivers blood into the systemic circulation, where blood moves to the entire body. The systemic circulation delivers blood rich in oxygen and nutrients to the organs of the body.

TABLE 4–2 Normal Ranges for Blood Cells

	Adult Dog	Adult Cat	Adult Cattle	Sheep	Goats	Swine	Horses
Red Blood Cells (millions/μl)	5.5–8.5	5.5–10.0	5–10	9–15	8–18	5–8	7–13
Packed Cell Volume (%)	37.0–55.0	24.0–45.0	24–46	27–45	22–38	32–50	32–53
White Blood Cells (cells/μl)	6,000–17,000	5,500–19,500	4,000–12,000	4,000–12,000	4,000–13,000	11,000–22,000	5,400–14,300
Neutrophils Mature	3,000–11,500	2,500–12,500	600–4,000	700–6,000	1,200–7,200	3,080–10,450	2,260–8,580
Neutrophils Bands	0–300	0–300	0–120	rare	rare	0–880	0–1,000
Lymphocytes	1,000–4,800	1,500–7,000	2,500–7,500	2,000–9,000	2,000–9,000	4,300–13,600	1,500–7,700
Monocytes	150–1,350	0–850	25–840	0–750	0–550	200–2,200	0–1,000
Eosinophils	100–1,250	0–1,500	0–2,400	0–1,000	50–650	55–2,420	0–1,000
Basophils	Rare	Rare	0–200	0–300	0–120	0–440	0–290

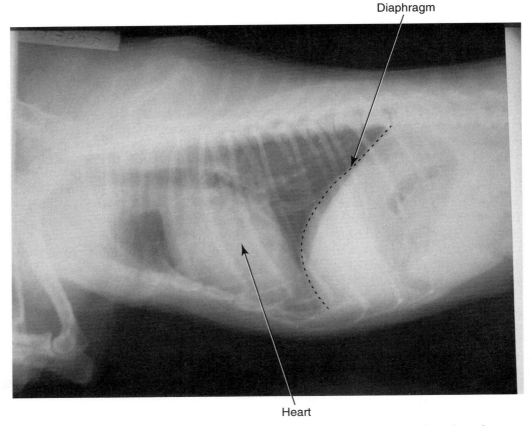

Diaphragm

Heart

FIGURE 4–5 Radiograph of a chest showing the position of the heart within the thoracic cavity.

The heart is located low in the chest between the two lungs (Figure 4–5) and is contained within a fine membrane, called the pericardium. This smooth-lined sac helps to protect the heart and allows it to beat with little resistance. The space between the pericardium and the heart is called the pericardial sac. In a healthy animal there is very little space in the pericardial sac.

The heart, a hollow muscular organ, divides into four chambers (Figure 4–6). The wall of the heart is mainly composed of cardiac muscle, the myocardium. Smooth epithelium lines both the inside and outside of the myocardium.

Valves, present in the heart, keep blood flowing in one direction. These valves separate the chambers on each side of the heart. They are also present in the major vessels leaving the heart. There is a muscular septum that separates the right and left sides of the heart.

BLOOD VESSELS AND BLOOD FLOW

OBJECTIVE

■ *Trace the Flow of Blood through the Heart and Body While Detailing the Parts of Blood Vessels and their Structural Significance*

Understanding the pathway of blood flow helps students understand the structure of the heart and circulatory system. By learning how the blood flows, pupils

gain knowledge of the two halves of the circulatory system and also the anatomy of the heart and vessels.

The blood returning from the systemic circulation to the heart has delivered oxygen and nutrients and picked up carbon dioxide and other waste products. Blood returns to the heart in the large vessels called the vena cava (Figure 4–7). The **cranial** (relating to the head) vena cava brings blood from structures in front of the heart. The **caudal** (relating to the tail) vena cava returns blood from structures behind the heart (Figure 4–8). Students familiar with human anatomy may recognize the terms *inferior* and *superior vena cava*. The inferior vena cava corresponds to the caudal vena cava, delivering blood from organs below the heart. The superior vena cava delivers blood from organs above the heart. The terms *inferior* and *superior* are used because of the human upright posture.

Blood from the vena cava flows into the right atrium, a relatively thin-walled chamber. The right auricle is the small blind pouch of the right atrium. The terms *atrium* and *auricle* are often wrongly used interchangeably. The auricle is a portion of the atrium. As the blood fills the right atrium, much of it passively flows into the right ventricle. Once the atrium contracts, it forces the remainder of the blood into the right ventricle.

The right ventricle has a thicker muscle wall than the atrium. The right ventricle pumps the blood to the lungs. Separating the atrium from the ventricle is

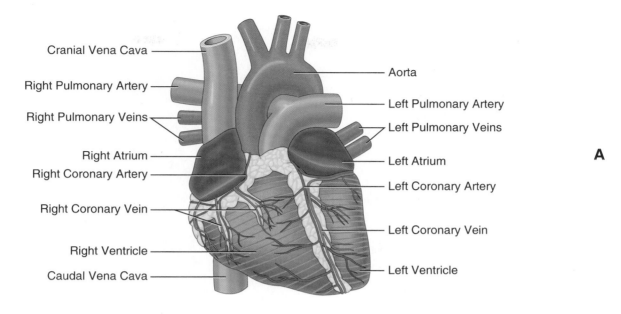

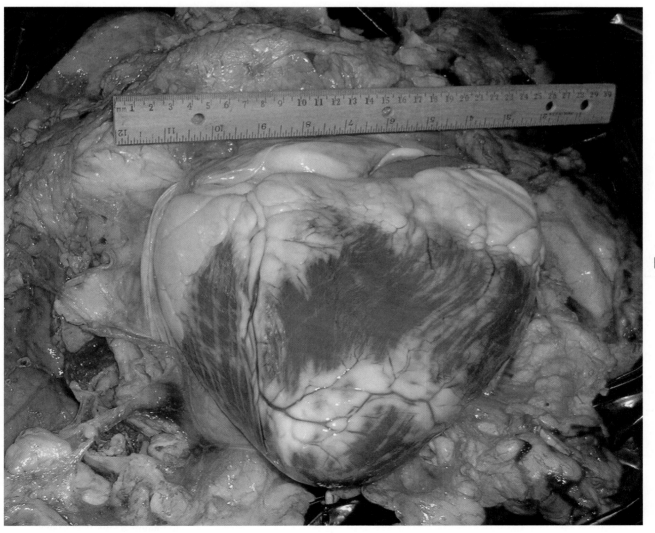

FIGURE 4–6 *A.* The external structures of the heart. *B.* A gross specimen of the heart of a cow.

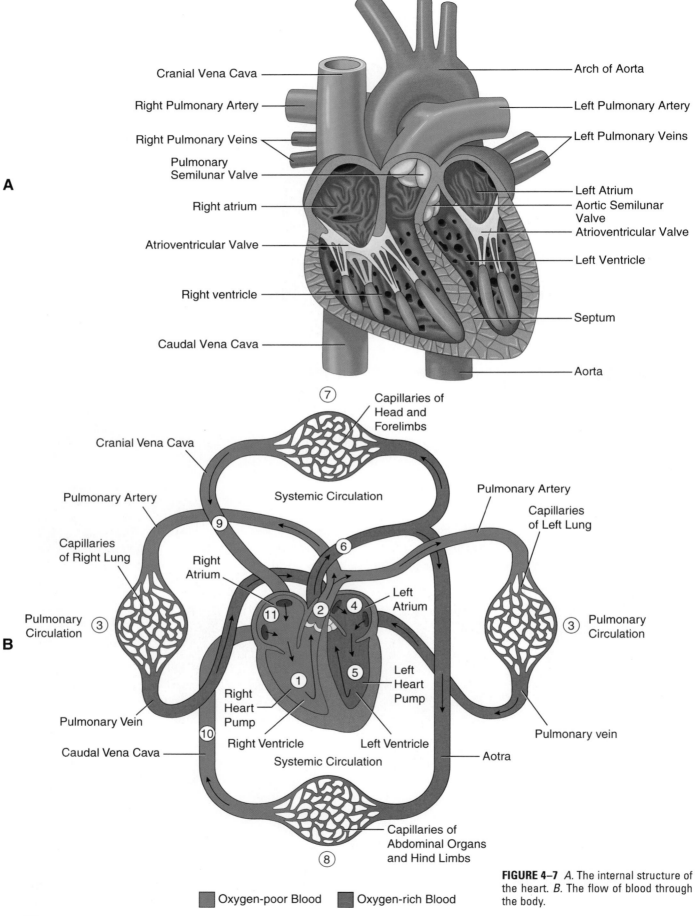

A

Cranial Vena Cava

Right Pulmonary Artery

Right Pulmonary Veins

Pulmonary Semilunar Valve

Right atrium

Atrioventricular Valve

Right ventricle

Caudal Vena Cava

Arch of Aorta

Left Pulmonary Artery

Left Pulmonary Veins

Left Atrium

Aortic Semilunar Valve

Atrioventricular Valve

Left Ventricle

Septum

Aorta

B

Capillaries of Head and Forelimbs

Cranial Vena Cava

Systemic Circulation

Pulmonary Artery

Pulmonary Artery

Capillaries of Left Lung

Capillaries of Right Lung

Right Atrium

Left Atrium

Pulmonary Circulation

Pulmonary Circulation

Left Heart Pump

Right Heart Pump

Pulmonary Vein

Pulmonary vein

Caudal Vena Cava

Right Ventricle

Left Ventricle

Aotra

Systemic Circulation

Capillaries of Abdominal Organs and Hind Limbs

Oxygen-poor Blood   Oxygen-rich Blood

FIGURE 4–7 *A.* The internal structure of the heart. *B.* The flow of blood through the body.

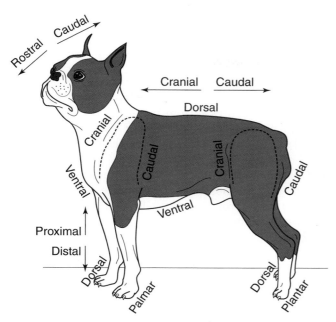

FIGURE 4–8 Directional terms used in describing relative position. Note the relationship of cranial and caudal with its association to the vena cava.

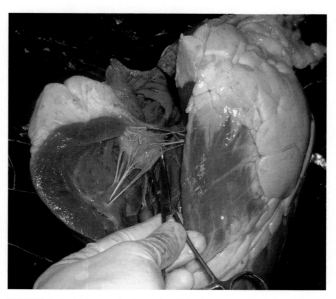

FIGURE 4–9 Atrioventricular valve. The hemostat is under the leaf of the valve. Note the chordae tendineae that secure the edge of the valve.

the right atrioventricular (AV) valve (Figure 4–9). The valve opens easily as the blood flows into the ventricle. Once the ventricle contracts, the pressure from the blood on the valve automatically forces it to close. These valves have flaps that are made of a thin layer of connective tissue, which are pushed by the blood. Tough fibrous cords, chordae tendineae, are attached to the flaps of the valves. These cords keep the flaps of the valves from going too far and effectively sealing the opening between the two chambers. The AV valve therefore is necessary to keep blood from going back into the atrium. The blood is then efficiently pumped into the pulmonary circulation. The right AV valve has three flaps (also called cusps) and is often referred to as the tricuspid valve.

As the right ventricle contracts, the blood flows into the pulmonary **arteries**. Arteries carry blood away from the heart. In an adult animal the pulmonary arteries are the only ones that carry oxygen poor blood. The right ventricle separates from the pulmonary arteries by the pulmonary valve. The pulmonary valve keeps blood that has entered the pulmonary artery from flowing back into the heart.

Arteries have both smooth muscle and elastic connective tissue in their walls. The muscle can contract in vasoconstriction, decreasing the size of the opening. The muscle may relax, causing vasodilation or an increase in the size of the vessel.

The muscles influence the blood pressure by controlling the size of the vessels. The artery can also open to direct blood to certain areas. For example, a muscle

being used actively requires a greater amount of blood.

As the heart pumps blood into the artery, the elastic tissue stretches. Once the heart relaxes, the elastic tissue begins to shrink to its original size (just as in releasing a rubber band). This elastic tissue helps to force the blood through the artery. The pulmonary valve is necessary to keep the blood flowing toward the lungs.

The arteries branch into smaller arterioles as they distribute through the lungs. The arterioles further branch into microscopic capillaries. The capillaries are the smallest vessels. These thin-walled capillaries have an opening so small that only one red blood cell can pass through at a time. Transfer of nutrients and gases occurs in the capillaries. For instance, in the lungs, carbon dioxide is exchanged for oxygen.

The oxygen-rich blood then gathers into slightly larger venules. Many venules join into **veins**. Veins carry blood toward the heart. The veins then join into the main pulmonary veins that return the blood to the heart. The pulmonary vein, the only vein in adult animals that carries oxygen-rich blood, delivers blood into the left atrium. The veins possess a much thinner wall than similar-size arteries. The size of the opening is related to the amount of blood flowing within the vessel.

The structure of the left atrium is very similar to that of the right side. An auricle is also present on the left atrium. As the atrium fills, much of the blood passively flows into the left ventricle. When the atrium contracts, the remainder of the blood is pumped into

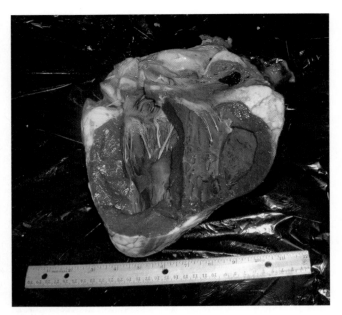

FIGURE 4–10 A sectioned heart showing left and right ventricles. Compare the thickness of the walls of the two ventricles.

the left ventricle. An AV valve also separates the chambers on the left side of the heart.

The left ventricle has the thickest muscular wall of all the four chambers (Figure 4–10). The left ventricle must force the blood through the entire systemic circulation. This requires higher pressure than the short pulmonary circulation. When observing an intact heart, the left ventricle makes up the point of the heart. As the ventricle contracts, the left AV valve prevents

blood from flowing back into the left atrium. The structure of this valve is similar to the right AV valve but only has two flaps to the valve (also called the bicuspid valve).

Blood is forced through the large elastic aorta by the left ventricle (Figure 4–11). The aortic valve separates the left ventricle from the aorta. Just as the pulmonary valve functions, the aortic valve prevents the blood from flowing backward into the heart. The main trunk of the aorta arches around the top of the heart and passes through the chest into the abdomen. Major branches of arteries separate from the aorta. Close to the heart, the coronary arteries branch to deliver blood to the cardiac muscle. The blood, within the chambers of the heart, does not supply nutrients to the heart muscle. Many other major branches derive from the aorta. These branches include the carotid arteries, supplying the head and brain; the mesenteric arteries, supplying the gastrointestinal tract; the renal arteries, supplying the kidneys; and the iliac arteries, supplying the hind legs.

Just as in the pulmonary circulation, the arteries branch into arterioles and then into capillaries. It is at this level that exchange with the tissues occurs. The capillaries organize into venules and then veins to eventually return the blood to the heart. In many locations there are arteries, veins, and nerves that all travel in the same bundle, a fact found very useful in many surgeries. Occasionally a need arises to amputate a leg. When amputating the hind leg at the middle of the thigh (mid-femur), the major vessels run down the inside of the leg, close to the femur. The femoral artery and vein run close together in a division between

FIGURE 4–11 Cross-section of a large artery and vein. Compare the differences between the thin-walled vena cava on the left and the thick-walled aorta on the right.

muscles. During the amputation it is necessary to identify these major vessels. The artery and vein are then tied off, or ligated, to prevent major blood loss during the surgery.

ELECTROCARDIOGRAMS, HEART SOUNDS, AND BLOOD PRESSURE

OBJECTIVE

■ *Use Knowledge of Heart Function and Control to Explain the Clinical Significance of the Electrocardiogram; Heart Sounds, Including Heart Murmurs; and Blood Pressure*

The rate at which the heart beats is controlled in part by the nervous system. The actual beat, though, begins within the heart itself, and will begin without any nerve supply. The cells that begin the heartbeat are called pacemaker cells. The **pacemaker system** is important in maintaining the heart's regular rhythm. The system must also function to make the heart contract in a highly organized manner. The atria must contract first to fill the ventricles. When the ventricles contract, they need to do so in a way that the blood is efficiently pumped out of the heart.

In mammals the pacemaker is called the sinoatrial (SA) node (Figure 4–12). This specialized group of muscle cells is found in the right atrium, close to the cranial vena cava. An electrical signal begins in the SA node and spreads to the surrounding muscle cells. The specialized connections, between cardiac muscle cells, allows for this signal to spread rapidly across the atria. The atria contract, forcing the blood into the two ventricles.

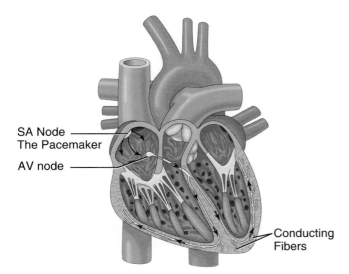

SA Node
The Pacemaker

AV node

Conducting Fibers

FIGURE 4–12 The heart conduction system. The signal is initiated in the SA node, or pacemaker. The impulse travels through the atria and is picked up by the AV node, which quickly distributes it through the ventricles via conducting fibers.

There is a slight delay as the signal is picked up by the atrioventricular (AV) node. As the name suggests, this node is found close to the junction between the atrium and the ventricle. The AV node then sends the electrical signal through conducting fibers to the ventricles. The contraction begins at the apex of the heart and then spreads across the ventricles. This provides an effective contraction for pumping the blood out through the major arteries.

A **cardiac cycle** includes one complete contraction and the following relaxation that occurs. During the relaxation the atria fill with blood in preparation for the next contraction. The rate of this cycle varies between species. The rate for an individual animal can vary with breed and fitness level. The contraction phase of the cycle is called **systole**. **Diastole** is the relaxation phase.

Blood pressure measurements are taken during both phases of the cardiac cycle. The higher number occurs during systole as the heart contracts. The pressure then declines during diastole as the heart relaxes. Blood pressure is measured in millimeters (mm) of mercury (Hg). In human medicine a typical number, such as 120/80 mm Hg, represents the systolic over the diastolic pressure.

The flow of ions that allows for the cardiac muscle to contract causes a small electrical current. The **electrocardiograph** is an electronic instrument that picks up this small electrical signal running through the body. The **electrocardiogram** (ECG) is the tracing made by the instrument (Figure 4–13).

A letter identifies the different peaks. Each peak can be related to activity within the heart. The P wave is formed as the SA node fires and the atria contract. The QRS complex is formed as the ventricles contract. The T peak forms as the ventricles prepare for the next contraction. The PR interval is the time delay that occurs between the beginning of the atrial contraction and the ventricular contraction. This interval shows the time necessary for the atria to contract and fill the ventricles.

The ECG identifies problems associated with the contraction of the heart. The size of the different peaks can be measured, and abnormalities indicate changes in the size of the heart or damaged portions. The shape and frequency of the peaks also detects problems associated with the pacemaker and conduction system.

The normal consistent rate and rhythm is called sinus rhythm. It tells the veterinarian that the SA node begins each signal and the rate is very consistent. An **arrhythmia** describes any change in the rate, rhythm, or conduction within the heart. Many changes can occur (Figure 4–14). Common descriptive terms include the following:

Sinus **tachycardia**: This indicates a heart rate that is faster than normal, with a normal rhythm.

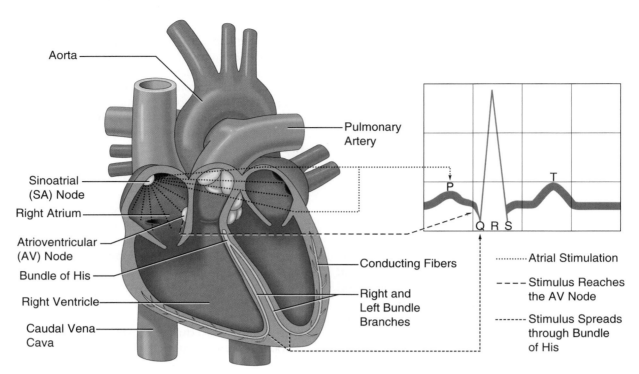

FIGURE 4–13 A typical ECG tracing and its relationship to the heart conduction system.

Sinus bradycardia: This term denotes a heart rate that is slower than normal, with a normal rhythm.

Sinus arrhythmia: In this condition the heart rate increases with inspiration and decreases with expiration. Every complex looks normal; there is just a change in the rhythm with every breath.

Atrial fibrillation: In this condition the SA node no longer acts as the pacemaker. Another site in the atria is irritated and fires very frequently. With this condition, the P waves occur very close together instead of in a clean peak. The atria no longer contract normally, but instead quiver. Only a small percentage of the P waves reach the ventricles to cause the contraction. The heart rate in this condition is extremely irregular and very rapid.

Ventricular fibrillation: In ventricular fibrillation, one of the most serious arrhythmias, a site within the ventricles fires extremely rapidly. The ventricles no longer contract normally, but quiver uncontrollably. This is the classic arrhythmia that requires the animal to be shocked with a defibrillator (commonly seen on medical television programs).

Asystole: The tracing in this condition is basically a flat line. The heart is no longer contracting. The animal is described as being in cardiac arrest. Depending on the cause, external stimulation may be able to start the heart. **Cardiopulmonary resuscitation (CPR)** can be used to stimulate the heart and deliver oxygen to the lungs.

The stethoscope is used to amplify heart sounds, allowing them to be heard much more clearly. The rate and rhythm of the heart can be determined with the stethoscope. Changes in the normal heart sounds can also be identified. Normal heart sounds have been described by the words *lub-dub,* heard when the valves close. The valves and surrounding heart wall vibrate during closing. These vibrations spread through to the chest wall and are picked up by the stethoscope. The first heart sound, *lub,* is created as the ventricles contract. During contraction, the AV valves close rapidly to prevent blood from flowing back into the atria. The closing of the AV valves creates the first heart sound.

The second heart sound, *dub,* occurs when the ventricles relax. At this point the pulmonary and aortic valves close. The valves and the surrounding vessels vibrate, creating the noise. When listening to the heart, veterinarians evaluate several features. First the heart rate in beats per minute (bpm) is counted. The heart rate is then compared with the normal rate for that species. Several factors may influence heart rate. Many diseases can cause an elevated heart rate (tachycardia). In the veterinary office it is important to realize that stress, fear, and nervousness can elevate the heart rate. It is very common for a healthy cat or dog to have an elevated heart rate just because it is afraid.

The rhythm of the heart can also be determined. In the normal sinus rhythm the *lub-dubs* occur at a very regular rate. In many disease conditions the heart may

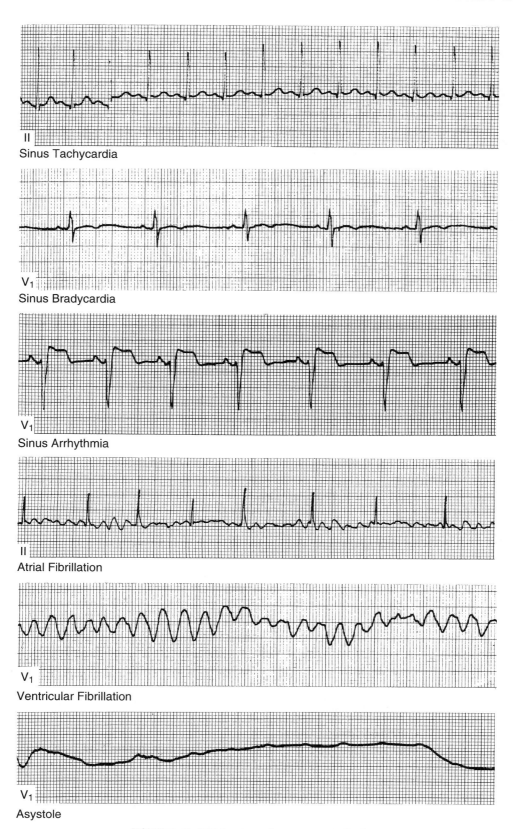

FIGURE 4–14 ECG tracings of common arrhythmias.

FIGURE 4–15 A diseased valve (endocarditis). This valve would not create a tight seal, resulting in a murmur. *(Photograph courtesy Dr. Arthur Hattel, Pennsylvania State University.)*

beat very irregularly, skipping beats or adding extra beats. When changes are heard, an ECG is often used to determine the exact cause. When listening to the heart, a pulse is often felt at the same time. It is useful to feel whether the pulse is strong and consistent with the heart sounds. The pulse is created as the blood surges through with higher pressure in systole and decreases during diastole.

Finally the stethoscope is used to detect any changes to the normal *lub-dub* sounds. **Heart murmurs** occur when there is a defective valve or an abnormal flow of blood (Figure 4–15). A murmur creates a swishing noise. If an AV valve leaks, the sounds become lub-swish-dub. This is called a systolic murmur, because the murmur noise occurs during the contraction of the ventricles. If the aortic valve leaks, the sounds become lub-dub-swish. This diastolic murmur happens when blood leaks back into the ventricle as it relaxes.

The pacemaker within the heart tissue establishes the basic heart rate (Table 4–3). The heart rate and

strength of contraction are controlled by other means. The volume of blood that the left ventricle pumps in one minute is called the cardiac output. The cardiac output can increase dramatically when the body needs it. For example, when a horse runs a race, the increase in muscular activity greatly increases the need for blood flow. The heart rate increases significantly, and the amount of blood pumped with each contraction increases as well.

The autonomic nervous system aids in this control of the heart rate. When exercise begins, the amount of carbon dioxide from the muscles increases in the blood. Special receptors detect this change, and the brain sends a signal, causing the heart rate to increase. Fear and stress can also cause the same type of increase in heart rate. In addition, the adrenal glands release the hormone epinephrine in response to fear and stress. The adrenal glands are small endocrine glands located very close to the kidney. The hormone is carried to the heart in the bloodstream, causing an increase in heart rate as well.

Another branch of the autonomic system causes the heart rate to slow. The signals for this branch are carried in the vagus nerve. Special receptors detect an elevated blood pressure, and the brain returns the signal through the vagus nerve. The vagus nerve causes the heart rate to slow and results in a decrease in blood pressure.

Blood pressure is maintained in a tight range. It is controlled by more than just the heart rate. Pressure receptors in the walls of some major arteries detect changes in blood pressure. As discussed earlier, when pressure increases, the vagus nerve causes the heart rate to decrease. In addition, nerves stimulate vasodilation (relaxation or increase in size of a vessel). The combination of these two events results in a decrease in blood pressure.

When pressure declines, the opposite events occur—heart rate increases and vasoconstriction (narrowing of the vessel as the muscles contract) occurs. In addition, hormones are used to control blood pressure (Figure 4–16). The kidneys produce an enzyme called renin. Renin releases when the kidney detects a decrease in blood pressure. Renin breaks down a protein found in the blood, forming angiotensin. Angiotensin causes vasoconstriction, resulting in a higher blood pressure. Angiotensin also causes the adrenal glands to produce a hormone called aldosterone. Aldosterone causes the kidney to retain higher amounts of sodium and water. The resulting increase in the amount of blood present causes the blood pressure to increase.

Blood pressure is much higher in arteries than veins. During surgery, when an artery is cut, the blood pumps from the vessel. The surgeon can easily tell if the cut vessel is an artery. The bright red blood (high in

TABLE 4–3 Typical Heart Rates (beats per minute)

Species	Typical Range
Cat	110–140
Cow	60–80
Dog	100–130
Goat	70–135
Hamster	300–600
Horse	23–70
Human	58–104
Sheep	60–120
Swine	58–86

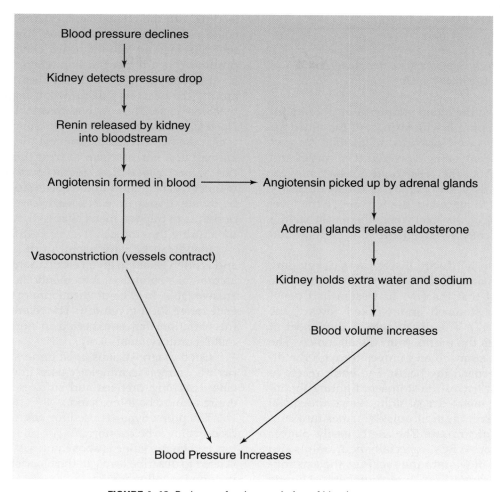

FIGURE 4–16 Pathways for the regulation of blood pressure.

oxygen) will shoot out, spurting forth with each contraction of the heart. When veins are cut, the dark red blood flows in a steady stream. (The description of the color of the blood is assuming that the surgery is not on the pulmonary artery and veins. The pulmonary circulation is the one instance where arterial blood is low in oxygen and venous blood is high in oxygen.)

Because of the lower pressure in the veins, mechanisms are in place to help fight gravity. With low pressure, it is difficult to pump the blood upward from the feet and legs. Movement can help with this. As the muscles contract in the legs, the veins are squeezed, forcing the blood upward. The veins that contain blood flowing against gravity have simple valves in them. These valves allow the blood to flow freely toward the heart but prevent the blood from flowing backward.

This system can be overcome when there is no movement. In soldiers forced to stand at attention, the blood can begin to pool in the veins of the legs. The veins overfill to the point that not enough blood returns to the heart. In turn, the heart cannot supply

enough blood to the brain. If the brain is deprived of blood, the soldier faints and subsequently lands flat on the ground. The heart is then no longer battling gravity and the blood supply is restored to the brain. Soldiers learn to prevent this by contracting and relaxing the muscles in the legs. Even without motion, the muscle contractions help to pump the blood back to the heart.

The mechanisms for controlling blood pressure usually respond quite quickly. However, there are instances where these measures are not quite adequate. Again consider using humans as an example. If one suddenly jumps up from a lying position, dizziness can occur. While lying, the heart is not working against gravity to pump blood to the brain. When the person jumps upright, there is an instantaneous need for higher pressure. The body quickly responds by constricting vessels, pushing more blood to the brain. Many students reading this will have experienced the mild dizziness associated with sudden changes in posture. Usually the effects are very short lived as the body reacts. In the most severe instances, fainting can occur.

CLINICAL PRACTICE

OBJECTIVE

■ *Discuss the Clinical Significance of the Academic Material Learned in this Chapter*

Sonic had stopped the blood supply to his foot by inadvertently wrapping it with string. All the functions of blood no longer were provided to his foot. No oxygen or nutrients were being delivered. The wastes and carbon dioxide from the cells could not be removed. As a result, the tissues in Sonic's foot had died. My only option was to amputate the foot above the level of the string. I selected a level where there was healthy tissue and surgically removed the foot and sutured the skin.

The cow with hardware disease was developing pressure as pus and fluid accumulated around the heart. The metal that the cow had eaten had penetrated from the stomach and worked toward the heart (Figure 4–17). The diaphragm, a thin sheet of muscle, separates the thorax from the abdomen. The metal leaves the stomach and moves through the diaphragm to approach the heart. The body reacts to the metal once it leaves the stomach, fighting the infection that the metal drags along. Once the metal punctures the diaphragm, it quickly comes into contact with the pericardium. The metal easily pierces the pericardial sac. WBCs, especially neutrophils and monocytes, also move into the sac. Pus, the accumulation of body fluids and huge numbers of white blood cells, eventually fills the sac, putting pressure on the heart.

The heart is less able to expand because of the outside pressure. This cow also had very large jugular veins. These large veins carry blood from the head,

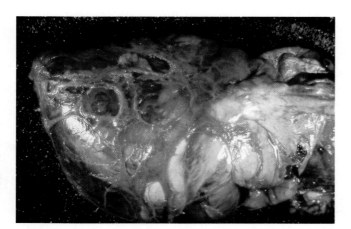

FIGURE 4–17 Diseased heart with pericarditis, which occurred in this cow due to hardware disease. A large amount of inflammation is present between the heart and pericardium. *(Photograph courtesy Dr. Arthur Hattel, Pennsylvania State University.)*

leading into the cranial vena cava. Because the heart could not expand easily, the pressure increased in all the veins leading into the heart. The jugular veins are positioned just under the skin where the enlargement is quite obvious. With every heartbeat, the fluid around the heart was disturbed, causing the washing machine noise. The cow was developing heart failure. *Heart failure* is a general term describing a condition in which the heart is unable to meet the demands of the animal. It is not practical to treat this condition, and this animal was culled (removed from the herd). A magnet can be placed in the cow's stomach to prevent hardware disease or to treat early cases. The magnet is designed to trap the metal against it, to prevent it from puncturing the stomach (Figure 4–18).

Blood can be very useful in diagnosing diseases and is often tested to give veterinarians information on an animal's condition. A tremendous number of tests are available. In general, the laboratory reports a reference range for a given test. The reference range gives the variation that is expected in normal animals for that laboratory (Table 4–4).

Good diagnosticians must understand the vast array of common terminology used in veterinary medicine. Learning prefixes and suffixes often allows an entire term to be interpreted.

The prefix **hypo–** is used for many items when the blood value is below normal. The prefix **hyper–** is often used when the value is above normal. The suffix **–emia** is used to describe levels in the bloodstream. Examples include the following:

Item Tested	Low	High
Calcium	Hypocalcemia	Hypercalcemia
Sodium	Hyponatremia	Hypernatremia
Glucose	Hypoglycemia	Hyperglycemia
Potassium	Hypokalemia	Hyperkalemia
Chloride	Hypochloremia	Hyperchloremia

Changes in blood cells are described differently. The suffix *–penia* is used when the white blood cell count is less than normal (for example, *leukopenia*, low total WBC count; *neutropenia*, low neutrophil count; *lymphopenia*, low lymphocyte count). The suffix *–cytosis* is often used for an elevated count (for example, *leukocytosis*, lymphocytosis). The suffix *–ia* is added to certain other cells when elevated (such as *neutrophilia*, *eosinophilia*).

Anemia is the term used to describe a low red blood cell count. With anemia, the blood supplies less oxygen to the tissues. The earliest signs include fatigue during exercise. As the anemia worsens the animal becomes very weak and lethargic (sluggish and inactive). Anemia is a sign of an underlying disease, not a disease itself. There are three basic causes of anemia: excessive blood loss, shortened life span of the RBCs, and decreased production of RBCs.

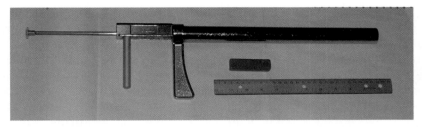

FIGURE 4–18 Balling gun used to administer oral medication to cattle. The rectangular gray magnet remains in the reticulum, where it can trap ingested metal.

The most obvious cause of blood loss is trauma. When a major vessel is cut, a large amount of blood can be lost in a short period. Sometimes this blood is external and very obvious. In other instances the blood loss can occur within the abdomen or thorax. Animals hit by cars (HBC) often present to veterinarians with internal injury. If the trauma causes an organ such as the spleen or liver to rupture, the animal can lose a tremendous amount of blood. HBC is a much too common presentation in private practice. Keep pets under close supervision to prevent such accidents.

The blood loss can also occur much more slowly. The intestinal parasite hookworm can cause a chronic (long duration) blood loss. The hookworm attaches to the lining of the intestine and feeds on the blood of the animal. Anemia is more common in young puppies with a high number of hookworms. The parasites can consume more red blood cells than the puppy can produce. If not treated in time, this chronic anemia can be fatal.

The life span of RBCs was discussed earlier in this chapter. If the RBCs are destroyed at a much higher rate than normal, the bone marrow cannot produce enough to maintain levels. Certain toxins and parasites can cause this increased destruction. The toxin can even be a drug. In Chapter 1 it was discussed that acetaminophen can be toxic in cats. One of the effects is the destruction of RBCs. There are several parasites that can specifically attack RBCs. The body then destroys these cells in attempt to eliminate the parasite.

The body's immune system can destroy its own cells. This condition is called an **autoimmune disease** (the prefix auto– is used to describe self). The red blood cells can be the target of such an autoimmune disease. Again in this condition the RBCs are destroyed faster than they can be produced.

In all the anemias just described, the body attempts to correct the low count. In these anemias the bone marrow releases as many RBCs as possible. This includes some immature RBCs (still containing a nucleus). Some anemias do not have these immature cells. In this situation the bone marrow is not producing enough RBCs. The bone marrow can be damaged with certain toxins and drugs, infections, and cancers. Animals in long-term kidney failure also develop anemia. As the kidneys fail, erythropoietin is not produced and the bone marrow slows down. Most commonly, correcting the underlying cause treats anemias. This includes treatments such as stopping the bleeding, eliminating the parasites, and eliminating any toxins. In the most severe cases, animals can receive a transfusion.

At times, blood loss occurs because animals cannot clot normally. A common reason for abnormal bleeding is consumption of rat poisons. Many of the commonly used mouse and rat poisons kill the rodents by causing bleeding disorders. Unfortunately, they are

TABLE 4–4 Normal Ranges for Blood Chemistry							
	Adult Dog	**Adult Cat**	**Adult Cattle**	**Sheep**	**Goats**	**Swine**	**Horses**
Albumin (gm/dl)	2.8–3.9	2.4–3.5	3.0–3.6	2.7–3.9	2.7–3.9	1.8–3.3	2.9–3.8
Calcium (mg/dl)	9.4–11.7	8.7–11.9	8.0–11.0	8.2–11.2	9–11.0	8.6–11.0	11.2–13.8
Carbon Dioxide (mEq)	11–27	10–21	24–33	20–27	—	18–27	22–32
Chloride (mEq/L)	108–131	115–125	95–110	95–103	99–110	94–106	99–109
Globulins (gm/dl)	2.6–4.4	2.9–5.5	3.0–3.5	3.5–5.7	2.7–4.1	5.2–6.4	.7–1.3
Glucose (mg/dl)	63–110	46.8–151	35–55	35–60	45–60	65–95	60–100
Phosphorus (mg/dl)	2.8–6.2	3.7–9.3	4.0–7.0	4.0–6.0	—	6.7–9.3	3.1–5.6
Potassium (mEq/L)	4.1–5.4	4.3–6.1	3.9–5.8	3.9–5.4	3.5–6.7	4.4–6.7	2.4–4.7
Total Serum Protein	5.7–7.6	5.3–8.5	6.0–8.5	6.0–7.5	6.4–7.9	6.0–8.0	5.5–7.5
Sodium (mEq/L)	143–168	147–161	132–152	139–152	142–155	135–150	132–146

quite dangerous in domestic pets as well. If detected early, these animals can be treated. Vitamin K is important in the whole process of blood clotting. The poisons hinder the function of Vitamin K. Animals that have consumed these poisons are often treated with Vitamin K at high doses for long periods.

During **shock** not enough blood is pumped to the vital tissues. This can occur when there is not enough blood volume. The blood loss described above from severe injury can cause shock. If the heart does not pump adequately, shock can also result. The third cause comes when too many blood vessels dilate or open. When this happens, there is not enough blood to fill every vessel in the body. In the normal animal the vessels dilate where the demand is the highest. A severe infection spreading through the body can be the cause of this type of shock.

Animals in shock are weak and depressed. Their heart rate is rapid and the pulses weak. One physical examination technique is to examine the gums. Normal gums (mucous membranes) should be a nice pink color. When the gum is pressed with a finger, it turns white. When the finger is released, the gum should return to the normal pink color within a second. This is called capillary refill time. Animals in shock have a much slower capillary refill time.

Treatment for shock is to stop the underlying problem. In addition, fluids are given intravenously (in the vein) to raise the blood volume and improve the flow of blood to the tissues. Other medications may be used to improve blood pressure as well.

Heart failure commonly affects elderly pets. Many of these older animals have developed a severe heart murmur. Remember, a leaky valve causes a heart murmur. Consider a dog with a heart murmur caused by a leaky left atrioventricular valve. As the ventricle contracts, a percentage of the blood leaks back into the atrium. To compensate and pump the same amount of blood, the heart beats faster and enlarges.

For mild leaks, the heart is able to compensate quite well. For a long time, the animal will show no clinical signs of the enlarging heart. As the murmur worsens, the heart continues to enlarge and beat rapidly. Often the animal develops exercise intolerance and a cough. The coughing occurs as the heart enlarges to the point that it puts pressure on the airways. As the heart failure progresses, fluid accumulates in the lungs or the abdomen. Eventually the condition is fatal.

The length of the disease varies dramatically among animals. There is no cure, but medications are used to improve the clinical signs. The medications are used to strengthen the contraction of the heart, lower the pressure in the arteries to lessen the work required by the heart, and remove excess fluid. The medications are used to make the animal more comfortable.

SUMMARY

Blood separates into fluid and cellular parts. Blood carries much-needed nutrients to the body and removes unwanted waste as well. Two circulatory pathways allow blood to exchange oxygen and carbon dioxide in the pulmonary system while the systemic circulation takes the oxygen-rich blood to the whole body. Knowledge of blood and blood pathways provides a deeper understanding of the four-chambered mammalian heart and its functions. Practitioners rely on this information to successfully identify and treat problems associated with the circulatory system.

REVIEW QUESTIONS

1. Define any 10 of the following terms:
   hardware disease
   centrifuge
   erythropoiesis
   cranial
   caudal
   arteries
   veins
   pacemaker system
   cardiac cycle
   systole
   diastole
   electrocardiograph
   electrocardiogram
   arrhythmia
   tachycardia
   cardiopulmonary resuscitation (CPR)
   heart murmur
   hypo–
   hyper–
   –emia
   autoimmune disease
   shock
   heart failure

2. True or False: The atrium is a portion of the auricle.

3. Blood can be divided into fluid and _____ parts.

4. Leukocytes help to fight _____.

5. Arteries carry blood _____ (direction) the heart.

6. _____ provide sites of nutrient/gas transfer in the circulatory system.

7. Tachycardia describes an _____ heart rate.

8. Name one of the three types of protein found in plasma.

9. What purpose do heart valves serve?

10. Is the cranial vena cava referred to as inferior or superior in humans?

11. Which of the four heart chambers has the thickest wall?

12. Give another name for the sinoatrial node.

13. What sound does a systolic murmur make?

14. Is blood pressure higher in the arteries or veins?

15. Describe the electrocardiograph during an asystole condition.

ACTIVITIES

Materials needed for completion of activities:
calculators
light microscope
blood smear slides from Chapter 1
stethoscope

1. Adult dogs have six to eight million RBCs per microliter of blood. A client brings in a 45-kilogram chocolate Labrador retriever and asks how many RBCs it has. Assume there are seven million RBCs per microliter. A dog's blood volume is about 80 milliliters per kilogram of body weight. How many RBCs does this dog have in its body?

2. If still available, examine the blood slides from Chapter 1. Following the illustrated lab guide and using the pictures in this text, identify the different white blood cells. To identify the cells, evaluate the shape of the nucleus, the color of the cytoplasm, and the appearance of any granules within the cells.

3. The veins of the arm can be used to demonstrate the presence of valves within the veins. Place arms at sides, allowing the blood to pool in the veins. The visibility of the veins will vary between individuals. Find a straight section of vein without branches entering it. Press down on the vein with a finger and push the blood out of the vein, moving toward the hand. The vein will collapse and not refill (assuming that the pressure is maintained and there are no deep branches entering this section). The blood does not fall down the vein because of the valves located within it. If the vein is pressed in the opposite direction, it immediately refills. The blood is returning from the hand and will fill the vein.

4. Stethoscopes are readily available at an economic price. Using the stethoscope, listen to your heart. Concentrate on identifying the two heart sounds present with each beat. Feel a pulse at the same time. The feel of the pulse should coincide with each heart cycle and will reinforce the action of the heart associated with each sound.

5. Capillary refill time (CRT) is typically tested on the gums in animals. In humans the fingernail provides a location where CRT can easily be tested. The tissue under the nail should be a rich pink color. Press the nail bed (the underlying tissue turns a very pale pink or white). Release the pressure and observe the time required to return to pink. In a healthy animal this should be less than one second.

6. A client brings in their six-year-old basset hound for you, the veterinarian, to examine. It has been weak and vomiting. You elect to do a blood chemistry to determine if there are any problems. You have plans to attend a conference and elect to refer this case to another veterinarian. You have received the following blood test results:

Albumin	3.2
Calcium	10.1
Carbon Dioxide	22
Chloride	115
Glucose	98
Phosphorus	3.4
Potassium	6.2 (high)
Serum Protein	7.0
Sodium	124 (low)

The veterinarian asks what abnormalities you found on the blood work. Using the correct technical terms, describe the dog's condition.

5

Respiratory System

Upon completion of this chapter, you should be able to:

- *Identify the basic components of the respiratory tract.*
- *List and discuss the function and control of breathing.*
- *Discuss the clinical significance of the academic material learned in this chapter.*

KEY TERMS

respiration	inspiration	pneumonia	roaring
palpated	expiration	pleural fraction rub	heaves
endotracheal tube	cyanosis	contagious	bronchodilators

INTRODUCTION

Chapter 4 dealt with the role of the circulatory system and included discussion on the transport of gases. **Respiration** is the exchange of gases between the animal and its environment. This chapter examines the function of the respiratory system.

A DAY IN THE LIFE

When Is an Emergency Really an Emergency?

Right in the middle of small-animal office hours, the receptionist slipped in and told me that an emergency case had arrived—a sneezing cat. My initial reaction was to question how much of an emergency sneezing could be. However, I finished the current appointment and then called in the emergency. Although the cat was in no immediate medical danger, it was really irritated. The cat sneezed frequently and was obviously quite annoyed. On physical examination, I found something in the cat's nose. I really cannot see very far into a cat's nose. I told the owner I would keep the cat and anesthetize it for further review.

Later I was called to a farm to examine a cow that labored to breathe and was not eating well. As soon as I saw her, I knew she was severely ill. The cow had been treated with antibiotics by the farmer but was not responding. This cow was particularly memorable because air had leaked out of her lungs and tried to escape from the body but became trapped in the fatty connective tissue, which lies under the skin. The skin on top of her back rose in a domelike fashion, making her look fat. When I pressed on this skin, I could hear and feel popping bubbles. It reminded me of popping packaging bubbles. When I listened to this cow's breathing with the stethoscope, I heard quite dramatic sounds. Each breath made a sound like a creaking saddle. No doubt this cow was in real danger.

FIGURE 5–1 A young kitten suffering from an upper respiratory tract infection. Note the sore eyes and the thick nasal discharge.

Every spring I see many kittens as I travel farm to farm. The young kittens play in the barn, anxiously awaiting their drinks of milk. During these months, my associates and I treat numerous respiratory infections in these young kittens (Figure 5–1).

THE RESPIRATORY TRACT

OBJECTIVE

■ *Identify the Basic Components of the Respiratory Tract*

The nose provides the opening for air to enter the inside of the animal. The nostrils then open into the nasal cavity, which divides by many scroll-like sheets of bone (Figure 5–2). Epithelium lines the bone. Mucus covers and protects the epithelial linings. The nasal anatomy increases the amount of surface area present and allows the air to come in contact with a large amount of epithelium before entering the lower airways.

The upper airways conduct several important functions. First the mucus-covered epithelium filters incoming air. Particles in the air, such as dust or dirt, become trapped in the mucus. The mucus accumulates

and eventually is swallowed, thus protecting the lower airways. Watch hunting dogs walk. The dogs constantly sniff the ground in an effort to detect odors. In addition to smelling, these animals inhale many dust particles that the nose helps to filter.

As mentioned, most of the trapped particles are eventually swallowed. If the particles cause irritation to the lining of the nose, the animal sneezes. Sneezing is a reflex action that occurs when there is irritation in the nose. During a sneeze, a large amount of air is rapidly forced through the nose and mouth. The sneeze physically forces the trapped and irritating particles from the nose.

The previously mentioned sneezing cat was obviously trying to clear its nose. Once the cat was asleep, I could see the edge of something green inside the nasal cavity. I reached in with a pair of forceps and

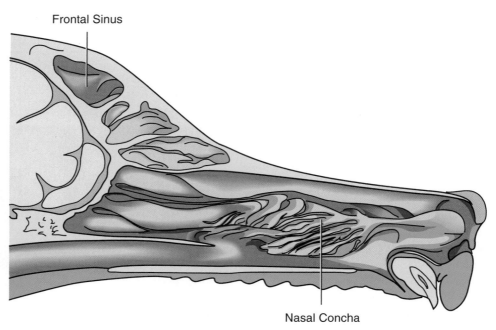

FIGURE 5–2 The internal structures of the nasal cavity of a dog.

removed a 3-inch piece of grass. The cat's sneezing was unable to dislodge such a long blade. The sneeze is protective. Had this blade of grass slipped into the lower airways, the cat could have choked. In addition to helping trap particles, mucus adds moisture to inhaled air, thus preventing incoming air from drawing too much moisture from the lining of the lungs. By passing over the large surface area of the nasal passages, the air is also brought close to body temperature, which protects the lungs from extremes of temperatures.

The air passes from the nasal cavity into the pharynx. The pharynx is the common area shared by the nose and mouth. Air then passes through the pharynx whether it is brought in through the nose or mouth. Food being swallowed also passes through the pharynx.

Because the pharynx functions both in swallowing and breathing, special structures are present to help ensure that food is not inhaled. The larynx is the firm cartilage structure at the opening to the major airways. This structure can be **palpated** (felt) at the top of the neck. In humans this structure is also called the Adam's apple. The larynx contains the vocal folds, which are the structures that allow vocalization in animals and humans. The opening in the larynx, between the vocal folds, is called the glottis (Figure 5–3). A cartilage flap, the epiglottis, protects the opening during swallowing. The epiglottis hinges at the base of the larynx and folds to cover the opening of the larynx during swallowing. When breathing occurs, the epiglottis

does not cover the opening, thus allowing for free exchange of air. This action is entirely an involuntary process, automatically protecting the airways.

The anatomy of the pharynx becomes significant when administering anesthesia. Many surgeries are performed using inhalant anesthetics in animals. This drug, delivered through the lungs, lowers the consciousness of the animal. An ensuing deep sleep keeps the animal from sensing the pain associated with surgery. To deliver the anesthetic, the drug is given through an **endotracheal tube** into the trachea. To place this tube, the technician or veterinarian opens the animal's mouth wide and then identifies the epiglottis. The epiglottis is pushed downward to expose the larynx and vocal folds. The tube inserts between the vocal folds. The tube connects to the anesthetic machine, which then sends the anesthetic through the airway of the animal.

Occasionally, foreign particles are able to slip into the larynx and trachea. Any particles or irritating gases initiate the coughing reflex. The cough is designed to further protect the airways. When coughing occurs, air is initially inhaled. Then the epiglottis and vocal folds close as the animal begins to exhale forcefully. The epiglottis and vocal folds open suddenly, allowing for air to rush out with great force. This reflex helps to force any irritant from the larynx and trachea.

The larynx leads into the trachea. The trachea consists of a series of cartilage rings joined by connective tissue. The cartilage rings are actually C shaped, not completely joining in a circle. Like the

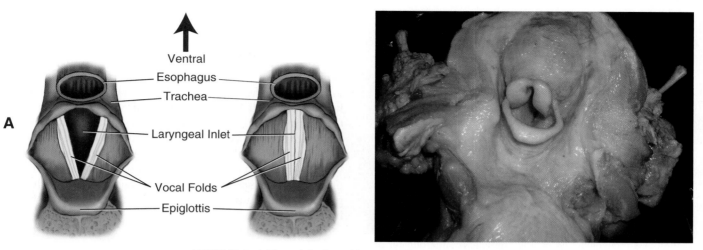

FIGURE 5–3 *A.* The epiglottis and larynx. *B.* The epiglottis and larynx.

larynx, the trachea can be palpated in the neck. If a finger is gently slid along the length of the trachea, the individual rings can be identified. This structure provides a rigid airway that also allows for movement. The neck can be bent sharply without pinching off the trachea.

The trachea is lined with a smooth epithelium that has surface cilia. Tiny particles that are able to pass through the nasal passages are caught in the mucus coating this epithelium. The cilia then work the mucus to the pharynx, where it is swallowed, hence providing for one more protective mechanism of the respiratory system.

The trachea enters the chest to about the region of the heart. At this point the trachea branches into two major bronchi. (The singular form of *bronchi* is *bronchus.*) These two bronchi each lead to a lung on opposite sides of the chest (Figure 5–4). These major bronchi branch into smaller bronchi, dividing and entering different areas of the lungs. The two lungs surround the centrally located heart.

The bronchi continue to divide into smaller and smaller airways, forming the bronchioles (Figure 5–5). The bronchioles have smooth muscle in their walls.

FIGURE 5–4 The lower respiratory tract.

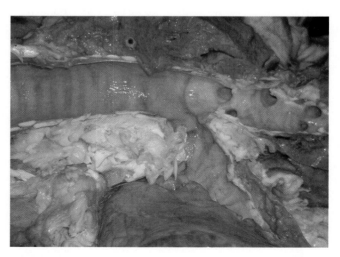

FIGURE 5–5 Trachea opening into the bronchi. The trachea and lung tissue have been incised to expose the openings. Note the tracheal rings and the openings into smaller bronchi.

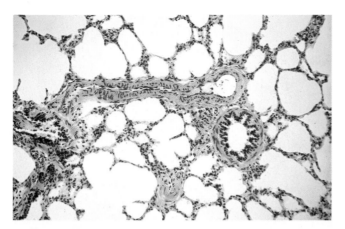

FIGURE 5–6 Photomicrograph of lung tissue showing a cross-section of a bronchiole and nearby arteriole. A large number of alveoli surround these structures. Note the thin wall of the alveoli, which is closely associated with a capillary. *(Photomicrograph courtesy William J. Bacha, PhD, and Linda M. Bacha, MS, VMD.)*

This smooth muscle can cause the airways to open further or close more tightly. Irritants, such as smoke, can cause the bronchioles to constrict. This protective reflex attempts to keep irritants of the lungs.

The bronchioles then form into the smallest openings, the alveoli. Gas exchange occurs in these microscopic pouches. A very thin simple squamous epithelium lines the alveoli (Figure 5–6). Capillaries, which also have a very thin epithelium, surround the alveoli. Therefore the inhaled air comes in close proximity to the circulating blood.

The blood being delivered in the pulmonary circulation is low in oxygen and high in carbon dioxide. The gases exchange readily with the alveoli delivering the oxygen and retrieving the carbon dioxide. At this location the hemoglobin absorbs the maximum amount of oxygen. The blood entering the pulmonary

veins has a high amount of oxygen and minimal carbon dioxide.

The lungs produce a product, called surfactant, which coats the alveoli. Surfactant keeps the alveoli from collapsing and makes them much easier to inflate. With surfactant, very little pressure is needed to keep the alveoli filled with air. Animals born prematurely may lack adequate amounts of surfactant. This makes the lungs very difficult to inflate, which eventually can be fatal because adequate oxygen is not provided to the tissues. In humans, premature babies are often put on respirators, which deliver oxygen under pressure to the lungs. These babies would be unable to inflate their own lungs and require the additional external pressure. As the babies mature, the lungs produce more surfactant. Eventually the lungs function without the respirator.

The lungs contain an enormous collection of alveoli, bronchioles, and bronchi. The lungs appear and feel quite spongy, with all the entrapped air. Although many tissues sink when dropped into a beaker of water, lung tissue floats (Figure 5–7). The entrapped air provides the buoyancy to keep the tissue floating. The spongy feel also occurs because of elastic tissue within the lungs. The elastic connective tissue holds all the enclosed airways together. The elastic tissue proves essential in lung function.

A smooth epithelium called the pleura covers the lungs. Pleura also line the inside of the thorax. A very small amount of fluid is present in the space between the lungs and the wall of the thorax. This fluid provides lubrication as the lungs expand and contract within the chest cavity. In a healthy animal the lungs come in close contact with the chest wall.

MECHANISMS OF BREATHING

OBJECTIVE

■ *List and Discuss the Function and Control of Breathing*

Breathing allows for air to be exchanged between the animal and the environment. This two-step process includes **inspiration,** where air is taken into the lungs, and **expiration,** in which the air is forced out. When inhaling, the chest cavity increases in volume. This occurs in two ways. First the muscles between the ribs (the intercostal muscles) contract to raise the ribs, expanding the chest. Also, the diaphragm contracts, forcing itself toward the abdomen (Figure 5–8). Again the movement of the diaphragm increases the volume of the chest. Increasing the volume of the chest effectively lowers the pressure within the lungs. This allows air from the environment to flow into the lungs. During this entire process the lungs maintain close contact with the lining of the thorax.

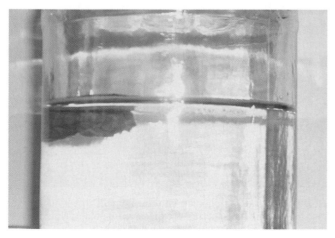

FIGURE 5–7 The buoyancy of a section of lung.

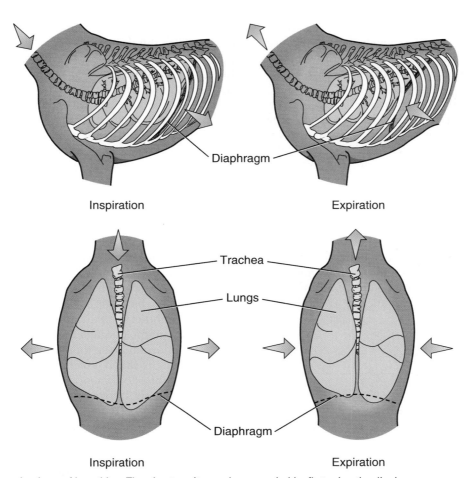

FIGURE 5–8 The mechanisms of breathing. The chest cavity can be expanded by flattening the diaphragm or expanding the rib cage.

When the intercostal muscles relax, the ribs lower. As the diaphragm relaxes, it arches forward toward the chest. The elastic tissue in the lungs recoils, driving out the air. Expiration can be a passive process. By relaxing, the process automatically forces air out. Contracting other intercostal muscles and the muscles of the abdomen can voluntarily increase expiration. This happens during periods of high activity, when the demand for oxygen increases. By initiating this muscle activity, air can be forced outward more quickly and more completely.

The brain controls the normal rate of respiration (Table 5–1). Many factors influence these rates. A hot day or a trip to the veterinarian raises the rates for many normal animals. Dogs rely on panting for cooling. The dog breathes very rapidly, which works to evaporate the moisture in the pharynx. Evaporation has a cooling effect. The blood in the vessels of the pharynx is cooled. This cooler blood then circulates through the body to help maintain a normal body temperature.

This basic rate changes in response to the demands of the body. When an animal is more active, the muscles produce more carbon dioxide. Carbon dioxide stimulates faster respiration. Special receptors in major arteries detect the increase of carbon dioxide in the blood. Signals are sent to the brain from these receptors. The brain then stimulates a faster respiration rate. The increase in carbon dioxide also stimulates the bronchioles to dilate. This opens the airways and improves the delivery of air to the alveoli.

TABLE 5–1 Normal Respiration Rates

Animal	Respiration Rate (breaths per minute)
Cat	26
Dog	22
Sheep	19
Cow	30
Horse	12
Human	12
Guinea Pig	90
Hamster	74

Surprisingly, oxygen plays only a minor role in the control of respiration. Oxygen levels must fall very low before they stimulate respiration. However, even minor changes in carbon dioxide levels are quite effective in changing respiration rates.

Breathing, an involuntary process, occurs without thought by the animal. Even so, the process can be consciously controlled. For example, humans can intentionally hold their breath. The longer it goes, the stronger the urge to breath. One evening during small-animal appointments, a young boy proved this point quite effectively. For some reason, he had become upset and held his breath. He overcame the urge to breathe until he passed out! The involuntary process of breathing then took over and he recovered. Fortunately the parents were used to such behavior and did not panic. They understood the natural protection that the body offered. The veterinarian's heart and respiratory rates both increased dramatically during that episode!

CLINICAL PRACTICE

OBJECTIVE

■ *Discuss the Clinical Significance of the Academic Material Learned in this Chapter*

In the last chapter, mucous membranes were discussed as a means to evaluate the circulatory system. The mucous membranes can also offer information on the respiratory system. If the level of oxygen falls too low, the blood takes on a much darker appearance. When observed through the tissues, it appears bluish. The blue color is termed **cyanosis**. Cyanosis is

a definite indication that inadequate oxygen is being delivered to the tissues.

Pneumonia is a disease with inflammation in the lungs (Figure 5–9). Usually this is caused by an infection, either bacterial or viral. The differences between these organisms are discussed later in the text. In cattle a number of viruses and bacteria commonly cause pneumonia. Once the infection invades the lungs, the body's immune system attempts to fight it. White blood cells move into the lungs to attack these invading organisms. Cells and tissue fluid accumulate in the alveoli. This decreases the amount of lung tissue that effectively transfers gases. To compensate, the animal breathes faster and harder.

The cow discussed in the beginning of this chapter had pneumonia. The infection had attacked the lungs and caused such damage that some of the airways had actually opened into the chest cavity. The air from the lungs entered the chest cavity and then dissected into the tissues under the skin. It was one more indication of how severe the damage had become.

The abnormal noise that I heard when I listened to her breathing is called a **pleural friction rub**. The creaking saddle noise told me that the pleura, the linings of the lungs and chest, were inflamed and irritated. Normally the smooth pleura slides silently without irritation. This cow had pleura that rubbed painfully with each breath. I treated this cow with medicine to reduce inflammation and switched to a different antibiotic. The last time that I spoke with the farmer, the cow was still alive but her productivity remained quite poor.

The stethoscope is used to evaluate breathing. The normal lung sounds are quiet, wispy noises. The changes that occur with pneumonia can produce abnormal lung sounds. In addition to the pleural friction

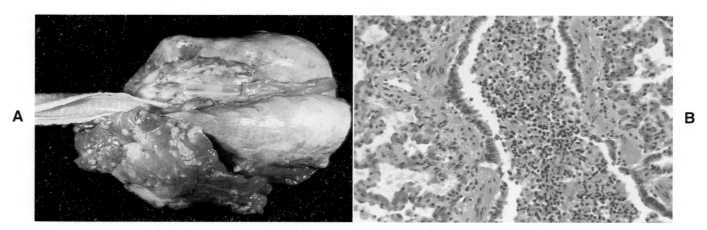

FIGURE 5–9 *A.* Cow lung displaying severe pneumonia. *B.* Photomicrograph of a cow lung with severe pneumonia. Large numbers of white blood cells are seen in a major airway. *(Courtesy Dr. Arthur Hattel, Pennsylvania State University.)*

rub, there are sounds described as crackles and wheezes. These sounds must be heard to truly appreciate the specific noises. Crackles sound almost like crinkling paper. The wheeze sounds more like a musical note. Both these noises are very quiet and do require the stethoscope to detect. Both sounds indicate that there is inflammation, fluid, or both within the lungs themselves.

Much like cattle, cats have their own group of disease-causing organisms. I commonly see cats and especially kittens that develop runny eyes, nasal discharge, sore throat, coughing, and sneezing. These animals often spike a high fever and do not eat well. The organisms causing these signs can be highly **contagious**. It is not unusual for all the cats in the household to develop these signs. Most of these kittens are able to survive with treatment. (It is interesting to note that these signs are quite similar to the common cold in humans. Humans have their own organisms that also attack the respiratory tract and are highly contagious.)

It is not surprising that clinical signs often start in the nose. The nose's ability to filter the incoming air means that it is susceptible to incoming organisms as well. The organism can become trapped in the mucus and then invade the epithelium lining the nasal passages. The hope is that the organism remains localized to the upper airway and does not move into the lungs. Infections in the upper airways can be quite severe at times, but pneumonia is in general a more life-threatening problem.

Breathing occurs with the diaphragm contracting and expanding the volume within the chest. The diaphragm is a relatively thin sheet of muscle that can be damaged by trauma (for instance, being hit by a car). The muscle can be torn, allowing organs from the abdomen to enter the chest (Figure 5–10), a condition termed diaphragmatic hernia. These abdominal organs, such as the stomach, intestines, and spleen, take space within the chest. This prevents the lungs from expanding fully, and the animal has difficulty breathing.

This problem requires surgery to repair the tear in the muscle. An incision made in the abdominal wall helps to locate the tear. The abdominal organs are moved back into the abdomen to their normal location, and the diaphragm is sutured. Once the abdomen is opened, the animal cannot breath on its own. As the animal tries to move its diaphragm, the lungs do not expand, because air is drawn into the chest cavity. The vacuum effect that allowed the lungs to expand is not possible when the chest cavity is open to the outside air. Once the incision is made, the anesthetist must do the breathing for the animal. Applying pressure to the air entering the lungs does this. The anesthesia machine allows the air to be forced into the lungs (Figure 5–11).

Horses have a well-developed respiratory system that allows them to be excellent athletes. Proportionately, horses have much larger lungs than do similar-size cattle. To be a productive racehorse, the entire respiratory system must work efficiently.

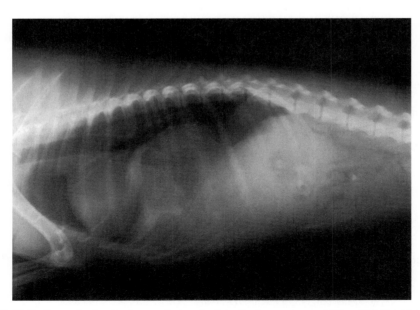

FIGURE 5–10 Radiograph of a diaphragmatic hernia. Intestinal organs are visible within the chest cavity.

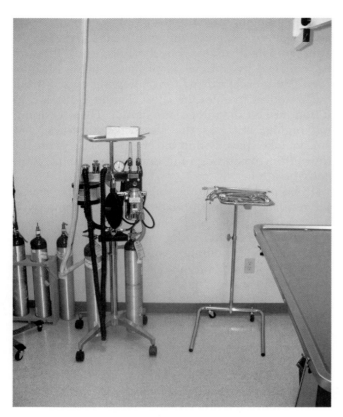

FIGURE 5–11 An anesthesia machine.

In a normal breath the vocal folds open wide to allow the air to flow into the trachea without obstruction. Horses can develop a problem where one of the vocal folds fails to open. This condition is called **roaring**. As the horse exerts itself, the air being forced through the half-open larynx creates a roaring noise. This condition develops when the nerve stimulating that side of the larynx stops working properly. Occasionally the cause can be traced to some sort of trauma. However, the cause often goes undiscovered. Surgery is required to hold the vocal fold open, once again allowing the air to travel without obstruction.

Horses may also develop a disease called **heaves**. Affected horses have coughing, nasal discharge, and labored breathing and tire easily. This could be confused with pneumonia, but it is not infectious. Heaves begins very slowly and is a long-lasting problem. The exact cause is not clear, but exposure to dusts and molds often brings on the clinical signs. Medications to open the airways **(bronchodilators)** and decrease inflammation are often quite helpful. It is essential to minimize the exposure of the horse to the potential dusts and molds that initiate the problems. Keeping a horse on pasture can be very effective when attempting to keep it away from the aforementioned irritants.

SUMMARY

Being able to identify respiratory structures and their associated functions, from the nose to the lungs, allows veterinarians to diagnose and treat such disease conditions as pneumonia and roaring. Moreover, respiratory rate provides a key piece of information to practitioners when accessing the overall health of animals. The status of the respiratory system affects the breathing and therefore the total health of animals.

REVIEW QUESTIONS

1. Define any 10 of the following terms:

   respiration
   palpated
   endotracheal tube
   inspiration
   expiration
   cyanosis

   pneumonia
   pleural fraction rub
   contagious
   roaring
   heaves
   bronchodilators

2. True or False: Mucus lines the epithelial tissue in the nostrils.

3. True or False: The cartilage rings of the trachea are shaped like an O.

4. The _____ is the common area shared by the nose and throat.

5. The human larynx is sometimes called the _____.

6. The trachea branches into two _____.

7. Gas exchanges in the smallest openings of the respiratory system. These openings are called the _____.

8. The muscles between the ribs are called the _____.

9. Name the reflex action that occurs when there is an irritation in the nose.

10. What substance lines the lungs, making them easier to inflate?

11. What controls the rate of respiration?

12. What is the normal respiration rate for a dog?

13. What plays a more significant role in the control of respiration, oxygen or carbon dioxide?

14. What medical tool is used to evaluate breathing?

15. What species can develop a condition referred to as roaring?

ACTIVITIES

Materials needed for completion of the activities:
   stethoscope
   balloons
   Y-shaped polypropylene connecting tubes

1. Use the stethoscope to listen to normal lung sounds. Have the "patient" take deep, slow breaths. The patient should breathe quietly, making noise through the nose and mouth. The stethoscope can detect these noises. Listen to different areas on the chest, from both the front and the back.

2. Take two identical balloons and inflate them to different sizes. Slip a balloon onto an end of Y-shaped polypropylene connecting tubes. Do not release the balloons yet. Plug the third opening of the Y piece. Hypothesize what will happen when the balloons are released. Will the large balloon deflate and fill the smaller balloon to equalize the size? Or will the smaller balloon deflate into the other balloon? Surfactant prevents this problem from occurring between alveoli. Even though the alveoli may be of different sizes, the pressure in each is similar. Without it, the small alveoli would deflate.

6

Renal System

OBJECTIVES

Upon completion of this chapter, you should be able to:

- *Identify and name the basic structures in the renal system.*
- *Name and explain the functions of the renal system.*
- *Identify structures within the kidney and detail the formation of urine and its regulation.*
- *Evaluate urine and blood as a measure of the health of the animal and the urinary system.*
- *Discuss the clinical significance of the academic material learned in this chapter.*

KEY TERMS

bucks	ventral	urinalysis	uremia
does	serum	specific gravity	acute
dorsal	gout	refractometer	chronic
retro	mastitis	free catch urine	subcutaneous
urinary incontinence	intravenous	azotemia	skin turgor
spayed	isotonic	parvovirus	

INTRODUCTION

The byproducts of metabolism are eliminated from the animal through excretion. In this chapter we investigate the renal system. In the renal system the kidneys produce urine as a means of this elimination.

A DAY IN THE LIFE

Some Procedures Are a Group Effort...

It was Friday afternoon. I felt very good as I finished my farm calls. By mid-afternoon I was done for the weekend. I sure was happy to be done so early in the day, or so I thought. Just before I got back to the office, I received a call on the two-way radio. The message stated that Rufus had returned to see us and was having difficulty urinating again. Everyone at the office knows Rufus. Rufus is a 9-year-old Dalmatian who has had multiple surgeries for stones, which form in his bladder (urinary calculi). The stones prevent him from urinating properly. My partner, Dr. Brofee, had examined Rufus and found that he had bladder stones once again (Figure 6–1). Dr. Brofee would be busy with other appoint-

ments for the rest of the day, and the staff wanted to let me know that Rufus might need surgery.

When I arrived at the office, Dr. Brofee and I examined Rufus again. We both agreed that Rufus needed surgery. Rufus had suffered from this condition before and had had two prior operations. With the aid of my assistants, I opened Rufus's abdomen and exposed the urinary bladder. I then incised the bladder and removed three large stones and many tiny pebbles (Figure 6–2). Fortunately, Rufus has been doing well since this last surgery. Dalmatians happen to have a much higher incidence of certain stones as a result of their metabolism. Understand-
continued

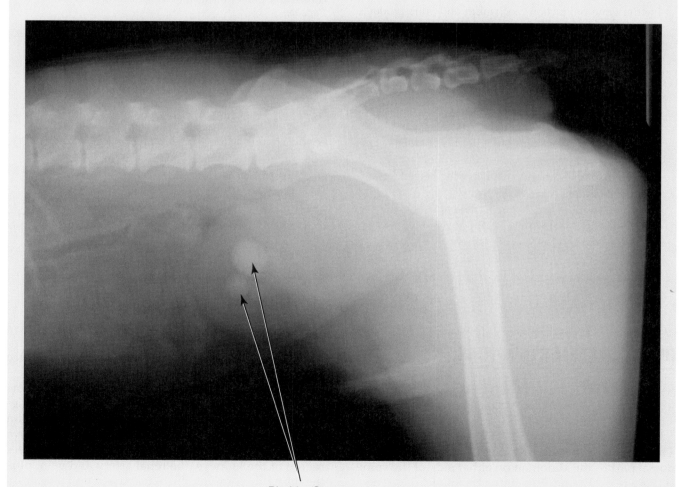

Bladder Stones

FIGURE 6–1 Radiograph of Rufus, showing bladder stones.

FIGURE 6–2 Bladder stones removed from Rufus. The rough texture of the stones causes severe irritation to the lining of the bladder.

FIGURE 6–3 Barney the goat. *(Photograph courtesy Dr. Debra Deppen.)*

ing the urine production mechanism and urine products aids in understanding how Rufus's problem developed.

Veterinarians' animals are not exempt from having problems. Several years ago, Barney, Dr. Deppen's pygmy goat, was straining to urinate (Figure 6–3). Goats can also develop bladder stones, causing urinary tract obstruction. **Bucks** (male goats) are more susceptible to obstruction than **does** (female goats). The anatomy of these animals makes them prone to this condition. Understanding the differences in the anatomy between males and females helps to explain why the genders are affected differently.

Last evening, Vanna came to the clinic because she was having accidents in the house. Vanna is a 10-year-old Basset hound. I can understand why the owners were so upset, because Vanna is such a sweet dog. My initial thought was that because Vanna was an old spayed female dog, she was losing control of her ability to hold

urine. But as I talked with the owner and examined her, I discovered that Vanna had developed a tremendous thirst. Any time the owner would fill the water dish, Vanna would drink until she emptied the bowl. I became concerned that there might be a more serious health condition causing the problem. I took a sample of blood from Vanna and submitted it to the lab for analysis.

Blood tests and urine tests are common methods used to investigate disease affecting the urinary tract. Understanding how the kidneys function and are controlled aids in interpreting lab test results. This chapter examines the normal function of kidneys and discusses many of the problems that can result from disease conditions. Fortunately, the blood tests did not show any significant problems for Vanna.

RENAL SYSTEM STRUCTURES

OBJECTIVE

■ *Identify and Name the Basic Structures in the Renal System*

The kidneys produce urine and are present in the **dorsal** part of the abdomen on either side of the spine (Figure 6–4). The peritoneum, a thin sheet of connective tissue, lines the entire abdominal cavity. The kidneys actually lie between the peritoneum and the muscles that are adjacent to the spine. Technically the

kidneys are located in the retroperitoneal space (**retro** = behind).

The kidneys are reddish brown bean-shaped organs. The surface of the kidney is very smooth (Figure 6–5). The kidney of a cow serves as an exception because it has multiple lobes, separated by a groove in the surface of the kidney (Figure 6–6). In the center region of the kidneys, the renal artery and vein, plus the ureter, are present. The renal artery branches from the abdominal aorta and supplies blood to the kidney. As much as 20% to 25% of the blood pumped from the heart each minute travels through the kid-

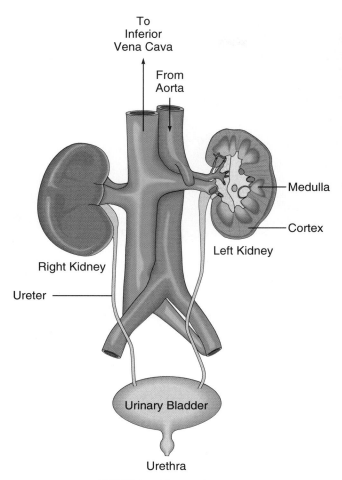

FIGURE 6–4 The renal system.

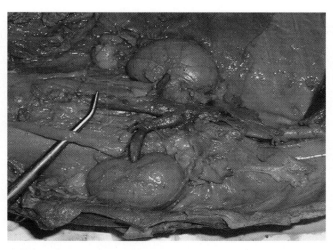

FIGURE 6–5 External appearance of feline kidney. This photograph is of a prepared specimen. The blue vessels are the renal veins and caudal vena cava injected with latex. The ureter passes over the probe.

ney. The renal vein returns the blood that has traveled through the kidney to the caudal vena cava.

A ureter emerging from each kidney is the tubular structure that carries urine to the urinary bladder. The wall of the ureter contains smooth muscle. The urine is actually pushed from the kidney to the bladder by an organized series of contractions. Peristalsis describes this type of contraction.

The urinary bladder, a hollow organ, has a great ability to expand for storing urine. Transitional epithelium lines the bladder. This epithelium provides the bladder with the ability to stretch as it fills with urine. In the empty bladder, this epithelium appears to be at least six or seven cell layers thick. In the fully distended bladder the same epithelium appears to be only a couple of cells thick. This epithelium also prevents any urine from penetrating into the underlying tissues.

The wall of the bladder contains layers of smooth muscle and surrounding connective tissue. This muscular wall also provides the bladder with the ability to stretch and is responsible for providing the contraction force necessary to empty urine from the bladder.

The bladder of an 11-kg dog can easily hold more than 100 ml of urine.

From the bladder the urethra carries urine to the outside of the body. The urethra also has a muscular wall that allows for control of urine flow. The urethra of a male is significantly longer than that of a female, which plays an important role in practice. Females have a higher incidence of **urinary incontinence**. Animals with this condition leak urine at inappropriate times. One common example occurs in elderly **spayed** (the ovaries and uterus have been removed) female dogs. These dogs often leak urine while sleeping. Fortunately they respond to a supplement of estrogen, a hormone normally produced by the ovaries. Vanna responded very well to this medication. Because she wasn't causing accidents in the house, she returned to the owners' bedroom to sleep. Likewise females are more prone to developing bladder infections. Bacterial

FIGURE 6–6 External appearance of bovine kidney.

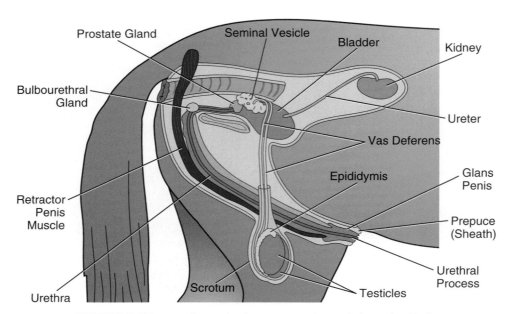

FIGURE 6–7 Urinary and reproductive structures in a male horse (stallion).

contamination on the outside skin can migrate into the urethra to develop infection in the bladder.

Males, with their longer and narrower urethra, are much less likely to have bladder infections (Figure 6–7). But as Rufus and Barney illustrate, males are much more prone to having urinary obstruction from small bladder stones. Male dogs have a bone in the penis (called the os penis), through which the urethra passes. This portion of the urethra is the narrowest. A small stone can travel from the bladder and through the urethra until it reaches this point. Then the stone becomes trapped, preventing the dog from emptying the bladder. The first surgery that Rufus had was to correct this problem.

Bucks do not have an os penis, but have two regions that may obstruct a stone. The end of a goat's penis has a narrow extension of the urethra. This narrowest portion of the goat's urethra often traps stones. Fortunately, this urethral process can be surgically removed and may relieve the obstruction. Unfortunately, this did not work for Barney.

As the urethra exits the pelvis in goats, it makes a sharp bend in the form of an S curve. The sharp curve narrows the urethra to an extent that it too may obstruct a bladder stone. A radiograph showed that Barney had a stone lodged in this location. In a team effort, Dr. Griswold and I made an incision directly above where the stone was lodged. This incision was **ventral** to (below) the anus and directly over the penis. The surgery became relatively bloody as we opened the penis and cut into the urethra to remove the stone. Dr. Deppen then sutured the lining of the urethra directly to the skin. Barney was quite fortunate that the incision healed well. Since that day, Barney has been urinating from the surgical site directly below his anus.

RENAL SYSTEM FUNCTIONS

OBJECTIVE

■ *Name and Explain the Functions of the Renal System*

The renal system primarily functions to excrete nitrogen-containing wastes from the body. The majority of these wastes arise from the breakdown of proteins and amino acids. As the amino acids are metabolized, ammonia (NH_3) is produced. Ammonia is extremely toxic to tissues. The liver converts ammonia to urea, which is much less toxic to the animal (Figure 6–8). In this process, ammonia combines with carbon dioxide to produce urea.

Nucleotides are also nitrogen-containing compounds. The breakdown of nucleotides results in the formation of uric acid. Mammals have an enzyme that

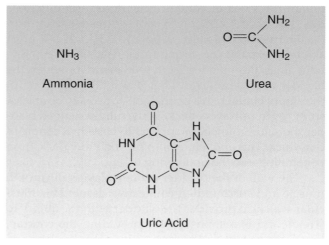

FIGURE 6–8 Chemical structure of nitrogenous waste.

further breaks down the uric acid. Dalmatian dogs have much less ability to metabolize uric acid. As a result, the kidney excretes much higher amounts of uric acid. The higher amount of uric acid in the urine can crystallize and form a bladder stone. Rufus suffered from this problem. The stones that I removed from his bladder were sent to a laboratory and determined to be urate stones. As a follow-up, Rufus was placed on a special diet that limits the amount of uric acid produced. In addition, Rufus is taking a medication daily that limits uric acid production.

It is interesting to note that although Dalmatians may have **serum** levels up to four times that of a non-Dalmatian dog, human levels reach even higher. Uric acid is relatively insoluble, and when concentrations reach adequate levels, crystals precipitate. In the urine this results in stone formation. In humans, serum levels can be so high that the crystals deposit in joints. The resulting painful condition is called **gout**.

A third waste product is creatinine, which results from the daily activity of muscles. Creatinine is not significantly affected by dietary intake. Muscle makeup of the animal does influence the concentration. Males and well-muscled active animals tend to have higher levels of creatinine. Creatinine is almost entirely filtered from the blood in the glomerulus. The level of creatinine in the blood can be a very helpful indicator of kidney function.

In addition, the kidneys help to regulate the water balance in the body. In times of need, the kidneys conserve water, making concentrated urine. This need to conserve water may arise because of limited intake or disease condition. Veterinarians often examine animals that are suffering from vomiting, diarrhea, or both. Many of these sick animals take less water into their system and have much higher fluid losses through the vomiting and diarrhea. In these situations the kidneys work to conserve a maximum amount of water.

The kidneys also are essential in regulating the amount of sodium, chloride, and potassium in the bloodstream. Again, vomiting and diarrhea can increase the loss of these electrolytes and the kidneys can work to conserve the electrolytes in the blood. In times when intake is excessive, the kidneys allow higher amounts to be excreted in the urine.

Along with the respiratory system, the kidneys aid in the control of the pH in the blood. When the pH of the blood declines (becomes more acidic), the kidneys excrete a higher amount of hydrogen ions. The pH of the urine can vary from acidic to basic, as the kidneys help to maintain the blood pH in a normal range.

In addition to producing urine, the kidneys produce hormones that are responsible for controlling blood pressure and red blood cell production. A more complete discussion of this control was presented in Chapter 4.

KIDNEY STRUCTURES AND URINE FORMATION AND REGULATION

OBJECTIVES

■ *Identify Structures within the Kidney and Detail the Formation of Urine and Its Regulation*

Sectioning through the kidney shows three distinct regions (Figure 6–9). The kidney visually divides into the outer cortex, the medulla, and the innermost renal pelvis. The cortex and medulla of the kidney consist of microscopic nephrons. The nephron, a structural unit, produces the actual urine. The kidney makes urine by filtering the blood in the nephron. Small molecules are passively forced into the filtrate. The kidneys then recapture useful molecules, returning them to the bloodstream.

The nephron is a tubular structure that is closed at one end and open at the opposite end (Figure 6–10).

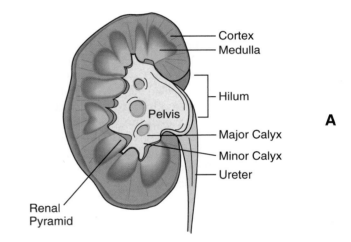

A

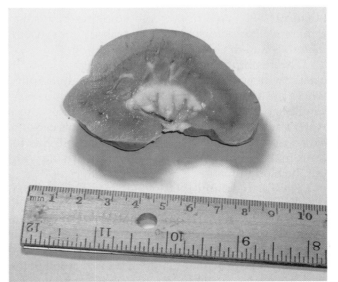

B

FIGURE 6–9 *A.* Structures of the kidney. *B.* Sectioned sheep kidney.

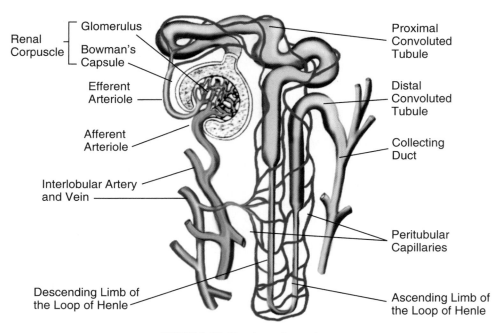

FIGURE 6–10 Structure of a nephron.

The closed end of the nephron in the cortex is folded into the structure called Bowman's capsule. Bowman's capsule is wrapped around a bundle of capillaries called the glomerulus. The nephron continues as a tubular structure that divides into three sections.

The first section is called the proximal tubule. This leads into the long, thin loop of Henle, which reaches into the renal medulla. The loop of Henle makes a sharp turn and returns to the region of Bowman's capsule in the cortex. The tubule then expands into the highly coiled distal tubule. Many nephrons empty into a collecting duct, which drains the final urine into the renal pelvis.

The entire nephron is tightly associated with blood vessels. In addition to providing the blood that is filtered, these vessels help to reabsorb any useful materials into the bloodstream.

The nephron produces urine through a complicated process. The process begins in the glomerulus, where an afferent arteriole delivers blood into these capillaries. The pressure within these capillaries forces water and small molecules into the Bowman's capsule. Proteins within the blood are not passed into this filtrate. In this respect the fluid is much like plasma without the blood proteins.

The filtered fluid then passes into the proximal tubule. The proximal tubule reabsorbs many of the essential molecules. The epithelial cells lining the proximal tubule are rich in mitochondria. These cells use a large amount of energy in the process of active transport. The proximal tubule reabsorbs much of the glucose, amino acids, vitamins, and ions that have

made it into the filtrate. These products are returned to the capillaries that surround the proximal tubule (Figure 6–11).

As the solutes return to the bloodstream, water follows by osmosis. More than 80% of the water is reabsorbed at this point. Not all of the solutes within the filtrate are removed. Much of the sodium, chloride, and urea (a form of nitrogen-containing waste) remain in the tubules.

The fluid then passes into the loop of Henle, where much more sodium is pumped out of the solution. The epithelium lining the loop of Henle does not allow water to follow the sodium. The extracellular fluid (ECF) in the medulla of the kidney, surrounding the loop of Henle, becomes very salty. The fluid within the tubules, however, becomes highly diluted.

The fluid continues into the distal tubules, where even more sodium is pumped into the ECF. The cells lining the distal tubules and collecting ducts control how much water may leave (Figure 6–12). The ECF in this region exerts a high osmotic force, trying to draw water out of the tubules. The epithelium is made more permeable when the body needs more water conserved. The draft horse working all afternoon plowing in the hot sun attempts to conserve a maximum amount of water. In this horse the walls of the distal tubules and collecting ducts become very permeable to water, allowing osmosis to pull water into the ECF. In these animals the urine becomes very concentrated, having a minimum of water. The kidneys are capable of making urine that is almost four times as concentrated as blood.

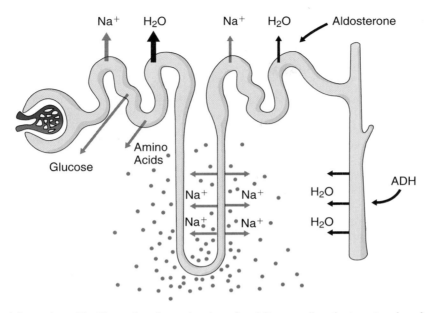

FIGURE 6–11 Function of the nephron. The illustration shows the sites of activity, as well as the target regions for the hormones aldosterone and ADH.

The discussion of urine formation to this point has been of filtration followed by reabsorption. The kidney also has the capability of secretion. The distal tubules help to control the pH of the blood by secreting either hydrogen ions (H^+) or ammonium ions (NH_4^+). The pH of the blood is kept in a very tight range, usually 7.3 to 7.4. In situations where the blood pH begins to decrease, the distal tubules secrete additional H^+. For each ion that is secreted, a sodium ion (Na^+) is absorbed. Likewise, if the blood becomes more basic (increases in pH), the kidneys secrete additional NH_4^+. Although the pH of the blood stays in the very tight range, the pH of the urine often ranges from 5 to 8.

The distal tubules also secrete excess potassium ions and another form of nitrogen-containing waste, creatinine. In addition, many drugs and toxins are removed from the bloodstream by secretion in the kidneys.

The collecting ducts empty into the renal pelvis. The pelvis collects the final urine produced and allows it to drain into the ureter. At this point the urine no longer changes in composition. It then moves to the urinary bladder for storage.

As mentioned earlier, more than 80% of the water produced in the filtrate is reabsorbed by the proximal tubule. The amount that is absorbed is selectively controlled in the kidney. Antidiuretic hormone is responsible for this control. Receptors in the brain detect when an animal begins to dehydrate (lose too much fluid) by detecting an increase in concentration in the blood. When this occurs, antidiuretic hormone (ADH) releases from the pituitary gland, which is located at the base of the brain. ADH is carried in the bloodstream to the kidneys, causing the distal tubules and collecting ducts to become more permeable to water. As a result, more water is reabsorbed and the urine produced becomes much more concentrated.

In addition to the control of the kidney, receptors within the brain also stimulate the sensation of thirst. When sodium levels increase, the animal is stimulated to drink. The thirst center can be used to aid in the treatment of animals. Veterinarians may use this technique in cows with severe **mastitis** (infection of the mammary gland). These cows would benefit from

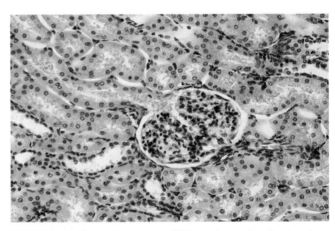

FIGURE 6–12 Photomicrograph of kidney tissue showing sections through many of the tubules of nephrons and a region of glomerulus. *(Photograph courtesy William J. Bacha, PhD, and Linda M. Bacha, MS, VMD.)*

large volumes of **intravenous** (in the vein; IV) fluids. Unfortunately, that administration is time consuming and often very difficult to do on a farm. As a substitute, a much smaller volume of a product called hypertonic saline (salt solution; NaCl) may be administered. Normal saline is 0.9%, which is **isotonic** with (the same concentration as) the blood. When hypertonic saline (7.2%) is administered, the brain detects the increased sodium concentration. The animal is then stimulated to drink. These cows often drink large volumes, which then is absorbed into the bloodstream. This procedure lets the cow do the work for the vet!

In Chapter 4 the renin-angiotensin system was discussed as a mechanism to help control blood pressure. As blood pressure declines, which might occur with dehydration, the kidney releases renin. This causes angiotensin to be formed and aldosterone to be released from the adrenal glands. (Review Figure 4–16.) The overall result is that the kidneys reabsorb more water and sodium. This increases the blood volume and, when combined with vasoconstriction, results in increased blood pressure. The net effect is also less urine production.

URINE AND BLOOD EVALUATION

OBJECTIVE

■ *Evaluate Urine and Blood as a Measure of the Health of the Animal and the Urinary System*

Veterinarians can evaluate urine (**urinalysis**) as a guide to assessing many aspects of the health of the animal. Obviously the urine reflects the health of the urinary system. Remember that products of metabolism by other organ systems are also excreted in the urine. Therefore urinalysis can provide many clues about the overall health of the animal.

Table 6–1 shows a list of the common characteristics of urine that are evaluated and the normal results expected for dogs and cats. Visually urine can be judged on color and appearance. Urine is typically yellow and should be clear. The darkness of the yellow can vary with the concentration of the urine, but the urine should always be clear. Cloudy urine can be an indication of a urinary tract infection.

Specific gravity is the measure of urine concentration. Technically, specific gravity is the weight of a liquid as compared with distilled water. The more concentrated (the more solute in the urine) the urine becomes, the higher the specific gravity. A special tool called a **refractometer** is used to measure specific gravity (Figure 6–13). A drop of urine is placed on the refractometer. The technician holds the refractometer to the light and looks through the viewer. A shadow

TABLE 6–1 Normal Urinalysis

	Dog	Cat
Color	Yellow	Yellow
Transparency	Clear	Clear
Specific gravity	1.015–1.045	1.035–1.060
Volume (ml/kg body wt/day)	20–40	20–30
Glucose	Negative	Negative
Ketones	Negative	Negative
Bilirubin	Negative to trace	Negative
Protein	Negative to trace	Negative to trace
Blood	Negative	Negative
pH	5.0–7.5	5.0–7.5

line is created across a scale that is read as the specific gravity. The number for dogs typically ranges from 1.015 to 1.045. A specific gravity of 1.025 is read as "ten twenty-five."

Figure 6–13 also shows a standard urine test strip. The strip is dipped into a urine sample, and at the appropriate interval (usually 30 to 60 seconds) the color of the small blocks is compared with a scale on the container. Table 6–2 shows the tests that are evaluated with strips at our clinic. A brief description of the significance of each test is included.

Recognize that testing urine reflects the condition of the entire urinary tract and body. The presence of blood, protein, or both in the urine is a very common sign of a bladder infection. However, a positive test does not localize the source to the bladder. The blood could be a result of kidney disease or a clotting disorder in the circulation.

Microscopic evaluation of urine is also performed. The urine is placed in a centrifuge tube and spun for 5 minutes at 1500 to 2000 rpm. When removed, a small plug of sediment rests on the bottom of the tube. The liquid portion is removed, and a drop of stain is added to the sediment. A drop of this stained sediment is then applied to a microscope slide and examined. A normal urine sample contains a few red and white blood cells, epithelial cells, and a few crystals (Figure 6–14).

The method in which the urine is collected can influence the number of cells present in the urine. **Free catch urine**, collected while the animal is urinating, may easily have an increased number of bacteria. These bacteria are added to the urine by contamination with the skin of the genital region. To avoid this problem, urine may be collected by inserting a hypodermic needle into the bladder and aspirating urine with a syringe. This technique avoids the problem of

FIGURE 6–13 *A.* Equipment used in urine analysis. *B.* A refractometer scale showing a urine specific gravity of 1.034 (far left scale). When measuring protein in serum or plasma, the center scale is used.

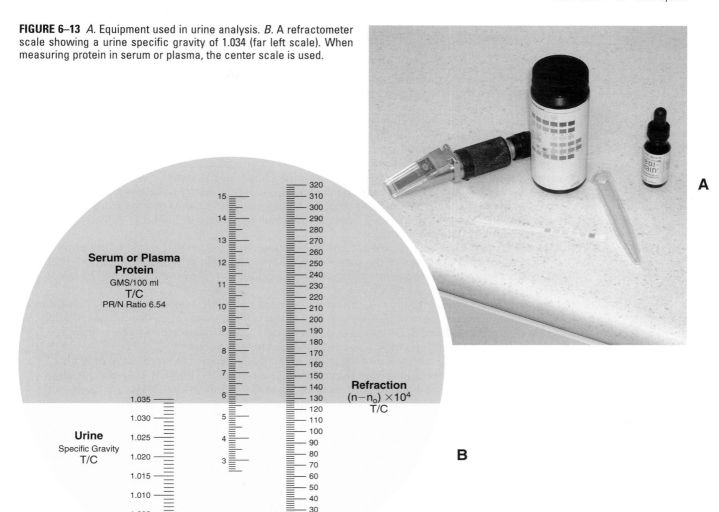

bacterial contamination but makes an increased red blood cell count much more likely.

A small number of crystals can be a normal occurrence in urine. Excessive numbers of crystals can indicate that bladder stones are much more likely to form. The crystals can be made of molecules of calcium oxalate, ammonium urate, and magnesium ammonium phosphate (struvite). Crystals form when the concentration of a molecule is too high to stay dissolved. Numerous crystals can form.

The presence of crystals in the urine can be helpful in identifying animals at higher risk for bladder stones or aiding in the identification of the type of stones present. Some animals with bladder stones do not have serious clinical signs. Certain stones may be able to be dissolved by using special diets and medications. These diets restrict the mineral found in the

stone and also adjust the pH of the urine. The pH of the urine influences the solubility of crystals. Proper identification of the stone is important to correctly adjust the urine pH. Evaluating the urine while receiving treatment can be helpful in monitoring the success of that treatment.

In addition to urinalysis, many blood tests are available to test kidney function. Both urine and blood tests can be significant in evaluating the condition of the animal. Many laboratories offer a group of tests as a standard evaluation. My practice utilizes a commercial laboratory that calls this group of tests a chemistry profile. Many practices have this type of testing equipment available at their facility.

Table 6–3 shows a partial list of the chemistry profile results for Symba, a patient seen at our practice. Laboratories report a reference range with the

TABLE 6–2 Urine Test Strips: Common Tests and Significance

Test	Significance
Urobilinogen	Used more commonly in human medicine to evaluate liver disease or the breakdown of red blood cells. This test is less useful in veterinary medicine. (The test strips used are designed for human urinalysis.)
Glucose	Used to screen for diabetes. Diabetics have elevated blood sugar. The kidneys are unable to conserve all the sugar once the level becomes too high. The test is also used to monitor diabetics once treatment has been started.
Ketones	In dogs and cats, the presence of ketones is typical of an animal with uncontrolled diabetes. Ketones result when metabolism is shifted from carbohydrates to lipids.
Bilirubin	Aged red blood cells are removed from the circulation in organs such as the spleen. Bilirubin is formed in the breakdown process of hemoglobin. Normally bilirubin is cleared from the blood by the liver and excreted in bile (see Chapter 7). Bilirubin is found in the urine in liver disease or excessive blood cell breakdown.
Protein	Proteins are large molecules that are not normally filtered into the urine. Protein in the urine can be present with a disease of the glomerulus (making it leaky) or with inflammation of the urinary tract (such as a bladder infection). A small amount of protein may normally be detected in very concentrated urine.
Blood	The test strip detects occult blood (i.e., blood that cannot be visibly seen in the urine). Blood can be present in diseases that cause inflammation of the urinary tract, much like protein. Bladder infections, stones, tumors, and trauma (e.g., hit by car) can all cause blood to be present in the urine. Bleeding disorders may also cause the test to be positive.
pH	Urine pH is influenced by diet and disease states in the body. Acidic pH is typical in animals with a meat diet or with acidosis (the kidney attempting to rid the body of excess acid). Basic or alkaline pH is typical in animals with a cereal grain diet, some urinary tract infections, and alkalosis in the body.

test results. A reference range reports the results that are normal or average for healthy animals evaluated with that test equipment. A reference range is important because variation occurs with the equipment used and results may vary slightly among laboratories.

With an understanding of how kidneys function and what role they play, interpretation of blood values is possible. Notice that many of the tests have a direct relationship to kidney function. Urea and creatinine are both compounds cleared from the bloodstream by the kidney. Urea nitrogen is measured as the means to determine the urea level in the blood. The kidney also plays a critical role in maintaining a normal blood level for the electrolytes (Na^+, K^+, Cl^-, bicarb, Ca^{++}, $PO4^{--}$).

Symba's blood report shows an elevation of both urea nitrogen and creatinine. This is defined as **azotemia**. Azotemia does not necessarily mean that the kidneys are damaged. The azotemia can result when adequate blood flow is not delivered to the kidneys. This prerenal azotemia can result when the animal is severely dehydrated. Consider animals with vomiting and diarrhea. These animals lose extra water and cannot take in water naturally, especially when vomiting is severe. The animals often have an increased concentration of protein in the blood. The total amount of protein has not increased, but because there is less water in the blood, the concentration is increased.

Azotemia can also result when the urine produced cannot be eliminated from the body. Animals with urinary obstruction develop this postrenal azotemia. We have already discussed dogs and goats with stones causing obstructions. These animals develop postrenal azotemia. As the pressure increases in the bladder, the kidneys are unable to continue producing urine. Eventually the kidneys themselves will be damaged from this pressure. In this situation, even when the obstruction is relieved, the animal will continue to be azotemic.

Kidneys that are not functioning normally often cause azotemia. This is called primary renal azotemia. The kidneys have a tremendous reserve of nephrons. For primary renal azotemia to occur, more than 75% of the nephrons must be damaged. This reserve allows for a kidney to be removed in an otherwise normal animal. A dog that has a kidney damaged by trauma or by a tumor can have that organ removed. This dog will function quite normally as long as the remaining kidney is healthy. Blood tests on such an animal would be normal. This also allows for kidney

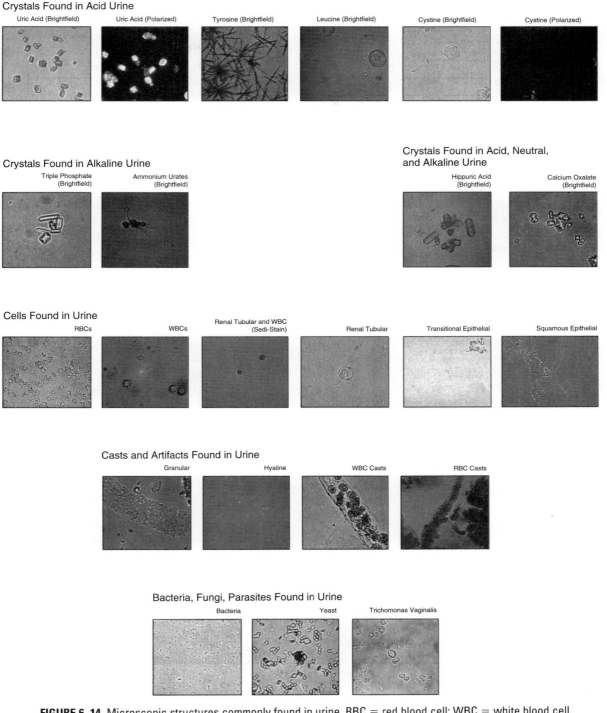

FIGURE 6–14 Microscopic structures commonly found in urine. RBC = red blood cell; WBC = white blood cell.

transplant patients to receive only one kidney. The newly transplanted kidney has adequate function to support the animal. Kidney transplants have increased in companion animals, especially cats, in recent years.

The kidneys play a critical role in regulating the mineral content of the blood. When diseases occur within the urinary tract, the concentration of these electrolytes is often disturbed. There is some variation in whether the concentration increases or decreases. The changes are influenced by how quickly the disease develops, where the disease is attacking, and other systems affected (such as vomiting and diarrhea).

CLINICAL PRACTICE

OBJECTIVE

■ *Discuss the Clinical Significance of the Academic Material Learned in This Chapter*

Symba's blood results show severe azotemia (Table 6–3). The urea nitrogen, creatinine, and phosphorus are extremely high. Symba, a 2-year-old male mixed breed dog, was presented to our clinic for vomiting and diarrhea. Symba was well vaccinated, and therefore infectious diseases such as **parvovirus** (a viral disease in dogs causing severe vomiting and diarrhea) were much less likely. The history showed that Symba had been free to roam the neighborhood, which raised the concern that Symba had eaten something that had upset his stomach or gotten into some sort of poison.

The blood tests were taken while we treated Symba for the vomiting and diarrhea. He did not respond to the medications, and the blood results helped to show why. **Uremia** describes the group of clinical signs associated with azotemia. Although Symba presented with what appeared to be a disease of the intestinal tract, the actual problem was being caused by kidney disease. Indications of uremia may include one or more of the following: vomiting, diarrhea, poor appetite, bad breath, and lack of energy. The signs that occurred in the intestinal tract were due to the increased concentration of waste products (urea and creatinine). These waste products are toxic to the lining of the entire gastrointestinal tract and may cause the development of ulcers (a defect in the surface), which may even be seen in the mouth.

Symba's presentation was classic for antifreeze toxicity. Many types of antifreeze contain the active ingredient ethylene glycol. The animal metabolizes the ethylene glycol in an attempt to eliminate it from the body. Unfortunately, these breakdown products are toxic to the kidney. Care must be taken when antifreeze is changed in a car or if there is a leak in the system.

The damage can be prevented if the ingestion is detected in time. If the animal is caught drinking the antifreeze, it can be made to vomit. Activated charcoal can also be administered to help prevent more antifreeze from being absorbed into the system. Two antidotes are also available that can be administered to the animal. These keep the ethylene glycol from being metabolized but are only helpful if the problem is detected before the damage to the kidneys is already done.

Symba was in renal failure. Renal failure happens when the kidneys are no longer able to maintain normal function. This function obviously presents as an increase in the nitrogenous wastes and electrolytes within the blood. In addition, the endocrine functions are also affected. Erythropoietin production may also decline, resulting in anemia.

Renal failure can be either **acute** (sudden onset) or **chronic** (long term). Symba was in acute renal failure. His condition developed very suddenly and was likely due to a toxin. The list of toxins that can cause kidney damage is quite long and includes medications such as acetaminophen. Any factor causing severe inflammation, infection, or loss of blood flow to the kidney can cause acute renal failure. Table 6–4 lists only a small portion of the causes of acute renal failure.

When detected early, the predisposing cause of the kidney damage can be corrected. This correction is specific for the cause. For example, with antifreeze poisoning, an antidote can be given to prevent further damage. If the underlying cause is corrected, the kid-

TABLE 6–3 Chemistry Profile Results for Symba

	Test Result	Reference Range	Unit*
Glucose	114	65–120	mg/dl
Urea Nitrogen	229	6–24	mg/dl
Creatinine	19.7	0.4–1.4	mg/dl
Na$^+$	144	140–151	mEq/L
K$^+$	5.7	3.4–5.4	mEq/L
Cl$^-$	95	105–120	mEq/L
Bicarbonate ion	12	15–28	mEq/L
Cholesterol	233	110–314	mg/dl
Total bilirubin	0.1	0.0–0.4	mg/dl
Total protein	6.6	5.2–7.2	g/dl
Albumin	3.2	2.4–4.3	g/dl
Globulin	3.4	0.9–4.0	g/dl
Ca^{++}	11.9	7.9–12.0	mg/dl
Inorganic phosphorus	24.1	2.1–6.8	mg/dl

*mg/dl = milligrams per deciliter; m/EqL = milliequivalent per liter; g/dl = grams per deciliter.

TABLE 6–4 Causes of Acute Renal Failure

Severe blood loss
Shock
Heat stroke
Trauma
Burns
Infections
Autoimmune disease
Urinary outflow obstruction (e.g., bladder stones preventing urination)
Toxins (certain antibiotics, cancer drugs, acetaminophen, lead, mercury, ethylene glycol, solvents, snake venom)
Tumors

ney may be able to heal. With acute renal failure, there is hope that the animal can recover with supportive treatment.

Chronic renal failure is much more common. The classic presentation is an elderly pet that has begun to lose weight and has become lethargic. The owners often report that the pet's appetite has been poor but that it drinks a large amount of water. Along with the excessive thirst, the animal then has to urinate frequently. This is often the first hint to the owners that a problem is developing. A pet that has been housebroken for years and then begins to have accidents in the house or has to get the owners out of bed several times a night may be in the beginning stages of chronic renal failure. Many times, vomiting, diarrhea, or both are associated with the kidney failure as well.

Chronic renal failure develops over a period of months to years. The condition often occurs slowly without any visible indications. Although the animal would not be showing clinical signs, performing a chemistry profile would detect the earliest changes. Typically, urea nitrogen, creatinine, and phosphorus are the first items to increase above the normal reference range. With time, the animal becomes anemic due to the lack of erythropoietin production by the kidney. As time progresses, the values on the profile continue to worsen.

The failure of kidney function may come from old age, when the organ just fails to function properly. It is much less common, but animals can also be born with conditions in which the kidney is not formed normally at birth. These animals develop kidney failure at a young age.

With chronic renal failure, there is no specific medication to correct the damaged kidney. If detected early, the animal is often placed on a special diet. Remember that the urea nitrogen in the blood is formed from the breakdown of protein in the diet. Therefore these diets often restrict the amount of protein. Phosphorus is also limited because it is typically elevated in renal failure. However, protein and phosphorus are present in these diets; they are just kept to minimal levels to support the animal.

A complete cure of chronic renal failure requires a kidney transplant or dialysis. The use of transplants has proven easier in cats than in dogs. Both procedures are quite expensive and at this point are still being performed only in specialty practices.

Dehydration may occur in either acutely or chronically dehydrated animals. In companion animals the two most common treatments are **subcutaneous** (subQ) or IV fluids. Dogs and cats have loosely attached skin that allows for relatively large amounts of fluids to be given in one location. The fluid then slowly absorbs into the circulation. SubQ fluids are typically used when the treatment is expected to be of short du-

ration. When administering IV fluids, a catheter is used to enter a vein. The catheter has a needle to allow the vein to be punctured and a soft plastic catheter that gets threaded into the vein. The softer portion prevents any damage to the animal, because the catheter remains in the body for extended periods.

Veterinarians are often faced with the need to evaluate the dehydration status of an animal. Laboratory tests are very helpful. When on a farm, vets normally do not have the luxury of these tests and often have to make decisions based on the appearance of the animal. Three factors are typically evaluated on the animal: position of the eye in the socket, skin turgor, and mucous membranes.

As the animal dehydrates, the eye sinks farther into the socket. There is a fat deposit behind the eye, and as the water leaves the fat, the eye settles deeper into the socket.

Skin turgor is a measure of how quickly the skin returns to its normal position when pinched. Students can test their own skin turgor by gently pinching a fold of skin on their arm. Normally the skin snaps back into its normal position immediately. The more dehydrated the animal, the slower the skin returns to normal.

Feeling the mucous membranes on the gums is also helpful in detecting dehydration. Normal mucous membranes are moist. As dehydration worsens, they become tacky and then dry. Table 6–5 provides a guide to estimating dehydration.

Once the percentage of dehydration has been estimated, the amount of fluid to correct the problem can be calculated:

$$\text{Percent dehydration} \times \text{Body weight in kg} \times 1000 = \text{Replacement volume (ml)}$$

This volume is generally given over the first 2 to 4 hours after diagnosis. In addition, the fluid treatment must continue to meet the current demands of the animal. Maintenance in the normal animal is 20 to 25 ml/kg/day.

TABLE 6–5 Estimating Dehydration

Percent Dehydration	Clinical Presentation
6%–7%	Eyes slightly sunken, skin turgor slightly slower, moist membranes
8%–9%	Eyes obviously sunken, skin turgor obviously slower, membranes tacky
10%–12%	Eyes deeply sunken, skin very slow to return to normal position, membranes dry

Adapted from: Roussel Jr, Allen J., and Constable, Peter D., ed: *The Veterinary Clinics of North America, Food Animal Practice, Fluids and Electrolyte Therapy*, 15:3, Philadelphia, W.B. Saunders, Nov 1999, p. 547.

An example may help to illustrate the necessary calculations. Sparky, a 20-kg Australian shepherd, presents for a severe vomiting problem. The owners caught the problem early, and you estimate that he is 6% dehydrated. You decide that he should be kept from food and water for the next day to rest his stomach. How much fluid will he require over the next 24 hours?

$$\text{Replacement volume} = 20 \text{ kg} \times 6\% \text{ dehydration} \times 1000$$
$$\text{Replacement volume} = 20 \times 0.06 \times 1000$$
$$\text{Replacement volume} = 1200 \text{ ml}$$
$$\text{Maintenance volume} = 25 \text{ ml/kg/day} \times 20 \text{ kg}$$
$$\text{Maintenance volume} = 500 \text{ ml}$$
$$\text{Total first day requirement} =$$
$$\text{Replacement volume} + \text{Maintenance volume}$$
$$\text{Total first day requirement} = 1200 + 500$$
$$\text{Total first day requirement} = 1700 \text{ ml}$$

This amount may actually have to be increased if there is excessive loss of fluid in the form of vomiting or diarrhea.

Much of our discussion has revolved around bladder stones in dogs and goats. Male cats are commonly affected by urinary obstruction as well. Inexperienced cat owners often call complaining that their cat is constipated. They see the cat sitting in the litter pan straining without success. The owner assumes that the cat is having trouble defecating. The more common problem is that these cats are straining to urinate. The narrow urethra in the penis becomes obstructed with a plug of crystals, blood, cells, and matrix. If left unattended, this condition can be life threatening.

The diagnosis is usually quite straightforward. The cat will present with a very painful abdomen and a huge bladder. These cats are generally sedated or anesthetized, and the urethra is flushed until the obstruction is relieved. The length of time that the cat has been unable to urinate then guides further treatment. In severe cases, acute renal failure is developing and must be treated. These cats must be monitored for reoccurrence, which can be quite common.

Prevention is aimed at controlling the diet. The diets are designed to control the mineral content of the urine and also maintain an acid pH. Both features attempt to minimize how many crystals are formed. Many cats develop the problem again even with a special diet. A special surgery is available in these problem cats. The surgery is called a perineal urethrostomy. The basic procedure is to amputate the penis, eliminating the narrowest urethra. The remaining urethra is sutured to the skin, creating a much larger opening for the cat to urinate.

SUMMARY

The renal system, including the kidneys and bladder, allow for waste products of body functions to be eliminated in the form of urine. Evaluation of urine can provide valuable insight into the status of the renal system and the total health of the animal. Veterinarians use knowledge of the renal system to successfully treat such problems as dehydration, acute and chronic renal failure, and bladder stones.

REVIEW QUESTIONS

1. Define any 10 of the following terms:

   bucks
   does
   dorsal
   retro
   peristalsis
   urinary incontinence
   spayed
   ventral
   serum
   gout
   mastitis
   intravenous
   isotonic
   urinalysis
   specific gravity
   refractometer
   free catch urine
   azotemia
   parvovirus
   uremia
   acute
   chronic
   subcutaneous
   skin turgor

2. True or False: Urine becomes more concentrated during times of excessive hydration.

3. When sodium levels increase, an animal is stimulated to _____.

4. When an animal's kidneys no longer function, the animal is said to be in _____ failure.

5. What tubelike structure connects the kidney to the bladder?

6. Do males or females have a higher incidence of urinary incontinence?

7. What breed of dog has a history of difficulty metabolizing uric acid?

8. What tool measures specific gravity?

9. What is the name of the test that evaluates urine?

10. What dietary component may be limited in animals with chronic renal failure?

11. Can azotemia result from dehydration?

12. What is the primary function of the renal system?

13. Differentiate between acute and chronic.

14. Name the two systems that control pH of blood.

15. List two symptoms of parvovirus.

ACTIVITIES

Material needed for completion of activities:
   metric to English conversion (1 kg = 2.205 lb)
   English to metric conversion (1 oz = 29.57 ml)
   beaker, at least 100 ml
   urinalysis kit

1. Earlier in this chapter, it was stated that the bladder of an 11-kg dog can easily hold more than 100 ml of urine. Convert the weight of the dog from kilograms to pounds. Look at a 100-ml beaker to see the potential size of the dog's bladder.

2. Monitor your fluid intake for a 24-hour period. Calculate intake on a weight basis (that is, divide the total fluid intake in milliliters by body weight in kilograms). Does this number come close to the 20 to 25 ml/kg/day in animals? You may find you have consumed more or less fluid than the 20 to 25 ml/kg/day needed by the body. Voluntary consumption often exceeds this amount. The kidneys are able to excrete the excess volume. Remember that water is also taken in during consumption of food. For example, soup adds fluid to the diet, much more than French fries.

3. Perform a urinalysis.

4. You can test your own skin turgor by gently pinching a fold of skin on your arm. Normally the skin snaps back into its normal position immediately. The more dehydrated you are, the slower the skin returns to normal.

7

The Digestive System

OBJECTIVES

Upon completion of this chapter, you should be able to:

- Identify the basic structures of the digestive system.
- Explain digestion in monogastrics, including:
  exocrine secretions and function.
  digestive tract function.
  digestive tract absorption.
  role of the liver in digestion and metabolism.
- Compare and contrast the specialization of dentition and digestive tracts found in the various domestic species, and define symbiosis and its significance in the ruminant.
- Discuss the clinical significance of the academic material learned in this chapter.

KEY TERMS

intussusception	deciduous teeth	symbiosis	vestibular system
colic	peristalsis	rumination	bloat
carnivore	monogastric	eructate	
herbivore	phenobarbital	retching	

INTRODUCTION

Digestion is the process in which food is taken into the body and broken down into small molecules, which can be absorbed and utilized by the animal. The process that accomplishes this is complex. The remnants of nonnutritious portions of the diet are then eliminated from the body.

A DAY IN THE LIFE

Rocks, Socks, and Underwear...

It is funny how my mind works sometimes. I was getting ready to do surgery this morning and I had to think of my high school biology teacher. He introduced our class to the word *gravy-splasher*. When we were given a new large term, he called it a gravy-splasher. He said that we could tell our parents about it at supper and they would be so impressed that they would drop their spoons in their gravy!

The gravy-splasher that came to mind today is **intussusception**. Toby, a year-old Jack Russell terrier, was on my list for neutering. I had seen Toby almost six months earlier for a severe vomiting problem. Toby had been vomiting all day and had then become uncomfortable. He was acting like his belly was hurting. When I palpated his abdomen, he winced. I was able to feel a thickening in one region of his intestine.

I was concerned that he may have eaten something that was trapped in his intestine. I obtained a radiograph of his abdomen but could not identify a foreign body. The thickening felt like a firm tube in the middle of his abdomen. His clinical signs, the vomiting and pain, along with the feel of his abdomen made me suspect an intussusception. An intussusception occurs when a region of the intestine begins to telescope into itself (much like the toe of a sock turned into the remainder). I discussed the options with the owner, and we elected to do exploratory surgery on Toby.

Dogs are incredibly intelligent animals, but sometimes they just do some dumb things. In the few short months since I began writing this text, veterinarians at our office have removed rocks, wrapping off of a chicken, the tip of a nipple bottle, and a pair of underwear from the gastrointestinal tracts of pets (Figure 7–1).

Buck is a 2-year-old Labrador retriever with an unusual desire to eat clothing. Dr. Griswold had received the call that Buck had vomited a sock two days earlier but was still not eating well. Buck continued to vomit even after getting rid of the sock. Buck eventually required surgery to have the underwear that he had also eaten removed from his small intestine. In the two months since his surgery, we have received two more calls from Buck's owner. In spite of the owner's best efforts to keep all clothes away from him, Buck has managed to eat socks twice more. Once we were able to make him vomit the sock, and the second time it passed through in the stool.

Vomiting and diarrhea are extremely common problems encountered in small animal medicine. There are numerous causes for these problems. Parasites, infectious agents, and poor diets can all contribute to

continued

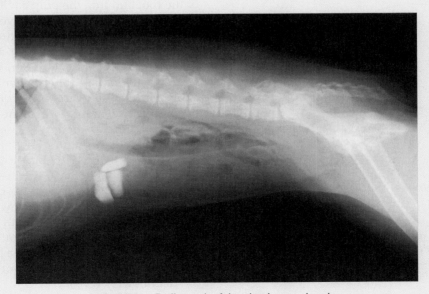

FIGURE 7–1 Radiograph of dog that ingested rocks.

vomiting and diarrhea. Dogs and cats often eat foods that upset their gastrointestinal tracts. A common source of stomach upset in pets is table scraps. This problem is especially common immediately after holidays, when families have big meals.

My practice does very little equine work. However, we still receive calls from owners about their horses suf-

fering from **colic**. *Colic* is a general term referring to abdominal pain. Colic generally reflects a problem with the gastrointestinal tract but is not a specific disease; it is a common problem in horses. Horses with colic kick at their abdomens, lie down, get up, or roll in attempt to relieve pain. Often these horses sweat profusely and have very high heart and respiration rates.

Carnivore

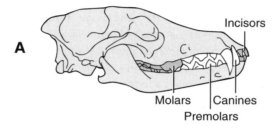

Herbivore

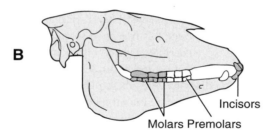

Omnivore

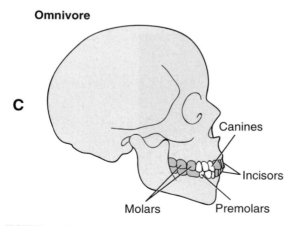

FIGURE 7–2 Comparison of skull and teeth. *A.* Carnivores (for example, dogs and cats) have pointed canines and incisors for obtaining and tearing flesh. Sharp molars and premolars are essential for shearing flesh. *B.* Herbivores have teeth adapted to biting off plant material and then grinding the food into smaller pieces. *C.* Omnivores, such as humans, have teeth adapted for eating a variety of foods.

DIGESTIVE SYSTEM STRUCTURES

OBJECTIVE

■ *Identify the Basic Structures of the Digestive System*

Teeth are used to gather, cut, and grind the food. Differences in teeth are quite noticeable when comparing a **carnivore** (such as a dog or cat) with an **herbivore** (such as a cow or horse) (Figure 7–2). The wild carnivore must capture prey and tear the flesh from the carcass. In contrast, the herbivore grazes on plant material that must be ground extensively. The teeth and jaw structure are adapted to their diet. Table 7–1 compares the number of teeth present in each species.

The crown of a tooth is the portion above the gum line (Figure 7–3). The root is the portion below the gum line anchoring the tooth in the bone. Many teeth have more than one root. The crown is covered with enamel, which is the hardest substance in the body. Dentin lies below the enamel and makes up the majority of the crown and root. The mineral content of dentin is very similar to that of bone. The nerve, artery, and vein that supply each tooth enter through the end of the root and into the pulp cavity.

Some newborn animals have teeth erupted from the jaw at birth. Other teeth enter the mouth as the animals mature. The initial set are called **deciduous teeth**. As an animal gets older, the deciduous teeth are replaced by larger permanent teeth. This process allows the animal to grow and therefore have room in the jaw for the larger permanent teeth.

Teeth are classified as incisors, canines, premolars, and molars (Figure 7–4). The incisors are the foremost teeth, used to bite into the food. Canine teeth, if present, are the longest teeth and are used for tearing the food. The root of the canine tooth is about twice as long as the exposed crown. This can make removal of a damaged canine tooth very difficult. When removing a lower canine from the mandible, care must be taken not to fracture the jawbone. The root of the tooth can involve a large portion of the depth of the mandible.

Premolars and molars are used for crushing and grinding food. In the dog and cat the upper fourth pre-

TABLE 7–1 Comparison of Dentition between Species*

		Deciduous				Permanent		
		Incisors	Canines	Premolars	Molars	Incisors	Canines	Premolars	Molars
Dogs	Upper	3	1	3	3	1	4	2
	Lower	3	1	3	3	1	4	3
Horse	Upper	3	0	3	3	1	3 or 4†	2
	Lower	3	0	3	3	1	4	3
Cattle	Upper	0	0	3	0	0	3	3
	Lower	3	1	3	3	1	3	3

*The chart lists the number of teeth present on one side of the mouth. The total number of teeth is double.
†The mare (female horse) often does not have the canine tooth, and the first premolar is often absent.

molar and the lower first molar are called the carnassial teeth. The shape and positioning of these teeth provide a shearing or cutting function. The muscles that close the jaws can exert a tremendous force across these teeth. Many people have seen dogs crush bone with their teeth. The carnassial teeth are the primary teeth used in this chewing in dogs and cats.

The premolars and molars in herbivores are much flatter and more tightly packed. These teeth are designed for repetitive grinding of plant material. The jaw moves in a circular fashion, as the teeth not only press together but also slide across each other. This motion provides a very effective means of grinding the plant material into tiny pieces that are more easily digested.

This grinding action has a wearing effect on the teeth. The teeth of horses continue to grow throughout their lives. The teeth move out of the jaw as they are worn down from the grinding action. As the teeth are worn down, the shape and appearance of the teeth also change. The appearance of the surface of the incisors can be used to approximate the age of a horse. The aging is not exact due to differences between individuals and diet. This process, though, can be very helpful in giving an estimate of the age of the horse, such as when a client is considering purchasing a new horse.

Saliva is produced by four salivary glands (Figure 7–5). Salivary glands are the first of several exocrine glands necessary for digestion. These exocrine glands produce a product that is carried by an epithelial lined duct into the digestive tract.

The tongue, made of skeletal muscle, helps to obtain food and moves it around in the mouth to aid chewing. Once the food is adequately ground, the tongue forms a bolus. This bolus of food is worked to the back of the throat to be swallowed. In the act of swallowing, the epiglottis covers the opening to the larynx. The food passes through the pharynx and into the esophagus.

The wall of the esophagus divides into four layers (Figure 7–6). As with all the tubular structures, there is an epithelial lining with an underlying connective tissue layer. The epithelial lining of the gastrointestinal (GI) tract is called the mucosa. There are two layers of smooth muscle in the wall of the esophagus. Finally there is a connective tissue covering, called the serosa. All the regions of the GI tract have this same basic structure. The appearance and structure of the epithelial layer varies significantly between each structure.

The food is propelled through the esophagus by **peristalsis** (Figure 7–7). This occurs with organized contractions of the muscles in the esophageal wall. The inner muscular layer is oriented around the esophagus in a circular manner. This layer contracts behind the bolus of food, forcing it toward the stomach. The outer layer of muscle is arranged lengthwise in the esophagus. This layer contracts, shortening the length of the esophagus. The inner layers of muscle contract farther

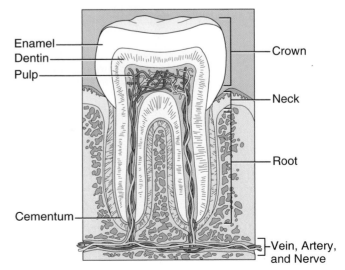

Enamel
Dentin
Pulp
Crown
Neck
Root
Cementum
Vein, Artery, and Nerve

FIGURE 7–3 The structure of a tooth.

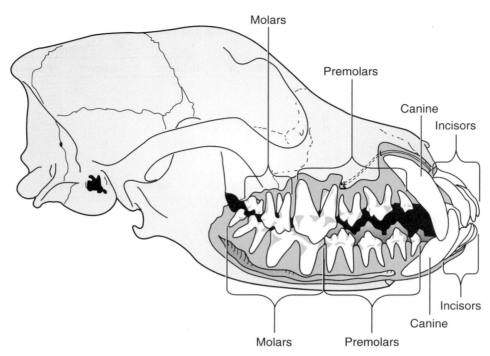

FIGURE 7–4 The types of teeth found in an adult dog.

down the esophagus, continuing to push the bolus of food into the stomach. Peristalsis describes this type of action, which occurs throughout the GI tract.

The esophagus delivers the food through the neck, chest cavity, and diaphragm into the stomach.

A sphincter controls the opening into the stomach. The sphincter is a circular muscle that remains closed until food is ready to enter from the esophagus, helping to prevent the acidic contents of the stomach from traveling backward into the esophagus. (If this backward

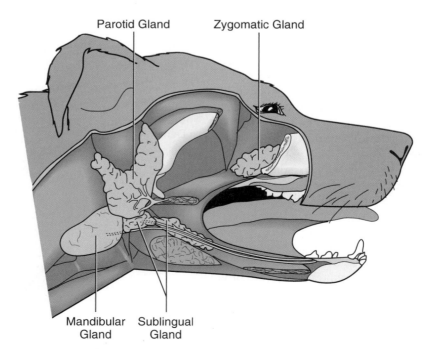

FIGURE 7–5 The salivary glands found in the dog.

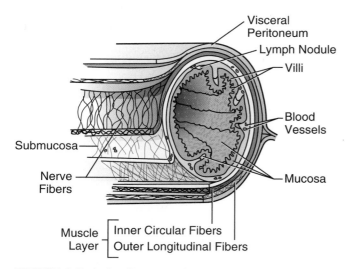

FIGURE 7–6 Typical wall structure found in organs of the digestive tract.

flow occurs, it causes the uncomfortable feeling of heartburn in humans.)

The basic structures to this point are very similar among species. The teeth are specialized based on the type of diet, but the other structures are similar. The following is a basic description of the gastrointestinal tract of dogs and cats. Dogs and cats are called **monogastrics,** meaning they have a single stomach. The variations in other species are discussed later in the chapter.

The stomach divides into three regions (Figure 7–8). The cardia is the inlet region associated with the esophagus. The major storage region is called the body of the stomach. When empty, the stomach is very small and has folds in the lining, called rugae, which allow for the stomach to expand when a large meal is eaten. The body of the stomach leads into the pylorus. The pylorus leads into the small intestine. Like the cardia, the pylorus has a sphincter that controls the flow of ingesta into the small intestine.

The stomach lies immediately behind the liver (Figure 7–9). The liver is important in digestion and also in processing the nutrients absorbed after digestion. The liver produces a secretion that is stored in the gall bladder. The liver and the gall bladder communicate with the small intestine through the bile duct (Figure 7–10).

The small intestine divides into three regions. The first relatively short section is called the duodenum. The duodenum is closely associated with the exocrine gland, the pancreas (Figure 7–11). This leads into the long middle region called the jejunum and then the final region, the ileum. (Note the difference in spelling from the ilium of the pelvis.) Much of the digestion

and absorption of nutrients occurs within the small intestine.

The ileum leads into the large intestine or colon (Figure 7–12). At this level there is a blind pouch off the colon, called the cecum. The name *large intestine* describes the diameter of the organ. The large intestine is much shorter than the small intestine. The large intestine divides into the ascending, transverse, and descending colons. The descending colon leads into the rectum. The rectum is the termination of the intestinal tract as it exits the body at the anus. The anus has a muscular sphincter that controls the act of defecation.

There are numerous connective structures within the abdominal cavity that help to support and protect the organs. The entire abdominal cavity is lined with the peritoneum. The peritoneum is a smooth thin epithelial lining with an underlying connective tissue. The smooth peritoneum allows the organs to move freely within the abdomen.

The mesentery is an extension of the peritoneum that carries the blood vessels and nerves to the small intestines. The omentum also carries blood vessels and surrounds much of the abdominal organs. The greater omentum is often the first tissue observed when the abdomen is opened. The omentum helps to minimize the spread of infection and inflammation within the abdomen.

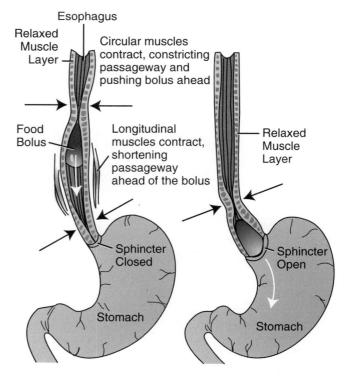

FIGURE 7–7 Peristalsis. Peristalsis actively propels ingesta through the intestinal tract.

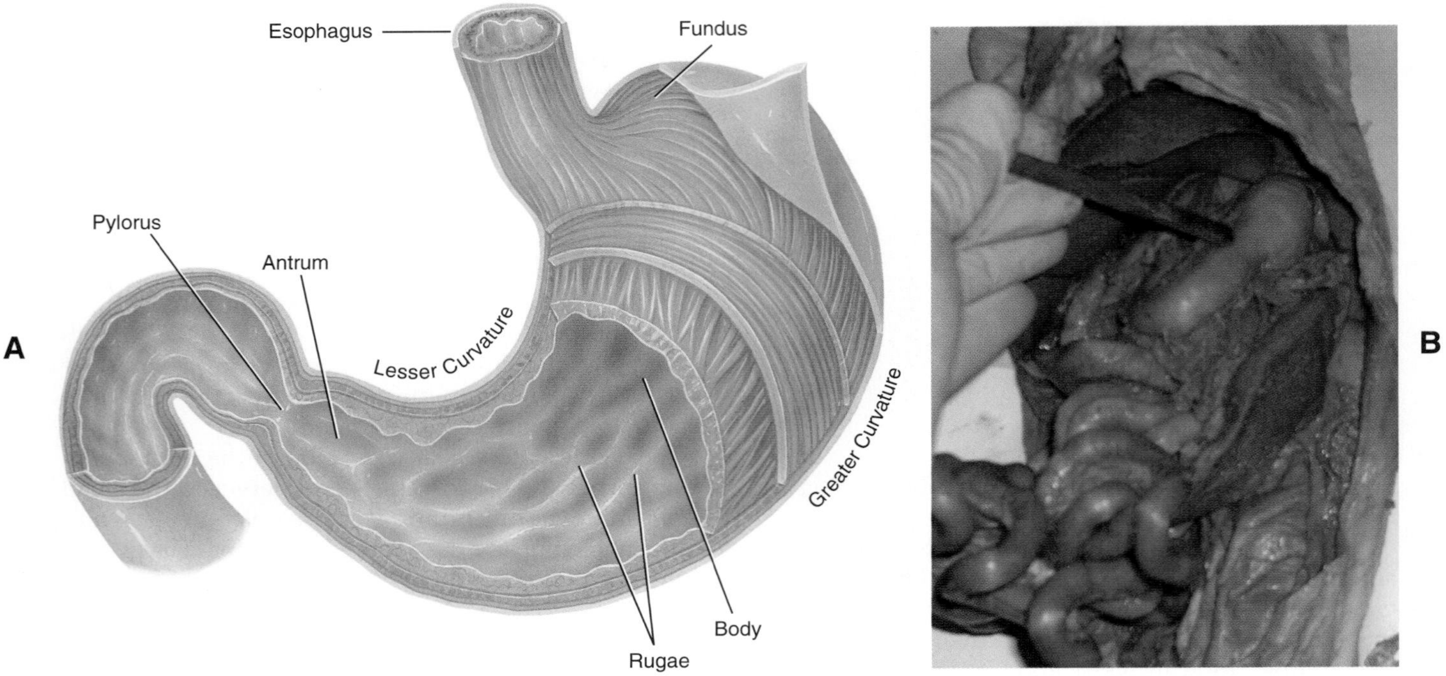

Esophagus

Fundus

Pylorus

Antrum

Lesser Curvature

Greater Curvature

Body

Rugae

A

B

FIGURE 7–8 *A.* The parts of the stomach. *B.* Normal positioning of the stomach in a cat.

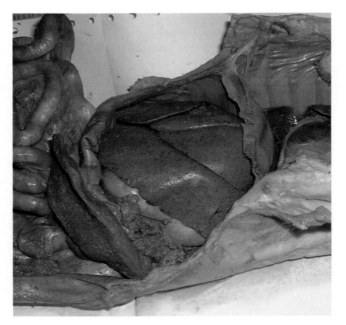

FIGURE 7–9 The liver of a cat.

MONOGASTRIC DIGESTION

OBJECTIVE

■ *Explain Digestion in Monogastrics, Including Exocrine Secretions and Function, Digestive Tract Function, Digestive Tract Absorption, and the Role of the Liver in Digestion and Metabolism (Table 7–2)*

The mouth helps to obtain food and break it down mechanically. The sight, smell, and taste of food stimulates a nervous reflex for a release of saliva. The liquid saliva mixes with the food, making it much easier to swallow. In addition, the saliva has a protective effect, coating the epithelium in the mouth and pharynx. Saliva contains sodium bicarbonate, which helps to maintain a stable pH in the mouth. Saliva also has antibacterial properties, helping to minimize the growth of bacteria. Digestion actually begins with the addition of saliva. Saliva contains the enzyme amylase, which begins to digest starch into the simple sugar maltose.

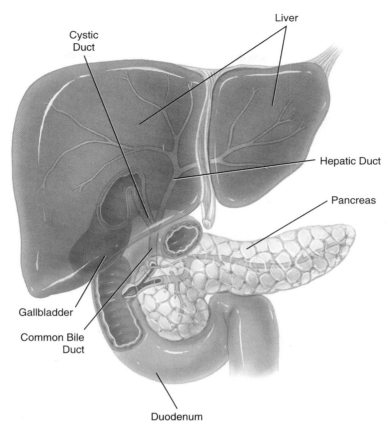

FIGURE 7–10 Gallbladder and bile ducts. Note the close proximity to the pancreas and duodenum.

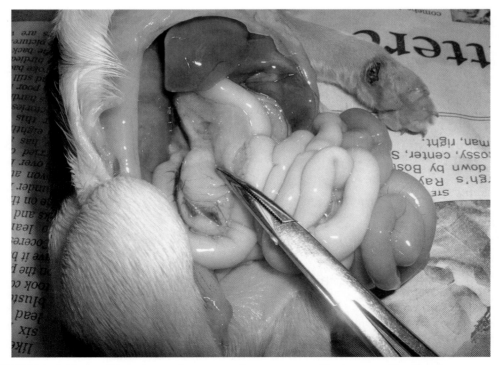

FIGURE 7–11 The small intestines and pancreas of a young puppy. The hemostat identifies the pancreas.

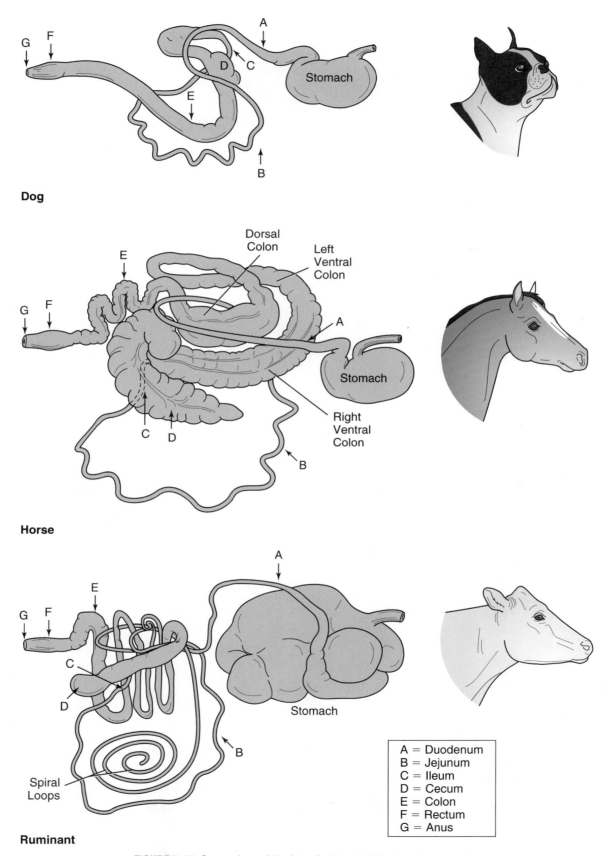

Dog

Horse

Dorsal Colon

Left Ventral Colon

Right Ventral Colon

Ruminant

Spiral Loops

A = Duodenum
B = Jejunum
C = Ileum
D = Cecum
E = Colon
F = Rectum
G = Anus

FIGURE 7–12 Comparison of the intestinal tract of the dog, horse, and cow.

TABLE 7–2 Summary of Digestive Enzymes and Hormones

DIGESTIVE JUICES

Secretion	Source	Location of Action	Action
Salivary amylase	Salivary glands	Mouth	Digests starch
HCl	Parietal cells (stomach)	Stomach	Degrades proteins Activates pepsin
Pepsin	Chief cells (stomach)	Stomach	Digests protein Activates pepsin
Trypsin, chymotrypsin, carboxypeptidase	Pancreas	Small intestine	Digests protein
Sodium Bicarbonate	Pancreas	Small intestine	Raises pH of chyme Inactivates pepsin
Lipase	Pancreas	Small intestine	Digests fat
Bile salts	Liver/gallbladder	Small intestine	Emulsifies fat
Amylase	Pancreas	Small intestine	Digests starch
Nuclease	Pancreas	Small intestine	Digests RNA and DNA

HORMONES (the hormones are released into the bloodstream; effect occurs when returned to target organ)

Hormone	Source	Target Organ(s)	Action
Gastrin	Stomach	Stomach	Increases HCl release
Secretin	Duodenum	Pancreas/liver	Increases bicarbonate release Increases bile production by liver
Cholecystokinin	Duodenum	Pancreas/gallbladder	Increases release of enzymes from pancreas Empties gallbladder
Gastric inhibitory peptide	Duodenum	Stomach	Decreases activity of the stomach

HCl = hydrochloric acid.

As food is swallowed, the bolus passes through the esophagus into the stomach. The smell, sight, and taste of food results in a nervous stimulation to the stomach. As food enters the stomach, the stretching of the stomach wall also stimulates activity. The stomach begins to contract and secrete gastric juices. The muscular contraction of the stomach wall helps to mechanically break down the ingested food.

Cells in the lining of the stomach begin to secrete gastric juices. Parietal cells, which secrete hydrochloric acid (HCl), contain many mitochondria, necessary to deliver the energy required for active transport of hydrogen ions into the stomach. The HCl lowers the pH within the stomach to 1 or 2. This very acidic pH helps break down protein and connective tissues. In addition, the low pH kills many of the bacteria that are ingested with the food. This is an important mechanism in helping to prevent disease.

The HCl also activates an enzyme called pepsinogen, which is secreted by chief cells that are also located in the lining of the stomach. The pepsinogen is converted to pepsin when it contacts the HCl. The pepsin enzymatically cleaves protein molecules and activates even more pepsinogen. Because pepsin digests proteins, it is secreted in an inactive form to protect the chief cells that secrete it.

The interior of the stomach is a harsh environment, with the low pH and the protein-digesting enzymes. Other lining cells in the stomach secrete a mucus that coats the epithelial lining of the stomach. The epithelial cells have a tight cell junction that prevents the stomach contents from penetrating between the cells. Even with the mucus protection and the tight cell junctions, the epithelial cells are very short-lived. Old cells are replaced every few days with new epithelial cells.

As proteins are broken down within the stomach, the resulting polypeptides stimulate the stomach to secrete the hormone gastrin, which is released into the bloodstream. As the blood returns to the stomach, the gastrin stimulates further HCl release.

Very little absorption occurs in the stomach. Only a small amount of water, electrolytes, simple sugars, and some medications are absorbed at this level. The remaining partially digested food, chyme, leaves the stomach through the pyloric sphincter, entering the duodenum.

The duodenum is essential for further digestion of the chyme. Ducts from the liver and the pancreas enter the duodenum. The liver and pancreas both secrete digestive enzymes that are necessary for the complete digestion of the chyme. The pancreas is a long, pale white gland that is closely associated with the duodenum. The pancreas secretes a variety of digestive enzymes into the duodenum through the pancreatic duct.

Sodium bicarbonate ($NaHCO_3$) is produced by the pancreas to neutralize the HCl of the stomach contents. The pH of the chyme is raised to a level of about 8 with the addition of the sodium bicarbonate. The elevated pH inactivates the pepsin from the stomach. The pancreas secretes several enzymes essential for the digestion of protein. Trypsin is one protein-digesting enzyme that is secreted in an inactive form. It is activated in the lumen when it contacts an enzyme (enterokinase) secreted by the epithelium of the duodenum. In the active form, trypsin begins to further digest protein. Trypsin also activates more trypsin, chymotrypsin, and carboxypeptidase. These three enzymes work to break down proteins into individual amino acids.

Salivary amylase is inadequate to completely digest all the starch. The pancreas also secretes amylase, which continues the digestion of starch and glycogen into smaller sugars. Lipase is another enzyme secreted by the pancreas. Lipase digests triglycerides into two fatty acids and a monoglyceride (glycerol with a single fatty acid). The pancreas also secretes nucleases to digest the nucleic acids, RNA and DNA.

The liver is constantly producing bile. The bile is transported to the gall bladder, where it is stored between meals. Bile salts act to emulsify fat. This works to create very small particles from larger fat globules. These smaller fat globules are more easily digested because they have a greater surface area than the large globules. The larger surface area allows for the lipases secreted by the pancreas to digest the fat. Pigments within the bile are produced from the breakdown of hemoglobin. The bile pigments are responsible for the coloration of feces.

Three hormones that control digestion are secreted by the duodenal mucosa. When the acidic contents of the stomach enter the duodenum, secretin is produced. Secretin stimulates the pancreas to release sodium bicarbonate and the liver to produce bile. As fats and proteins enter the duodenum, cholecystokinin (CCK) is released. CCK stimulates the pancreas to release its digestive enzymes and the gall bladder to empty. Gastric inhibitory peptide (GIP) is also released to decrease the activity of the stomach. All three of these hormones are released into the bloodstream. It is only when the hormone is returned to the target organ that its action occurs.

The small intestines are lined with tiny fingerlike projections called villi (Figure 7–13). Every villus is covered with even smaller microvilli. These villi and microvilli greatly increase the surface area of the intestinal lining. The villi are covered with a simple columnar epithelium. Within each villus is a rich blood supply to aid in the absorption of nutrients.

The partially digested food delivered into the small intestine is primarily in the form of short peptides, disaccharides, fatty acids, and monoglycerides. Many enzymes coat the microvilli that complete the digestion into individual amino acids and simple sugars. The amino acids, sugars (as well as many vitamins), electrolytes, and water are absorbed through the mucosal cells and delivered to the bloodstream. Much of the absorption occurs through diffusion. Active transport is necessary for much of the absorption. Active transport occurs when the blood level is higher than the concentration within the intestine.

The monoglycerides and fatty acids are absorbed into the lining cells. The cells reconstruct triglycerides within the smooth endoplasmic reticulum. The fats are then released into the center of the villus and enter a lacteal, rather than the capillaries. A lacteal is a duct associated with the lymphatic system. (A more complete discussion of the lymphatic system is presented in Chapter 11.) The lacteals join the much larger thoracic duct. The thoracic duct crosses the diaphragm into the thoracic cavity. The thoracic duct then empties into a major vein. If blood is collected from an animal shortly after a fatty meal, the serum will have a large concentration of fat. When separated from the blood cells, this type of serum has a cloudy white appearance, which is termed lipemia.

Blood from the villi collects into larger venules and veins, which then enter the liver. The liver processes much of the absorbed nutrients before they enter the systemic circulation. The vessels that carry blood from the intestinal tract to the liver make up the hepatic portal circulation. The liver has a wide variety of functions in the metabolism of the animal. The liver's production of bile salts to aid in fat digestion has already been discussed. The liver plays a major role in homeostasis. When excess nutrients are present, the liver removes them from the circulation. Likewise, when concentrations fall, the liver releases them back into the blood.

The liver helps to maintain a steady level of blood sugar. When excess glucose is present in the blood, the liver extracts and converts it into glycogen. The glycogen is basically used to store energy. When sugars other than glucose are present, the liver converts them to glucose and then into glycogen. When blood sugar levels decline, the glycogen is converted back to glucose. The glucose is then released into the bloodstream.

The pancreas helps to regulate this process of energy storage and release. In addition to being an exocrine gland, releasing digestive enzymes, it is also an endocrine gland. The pancreas has clusters of cells called islets of Langerhans. These islet cells produce the hormones glucagon and insulin, which are released into the bloodstream. Insulin causes the blood sugar to decline, whereas glucagon has the opposite effect, raising the blood sugar.

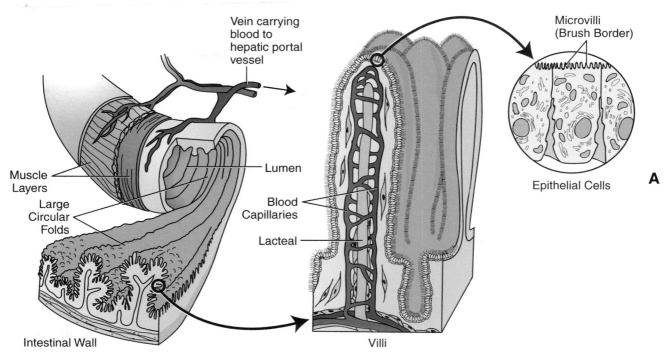

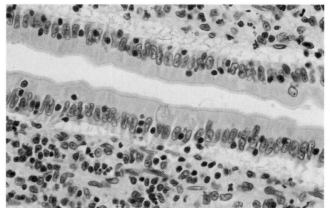

FIGURE 7–13 *A.* Linings of the small intestines showing villi and microvilli. These structures greatly increase the surface area of the intestinal lining. *B.* Photomicrograph of the mucosal lining of the jejunum of a cat. Note the very fine microvilli at the margins of the simple columnar epithelium. *(Photograph courtesy William J. Bacha, PhD, and Linda M. Bacha, MS, VMD.)*

In addition to controlling blood sugar levels, the liver removes excess amino acids from the bloodstream. The excess amino acids have the ammonia group removed, which is then converted into urea. Urea is a waste product produced by the liver and then excreted by the kidney. The remainder of the amino acid is then burned to produce energy.

The liver also stores certain vitamins and iron, acting as a reserve for when levels decline. The liver cells have many enzymes that help to break down toxins and drugs. If the liver is exposed to the same medication for long periods, the specific enzyme used to remove this drug increases. An example is **phenobarbital,** a drug used to help prevent seizures in dogs and cats. Many of these animals are given phenobarbital for many years. The liver becomes more and more efficient

at eliminating the drug from the blood. When this occurs, the blood level declines and the seizures may increase in frequency. This may require that a higher dosage be given to the animal to maintain the same concentration.

The small intestines absorb the vast majority of all the nutritious components of the ingested food. The residue that passes into the large intestine is primarily composed of indigestible products and water. Much of this residue is made of indigestible plant fiber. This fiber provides nourishment to the enormous population of bacteria present in the large intestine. The bacterial population is so large that up to half of feces can be bacterial cells. Some of these bacteria produce vitamins, which are then absorbed by the animal.

The main function of the large intestine is to absorb water from the feces. This water absorption occurs during the 12- to 24-hour period that the fecal material spends in the colon. Any disease that alters this timing can result in digestive problems for the animal. Diarrhea occurs when this time is shortened. The large intestine has inadequate time to absorb enough water, and the feces liquefy. In the opposite manner, constipation occurs when the transport takes longer than normal and the resulting feces become very dry.

As fecal material fills the rectum, the animal is stimulated to defecate. The act of defecation requires the anal sphincter to relax as the colon and rectum contract to force out the feces. Defecation requires both voluntary and involuntary muscular contraction to complete the act.

SPECIES VARIATION

OBJECTIVE

■ *Compare and Contrast the Specialization of Dentition and Digestive Tracts Found in the Various Domestic Species, and Define Symbiosis and Its Significance in the Ruminant*

In the discussion of the monogastric, it was mentioned that much of the fecal material that reaches the colon derives from indigestible plant fiber. Many of the domestic species, such as horses, cows, sheep, and goats, rely predominantly on plant fiber for their nutrition. Special adaptations are present in their digestive systems that allow these species to obtain their nutrients from plants.

As already mentioned, the teeth of herbivores (plant eaters) are adapted for their diet. The large flat molars and premolars are designed specifically for grinding the plant material. This mechanical processing is necessary to allow further digestion. The incisors of horses are aligned in a tight row that allows the grasses to be sheared off during grazing. Cattle, on the other hand, have no upper incisors. The lower incisors press the grass against the hard upper palate, and then the grass is torn loose.

Cattle and horses also eat loose feed such as grain, where the incisors are not necessary for gathering the food. Horses use their lips to gather the food and then the tongue to move it farther into the mouth. Cattle use their tongue much more to pull the loose food into the mouth.

Horses are monogastrics, like dogs and cats. The digestive process in the stomach and small intestines of the horse is identical to that discussed earlier. The large intestine is the region that distinguishes the horse's intestinal tract from other species.

Horses have a very well-developed large intestine (Figure 7–14). At the beginning of the horse's large intestine is a very long cecum. The cecum of an adult horse can be around 1.25 meters long and can hold 25 to 30 liters. The cecum leads into the beginning of the colon. The colon divides into four regions, each separated by a sharp bend or flexure. Table 7–3 shows a comparison between species on the relative sizes of the colon and cecum. Notice that the cecum and colon make up a much larger

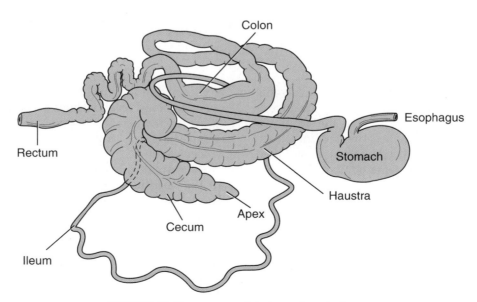

FIGURE 7–14 The anatomy of the horse intestinal tract.

TABLE 7–3 Comparison of Capacity within Digestive Systems*

Species	Cecum	Colon and Rectum	Reticulorumen
Dog	1	13	—
Horse	15	54	—
Rabbit †	43	8	—
Human	<1	17	—
Cattle	5	5–8	64
Sheep	8	4	71

Adapted from Van Soest, Peter, *Nutritional Ecology of the Ruminant, 2nd edition*
*The numbers in the chart represent the percent of the entire intestinal tract made up by the organ listed.
†Notice that the rabbit has a very large cecum in proportion to its intestinal tract.

portion of the horse's intestinal tract compared with the dog.

Plant fiber is digested in the cecum and colon by bacteria in a process called fermentation. These bacteria then release valuable nutrients for the host animal to absorb and use. The bacteria are provided with nutrients, fluid, and warmth by the horse, and in return they supply essential nutrients to their host. A close relationship between two organisms that benefits both organisms is called **symbiosis**.

In the horse, material is exchanged to some extent between the cecum and colon. A portion of the fermentation occurs within the cecum, but the majority of the fiber digestion occurs in the colon.

Cattle, sheep, and goats are called ruminants. Ruminants also rely on symbiosis and fermentation to digest plant material. The fermentation in ruminants occurs before food reaches the acid-secreting stomach and small intestines. It is generally said that ruminants have four stomachs (Figure 7–15), but technically these stomachs derive from the same region in the developing embryo and are four regions of one stomach. These

regions can be observed to be distinct in appearance, structure, and function. For simplicity, in our discussion we will call them stomachs.

The esophagus enters the stomachs at the junction between the rumen and reticulum. The rumen is considered the first stomach. The rumen is a large fermentation vat that occupies the majority of the left side of the abdomen in the cow. The lining of the rumen is covered with tiny fingerlike projections called papillae (Figure 7–16A). These papillae increase the surface area that allows for absorption of nutrients.

The rumen is not one large sac but has structures that divide it into a dorsal and ventral sac. The rumen regularly undergoes an organized contraction that helps to stir the contents within it. When performing a physical examination on a cow, the stethoscope is used to listen to these contractions. These contractions are easily heard as a rumbling noise. A healthy animal should normally have two to three rumen contractions every minute.

In the rumen the contents separate to a certain degree. The most liquid portion is in the lowermost region. There are many small particles floating in this liquid, above which is a fiber mat of large particles basically floating on the liquid. Above this is a gas that is continuously produced by the bacteria and must be belched off regularly by the cow.

The second stomach, the reticulum, is only partially separated from the rumen. The reticulorumen is often considered to be one functional unit that is the site of fermentation. The reticulum has a lining that is divided into regions, giving it the appearance of a honeycomb (Figure 7–16B). The position of the reticulum and the type of lining make it the site that traps any heavy objects ingested. See Chapter 4 for a discussion of a cow suffering from hardware disease. The reticulum is the stomach first affected by hardware. The metal settles out of the liquid-like contents and into the low pouch of the reticulum, becoming entrapped in the honeycomb lining. As the stomach contracts and moves, a sharp piece of metal can puncture through the wall of the reticulum into the abdomen. The reticulum is located immediately behind the diaphragm and liver. Often the metal continues to penetrate into the liver or through the diaphragm and into the heart. Placing a magnet in this stomach can prevent hardware disease. The heavy magnet settles into the stomach for the same reasons as the metal. The goal of the magnet is to hold the metal tight against it, preventing sharp ends from penetrating the wall.

The third stomach is called the omasum. The omasum is lined with long, thin folds that divide the stomach, preventing large particles from passing (Figure 7–16C). These folds greatly increase the surface area of the stomach and provide an area that absorbs water,

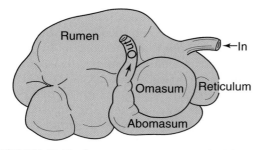

FIGURE 7–15 The four compartment stomach of the cow.

FIGURE 7–16 The linings of the four stomachs of a cow. *A.* Rumen. Note the small fingerlike projections, the rumen papillae, that increase the surface area for absorption. *B.* Reticulum. The distinct lining of the reticulum looks like *honeycomb*. *C.* Omasum. The lining has large sheets with ingesta between the layers. *D.* Abomasum. This smooth glandular lining appears very similar to the monogastric stomach.

electrolytes, and nutrients. The omasum then passes the ingesta into the fourth stomach, the abomasum.

The abomasum is often called the true stomach, because it is functionally identical to that of the monogastrics (Figure 7–16D). The abomasum contains acid and enzyme secretions just like the monogastrics. This stomach begins the digestion of all the materials that are not bound in the indigestible plant fiber. (The introduction of Chapter 2 discusses a common problem associated with the abomasum. Review the discussion of displaced abomasum and the treatments available.)

Ruminants eat very quickly. Food is taken into the mouth, chewed briefly, and then swallowed. Later the animal begins the process of **rumination**. In rumination the rumen and reticulum contract in a manner that forces some of the ingesta back through the esophagus and into the mouth (regurgitation). At this point the animal takes the time to chew the ingesta into fine pieces. This is also called cud chewing. Even more saliva is combined with the food, and then it is swallowed again. Cows at rest can often be observed to be chewing their cud. During this time the rumen continues to contract, keeping the contents stirred. After swallowing, another cycle of rumination can then occur.

Cud chewing breaks the ingesta into finely ground pieces that are able to be digested by the microorganisms within the rumen. The rumen houses a tremendous number of bacteria and protozoa that function to digest plant fiber. This is another example of symbiosis. These microorganisms release nutrients into the rumen that can be absorbed for the use of the cow. These organisms also produce gas as a byproduct of the fermentation. This gas accumulates in the dorsal sac of the rumen and is periodically **eructated** or belched.

The microorganisms of the rumen flourish in the warm, wet, nutrient-rich environment. There is such a tremendous growth that many of them pass into the abomasum. The bacteria and protozoa are then digested as a source of nutrients for the cow because they can be digested normally in the acid-secreting abomasum. When developing rations for feeding cattle, it must be kept in mind to feed the rumen organisms, which then feed the cow.

Once the ingesta reaches the abomasum, digestion occurs much the same as in other monogastrics. Cattle have a relatively large cecum, but the rumen is the primary site of fermentation.

CLINICAL PRACTICE

OBJECTIVE

■ *Discuss the Clinical Significance of the Academic Material Learned in This Chapter*

In the exploratory surgery I found that Toby did have an intussusception (Figure 7–17). At times the intestine can be pulled back into its normal structure. Unfortunately, I was unable to correct Toby's intussusception in that fashion. An intussusception eventually cuts off the normal blood supply to the intestine, and the tissue is no longer healthy. With Toby I had to remove a portion of his jejunum.

To accomplish this, I used a pair of clamps on each side of the intussusception. These clamps are gentle enough that they do not damage the tissue but firm enough to prevent any intestinal contents from leaking. Tying off the arteries and veins supplying that region stopped the blood supply to the damaged intestine, and I removed the diseased section. I then sutured the ends of the open intestine together until I was satisfied that I had a leakproof connection.

The important considerations in such an operation are similar to every surgery. The tissues that are sutured together are handled with care to prevent further damage. I remove a large enough section of intestine that I am sure I have healthy tissue at each end. I must also ensure that there is blood supply to the sutured tissues. When I am finished with intestinal surgery, I need to be comfortable that the contents do not leak from the incised intestine.

Not all foreign bodies are readily seen on a radiograph. Rocks in an intestine are easily detected on a radiograph, but materials such as a sock or underwear often are not so easily seen. To aid in diagnosis, I can give an animal oral barium, which shows up extremely well on a radiograph (Figure 7–18). By taking a radiograph immediately after administering barium, and then at regular intervals, the function of the gastrointestinal tract can be followed. When an obstruction is present, the barium stops progressing or at least slows down. This technique can be very helpful in deciding to perform exploratory surgery on an animal.

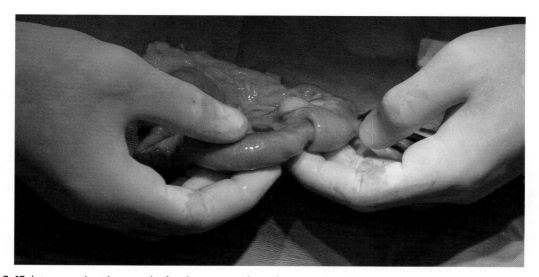

FIGURE 7–17 Intraoperative photograph of an intussusception, where a region of intestine has telescoped into another section.

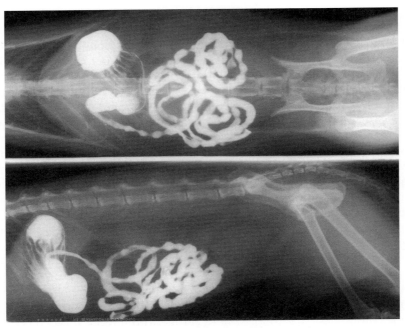

FIGURE 7–18 Radiograph of a cat during a barium series. The bright white areas are barium within the gastrointestinal tract.

Vomiting is a complex process that is not always a direct result of gastrointestinal disease (for example, it can occur with kidney failure). Vomiting begins with the nausea. Nausea makes an animal feel sick but also slows motility in the stomach and esophagus. Nausea proceeds into **retching,** or strong rapid abdominal contractions. Finally, vomiting occurs as the stomach's contents are forcefully driven out of the mouth.

Vomiting can be induced from several mechanisms. A signal is sent from the vomiting center in the brain. The vomiting center can receive stimulation from receptors in the stomach and abdomen. Inflammation or disease in the stomach or abdomen stimulates the vomiting center. In addition, there are receptors that pick up stimuli in the blood (such as azotemia or certain drugs) that can also induce vomiting. The balance center, called the **vestibular system,** also influences the vomiting center. You may have felt nauseous after an amusement park ride that makes you dizzy. This is a firsthand demonstration of the effect of the vestibular system on the vomiting center.

Diarrhea occurs when feces pass through the intestinal tract too quickly. The resulting feces has a high level of water. Many causes of diarrhea exist and include problems in either the large or small intestines. Diarrhea originating in the large intestine often results in a high frequency of defecation, often in small amounts and often with mucus and/or blood. Small-intestine diarrhea usually has an increased volume of very liquid feces without a tremendous increase in frequency.

The greatest risk associated with a sudden onset of vomiting and diarrhea is loss of water and electrolytes. Dehydration often kills animals that are experiencing severe vomiting and diarrhea. Vomiting prevents adequate intake of fluids. Diarrhea and vomiting both increase the loss of fluids. Animals with long-term or chronic vomiting and diarrhea usually present with severe weight loss. If these animals can maintain hydration, the problem develops as a loss of nutrients.

Many cases of vomiting, diarrhea, or both can be handled with symptomatic treatment. In cases of vomiting, the pet is taken off water and food for several hours (NPO). We often recommend up to 8 to 12 hours without food. However, very small pets or those with severe vomiting may lack the reserves to be kept off of fluids for that period. Each case must be evaluated individually and the animal monitored for dehydration. If the vomiting is under control when the animal is NPO, the pet is introduced to small amounts of water. If no vomiting occurs, the animal is then fed a bland diet. Bland diets, such as boiled chicken and rice, are very low in fat and high in fiber. Fiber in the diet helps to absorb water in the large intestine and slows down the transit of feces in the colon.

Animals that do not respond to such treatment must be treated for dehydration. Refer to Chapter 6 to review the calculations used in fluid therapy. Many medications are available to treat the symptoms of vomiting and diarrhea. It is also important to investigate the underlying cause and treat that condition if necessary.

In addition to diarrhea, animals can also suffer from constipation. One common cause of constipation in dogs results from the feeding of bones. The dog's powerful jaws crush the bones into pieces, which move into the colon and become tightly packed. The resulting feces can become extremely hard. Treatment is not fun for the animal or the veterinarian in these cases. These animals often require enemas (fluid is added into the rectum in attempt to soften the stool), manual removal, or even surgery to remove the firm feces.

Severe constipation is only one problem that veterinarians see as a result of dogs eating bones. Broken teeth, bones stuck in the roof of the mouth, bones stuck on the jaw, and vomiting can all result from dogs chewing bones. Many dogs eat bones without having problems; however, veterinarians see many of those that are not so fortunate. Because of all of the potential problems, it is strongly recommended that dogs should never be fed bones.

Horses show signs of colic when they have pain in their abdomen. This can result from any problem that causes pain. The most common causes are a direct result of intestinal pain. This pain can result in intestinal spasms with diarrhea, excess gas in the intestine, or severe constipation. The complete list is extremely long. Table 7–4 is a partial list of the causes of colic in horses.

Diagnosing the underlying cause of colic can be difficult. A complete physical examination, including an examination of the feces and rectum, are important. At times a problem can be felt within the abdomen with a rectal examination. Response to treatment is also used to evaluate many patients if a specific cause is not determined. Pain-relieving medications and laxatives are often used in nonspecific colic. Many horses recover with a single treatment.

Further blood testing and radiographic or ultrasound examinations may help to define a colic problem. If a horse does not respond to treatment and the condition worsens, surgery may be required. Loops of the intestines can become twisted upon themselves, cutting off blood supply (Figure 7–19). Surgery is required to correct such a condition. Such serious causes of colic can be life threatening for a horse. After surgery, these horses require intensive care, including antibiotics and intravenous fluids.

The condition of displaced abomasum has already been discussed in cattle. The rumen can also be a source of problems in cattle. The production of gas in the rumen is a natural event. The cows must periodically eructate or belch this gas to rid it from the rumen. If for some reason the gas cannot be eructated, the cow develops **bloat**. Bloat is a large gaseous distension of the rumen. This distension initially occurs high on the left side of the abdomen, where the dorsal sac of the rumen is located. The rumen can become so distended that the entire abdomen takes on a distended appearance.

There are two classifications of bloat: free gas or frothy. In frothy bloat the gas is trapped in many small bubbles (much like the appearance of soap bubbles), which prevents the animal from belching a larger pocket of gas. Classically this occurs after cattle are turned onto a lush pasture. The change in diet results in conditions in the rumen that create the small bubbles. Medications are available that break down the small bubbles, allowing the animal to eructate.

In free gas bloat the gas accumulates in a large pocket in the dorsal sac of the rumen. The gas is in a form that could be eructated, but for some reason the cow cannot. One cause can be choke, in which the

TABLE 7–4 Common Causes of Equine Colic

Accumulation of gas in intestines
Intestinal cramping or spasms
Impaction and constipation
Twisted intestine or colon
Abdominal tumors
Accumulation of sand in intestines
Parasitism
Intussusception
Peritonitis (inflammation of the peritoneum)
Stomach dilation
Hernia

FIGURE 7–19 A gangrenous (dark area) section of small intestine of a horse. The gangrene is due to lack of blood supply to the affected portion of the intestine and represents tissue damage at this site. *(Photograph courtesy Dr. Arthur Hattel, Pennsylvania State University.)*

esophagus becomes obstructed with ingesta. The rumen is functioning normally, but the gas cannot physically pass out of the esophagus. Other causes of free gas bloat result when the rumen is not functioning properly. The vagus nerve controls the normal contractions of the rumen. This nerve travels through the chest cavity to reach the rumen. In animals with pneumonia, this nerve can be damaged. The result is that the rumen stops contracting normally. Another common cause is low blood calcium (hypocalcemia or milk fever) that often occurs in older cows around calving. Calcium plays a critical role in the function of muscles, including the smooth muscle in the walls of the rumen. Low calcium causes poor tone in the rumen wall and results in bloat.

Supplementing calcium corrects bloat associated with milk fever very quickly. In other forms of free gas bloat, a stomach tube is passed through the mouth and esophagus into the stomach. When the tube reaches the gas pocket, the rumen deflates very quickly. Treating the underlying cause is then important to prevent recurrence. Bloat can be so severe that it is life threatening because of the pressure that it places on the diaphragm and blood vessels in the abdomen. Immediate treatment may be required in which a trocar is driven into the rumen through the abdominal wall. The trocar is a large, sharp probe with a metal sleeve around it (Figure 7–20). Once into the rumen, the probe is removed, leaving the metal sleeve as an opening to drain the gas. A plastic trocar

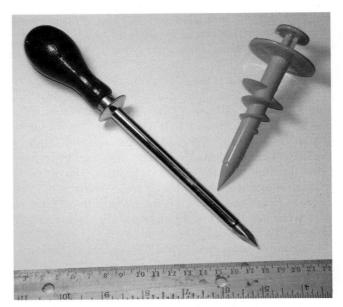

FIGURE 7–20 Metal and indwelling plastic trocar. Both trocars have a solid pointed spike in the center that is removed once the trocar is in place. With the spike removed, the tube portion of the trocar allows gas to escape.

is also available that can be placed in the rumen for a longer term. This can be useful in cattle that have developed the bloat secondary to pneumonia. The stomach stays deflated while the animal is being treated for the pneumonia.

SUMMARY

The basic process of digestion is the same for all animals. However, a variety of dentition and digestive tracts exists among differing species. Carnivores' teeth tend to be sharp to allow for tearing meat, whereas herbivores' teeth are smoother and more suited to grinding forage and grains. Swine, cats, and dogs are examples of animals possessing monogastric or simple stomachs. Cattle, sheep, and goats are all ruminants, which have multiple compartment stomachs. Horses are monogastric animals with specialized cecums that accommodate large volumes of forages. The differences among species challenge veterinarians to understand the disease conditions associated with each.

REVIEW QUESTIONS

1. Define any 10 of the following terms:

   intussusception
   colic
   carnivore
   herbivore
   deciduous teeth
   monogastric
   phenobarbital

   symbiosis
   rumination
   eructate
   retching
   vestibular system
   bloat

2. True or False: Looking at the teeth of a horse can help determine the animal's age.

3. True or False: Barium is distinctive on a radiograph.

4. True or False: Small-intestine diarrhea typically occurs very frequently.

5. The foremost teeth used to bite into food are called _____.

6. NPO (veterinary lingo) means _____.

7. The following symptoms are characteristic of which equine condition: kicking at abdomen, rolling, lying down and standing up repeatedly, sweating?

8. Are cats carnivores or herbivores?

9. Which portion of the tooth lies above the gum line?

10. What structure covers the opening of the larynx?

11. What are the small fingerlike projections that line the intestine called?

12. What portion of the ruminant stomach has a honeycomb-type lining?

13. What portion of the ruminant stomach is called the true stomach?

14. List the pH level of the stomach.

15. Name three ruminants discussed in the text.

ACTIVITIES

Materials needed for completion of activities:
    rumen fluid
    dropper
    beaker
    oil
    slides
    light microscope

1. The students can go to one of the following web sites to investigate how the age of horses can be assessed based on the appearance of their teeth: <http://www.equineestates.com> or <http://www.upperegypt.com>. Key search words are anatomy, teeth, and aging.

2. Contact a butcher shop to obtain a sample of rumen fluid. If a fresh sample can be obtained, special care should be used to keep the organisms alive. Place the liquid rumen contents in a glass beaker. Cover the surface of the fluid with a layer of mineral oil. The organisms in the rumen do not survive when exposed to air; the layer of oil floating on top of the liquid helps keep air out of the sample. Place the beaker in a warm water bath (37°C). Place a drop of the liquid on a microscope slide, and cover with a coverslip. Examine the sample under the microscope. Numerous bacteria and protozoa should be observed. If a fresh sample can be obtained, there should be a large amount of movement from the organisms.

3. Use Table 7–1 to investigate the total number of teeth in the various species. Remember that the table lists the number of teeth for one side of the mouth. Research the number of teeth in the human mouth.

8

The Reproductive System

OBJECTIVES

Upon completion of this chapter, you should be able to:

- *Identify male anatomy and relate associated hormonal function.*
- *Discuss female anatomy and the estrous cycle.*
- *List the steps in establishing pregnancy and identify the stages of parturition.*
- *Discuss the clinical significance of the academic material learned in this chapter.*

KEY TERMS

spay (ovariohysterectomy)	estrous cycle	pheromone	cesarean section
castration	puberty	parturition	ligated
prolapsed uterus	estrus	gestation	pyometra
epidural	polyestrus	weaned	cryptorchidism
lidocaine	seasonal polyestrus	obstetric	
	anestrus	whelping	

INTRODUCTION

The male produces sperm and delivers them into the female. The female then has the responsibility of providing the path and helping, through muscular contractions, to deliver the sperm to the location of the egg, which she produces. After the sperm and egg join, the female houses and nourishes the developing embryo until it is mature enough to survive on its own. At that point the female delivers the newborn. Appropriately functioning reproductive systems in livestock largely determine the economic success of the producer. Veterinarians assist farmers in caring for the reproductive health of pets.

Reproduction...

Preparing this chapter made me think about how much of my career revolves around the reproductive system. I just finished my morning surgeries and had a break until my afternoon office appointments. My morning's surgeries included a cat **spay** and front declaw, a dog spay, and two dog **castrations.**

In a spay, technically called an **ovariohysterectomy,** the ovaries and uterus are removed. In castration the testes are removed. Both spaying and castration prevent any unwanted pregnancies. In addition, the neutering procedure provides other health benefits.

Earlier in the text I mentioned that herd checks make up a large portion of my job. Dairy cattle begin producing milk after their first calves are born. Subsequently producers attempt to have the cattle calve every 12 to 13 months. At herd checks I help facilitate and check for pregnancy. I perform more herd checks than any other dairy-related task.

Yesterday afternoon I received a call from one of our farm clients. He had a down cow. The cow had begun to calve and did not have the strength to rise. She had developed hypocalcemia, or milk fever. (Review the role of calcium in the function of muscles in Chapter 2.) I gave her two bottles of calcium intravenously. Afterward the farmer and I examined another cow that was not eating well. When we returned to the down cow, the calcium had entered the muscles and the cow was able to rise.

I then thoroughly cleaned her vulva and reached into the uterus to find that the calf was ready to be born. Many cows with milk fever do not deliver normally because they lack muscle strength. Having been treated, this cow may have calved normally, but I could not be positive. I placed calving chains around the legs of the calf and, along with the farmer (with the help of the cow pushing), pulled the calf. Fortunately, the calf was born alive and the mother was standing (Figure 8–1). To this day, cases such as this continue to be the most rewarding in my profession. The results are immediate and dramatic. I still enjoy helping to bring a newborn into this world.

I received an emergency call as I was writing this chapter. The receptionist at the office called to tell me that one of our Amish clients had a cow that had thrown

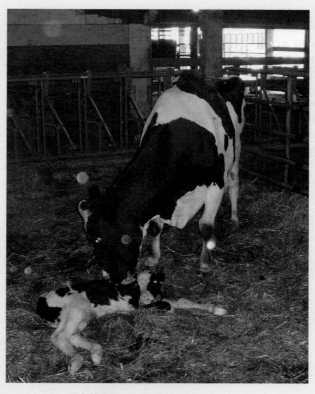

FIGURE 8–1 Newborn calf and dam.

her calf bed, or in other words, had a **prolapsed uterus.** This heifer had just delivered her first calf 2 hours earlier. The calf was quite large, and following the delivery, the cow continued to strain. Consequently, the uterus turned inside out and was pushed through the vulva.

When I arrived, the cow was down and her uterus protruded from out behind her. Because our Amish clients have no electricity on their farm, the farmer lit a lantern for me to see. I felt like an old timer, working in the barn by lantern light.

I gave the cow an **epidural,** in which I inject a local anesthetic (**lidocaine**) into the fluid around the spinal cord. The epidural relieves pain in the region around the vulva and rectum. Because the cow felt no pain, she resisted my efforts less. I then pushed her large, swollen uterus back through the vulva. Fortunately, this case went very well. The farmer and I hope for a good outcome.

MALE ANATOMY AND HORMONAL FUNCTION

OBJECTIVE

■ *Identify Male Anatomy and Relate Associated Hormonal Function*

In mammals the male produces sperm cells and provides a means to deliver them to the reproductive tract of the female. The male anatomy provides the organs and structures to accomplish both.

In the male the testes (singular is testis) lie external to the abdomen, housed in a skin-covered sac called the scrotum (Figure 8–2). The scrotum is lined by peritoneum, which stretches from within the abdomen. A muscle within the wall of the scrotum al-

lows for the testes to be pulled closer to the abdomen. For sperm production to occur optimally, scrotal temperature must be lower than body temperature. In warmer temperatures the muscle relaxes, allowing the testes to stay farther away from the heat of the body. As external temperatures decline, the muscle contracts, pulling the testes close to the body. This action helps to maintain a consistent temperature within the testes.

In the embryo the testes develop within the abdomen. As development progresses, the testes descend into the scrotum. The testis is basically oval with flattened sides. The testes produce sperm. A large number of seminiferous tubules fill the testes. Sperm production occurs within the seminiferous tubules (Figure 8–3). In addition to being a reproductive organ, the testis is also an endocrine gland. Cells found be-

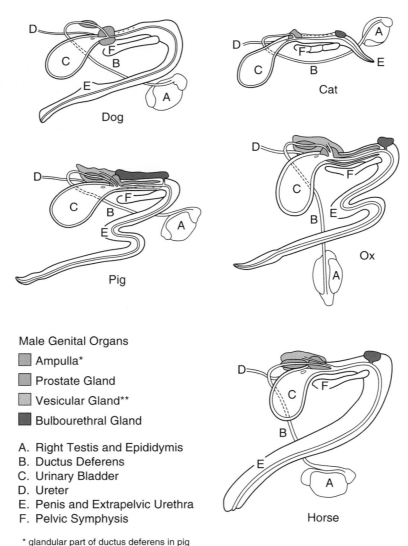

Male Genital Organs

⬜ Ampulla*

⬜ Prostate Gland

⬜ Vesicular Gland**

⬛ Bulbourethral Gland

A. Right Testis and Epididymis
B. Ductus Deferens
C. Urinary Bladder
D. Ureter
E. Penis and Extrapelvic Urethra
F. Pelvic Symphysis

* glandular part of ductus deferens in pig
** seminal vesicle in horse

FIGURE 8–2 Male reproductive tracts.

tween the seminiferous tubules produce a hormone called testosterone.

The testes closely adhere to the epididymis, a long, convoluted (highly folded) tube attached to the outer surface of the testes. The epididymis acts as a storage area for the sperm produced within the testes. In addition, the sperm go through their final maturation within the epididymis. This final development is necessary before the sperm become capable of fertilizing an egg.

On physical examination of a male, such as a bull, the testes and epididymis can be palpated. The spermatic cord leads from the testis toward the abdomen. The spermatic cord contains the testicular artery, vein, and nerve. The arterial blood supply originates from the descending aorta in the abdomen. In addition to the blood vessels, the spermatic cord also contains the ductus deferens. The spermatic cord enters the abdomen through a small slit called the inguinal canal, which is located in the abdominal muscles.

The ductus deferens carries the sperm from the epididymis to the prostate gland. The prostate gland is located at the base of the urinary bladder, surrounding the urethra. The prostate produces an accessory secretion, which adds to the sperm to produce semen. The prostate is the only accessory sex gland present in the dog. The secretions produced by the prostate are necessary for the survival and motility of the sperm. Cattle and horses have two additional accessory sex glands. They are the seminal vesicles and bulbourethral glands. Both these paired glands closely associate with the urethra. They too contribute to the fluid portion of the semen.

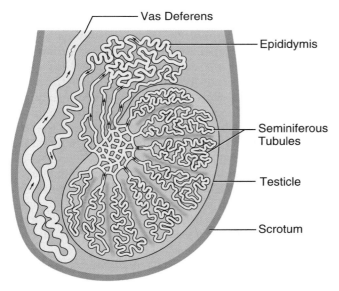

FIGURE 8–3 The structure of a testis.

Within the prostate, the ductus deferens joins the urethra. The urethra therefore carries the dual role of carrying urine for excretion and carrying semen for reproduction. The urethra then leads through the penis. The penis lies within a tubular, skin-covered sheath. This protective sheath is called the prepuce.

The penis has numerous blood vessels within its structure. Larger open areas called sinuses connect the veins. As sexual excitement occurs, the flow of blood out of the penis is restricted. As blood continues to flow into the penis through the arteries, the sinuses become filled with blood. As the pressure increases, the penis becomes erect, protruding from the prepuce. Through this mechanism, the penis is able to penetrate the vagina of the female and deliver the semen.

Precursor cells that line the seminiferous tubules produce sperm cells. Throughout the male's life, cells undergo mitosis, providing an unlimited supply of precursor cells. To produce sperm cells, the precursor cells undergo meiosis. The divisions of meiosis produce sperm cells with half the normal number of chromosomes. Review the process of meiosis in Chapter 1.

Sperm cells have flagella that provide motility but have very little cytoplasm. Rather, chromosomes are densely packed in the head of the sperm. In addition, mitochondria at the base of the flagellum provide the energy necessary for motion.

As mentioned earlier, the testes act as an endocrine gland producing the hormone testosterone. Testosterone stimulates the development of the male genitals. Testosterone also produces a male's distinguishing features. The powerful musculature of the bull and the associated deep bellow are both a result of the hormone testosterone. Testosterone is also essential for the production of sperm.

Two hormones released from the pituitary gland help control the function of the testes. Both males and females produce these two hormones. They are named based on the function that they play in females: luteinizing hormone (LH) stimulates the production of testosterone by the testes, and follicle-stimulating hormone (FSH) facilitates actual sperm production.

FEMALE ANATOMY AND HORMONAL FUNCTION

OBJECTIVE

■ *Discuss Female Anatomy and the Estrous Cycle*

The vulva, the external opening to the urogenital tract in the female, leads into the vagina. The opening of the urethra lies on the floor of the vagina. Urine exits through the urethra into the caudal portion of the vagina and then through the vulva. The vagina also leads into the cervix. The cervix, a portion of the uterus, is a firm structure that protects the opening of the uterus.

The uterus has a short body that joins the two horns of the uterus. The uterus takes on the shape of a Y, with the two horns leading from a common body and cervix (Figure 8–4). The uterus, a hollow muscular wall organ, is lined with a simple columnar epithelium and has the ability to expand dramatically to support the developing fetus.

The uterus tapers into a small oviduct that leads to the ovary. The ovary, the female gonad, produces eggs. The oviduct carries the egg from the ovary to the uterus. At the end of the oviduct lies a thin membranous infundibulum. The infundibulum wraps around the ovary to catch the egg after it is released from the ovary. The ovaries are located caudal to the kidneys.

Branches from the descending aorta supply blood to the ovary and uterus. A separate ovarian artery sends blood to each ovary and oviduct and to the cranial portion of the uterine horn. Caudally a uterine artery delivers blood to the cervix, uterine body, and each horn. The blood from the uterine and ovarian arteries then returns to the caudal vena cava by the uterine and ovarian veins. Knowledge of the reproductive tract's blood supply is critical during spays and castrations.

Ovaries produce eggs through meiosis. As the ovaries develop in the embryo, the precursor cells undergo the first step of meiosis and then stop. The number of eggs that can be produced is established at this point. This differs from the male, where the sperm precursor cells continue to undergo mitosis. The male basically has an unlimited supply of sperm cells. The number of eggs is established early in development but allows for a supply larger than the female would naturally use during her reproductive lifetime.

Females undergo an **estrous cycle.** This process begins at a point in the animal's development called **puberty.** The estrous cycle prepares the female to become pregnant. Hormones control the estrous cycle. Specific details associated with the estrous cycle vary among species. The bovine estrous cycle serves as an example and a species comparison follows.

The average estrous cycle in the cow is 21 days, although it may normally range from 17 to 25. The estrous cycle in cattle is described as **polyestrus,** meaning that cycling continues until the cow becomes pregnant. **Estrus** describes the animal when in a state of sexual excitement. In lay terms the animal is said to be in heat. In the animal kingdom, estrus tells the male of the species that the female is receptive.

An estrous cycle divides into four phases. The events of each phase are distinct. However, the cycle flows without pause. Proestrus in cattle describes the 3 days prior to heat. The ovary contains a corpus luteum (CL), which produces the hormone progesterone (Figure 8–5). At the beginning of proestrus, the hormone prostaglandin releases into the bloodstream from the uterus. The prostaglandin causes the CL to regress, lowering the level of progesterone (Figure 8–6).

The declining progesterone results in increased activity of the pituitary gland and release of the hormones FSH and LH. The rising level of FSH results in the development of a follicle in the ovary. The follicle, a fluid-filled structure, surrounds the egg. The cells lining the follicle secrete the hormone estrogen in response to the FSH. Estrogen is responsible for the behavioral changes that occur in the female during estrus. The day prior to estrus there is also a sharp increase or pulse in the release of LH.

The actual estrus in cattle lasts 8 to 30 hours, although most animals cycle toward the shorter end of that range. The most definitive sign of heat occurs when a cow stands when mounted (Figure 8–7). Other secondary signs of heat include increased excitability, restlessness, bellowing, and mounting of other cows. (Cows not in heat usually try to run when mounted by another cow.) Cows in estrus may also show a clear mucus discharge from the vagina and a reddened vulva.

The developing follicle usually releases the egg 10 to 14 hours after estrus. Release of the egg, or ovulation, occurs as a result of the surges in LH and FSH that occurred the prior day. Ovulation marks day 1 of the estrous cycle. The infundibulum captures the egg and transports it into the oviduct. If the cow was inseminated or bred, sperm is present in the oviduct to fertilize the egg.

After ovulation, the blisterlike follicle collapses, leaving a depression in the surface of the ovary (ovulation depression). Metestrus follows ovulation. At this point the level of estrogen begins to decline. The luteinizing hormone stimulates the former follicle to develop into a corpus luteum. *Corpus luteum* translates from Latin as "yellow body." The cells lining the CL begin to produce progesterone. The progesterone stimulates the uterus to prepare to nourish the embryo. Metestrus lasts 3 to 4 days, during which time the egg or embryo resides in the oviduct.

During diestrus the embryo moves into the uterus to establish a pregnancy; the corpus luteum produces a large amount of progesterone at this time. Progesterone stimulates the growth of the lining of the uterus and provides the necessary environment for the embryo to survive.

Diestrus lasts 12 to 15 days in an animal that does not become pregnant. Without a pregnancy the uterus releases prostaglandin and the cow enters into the

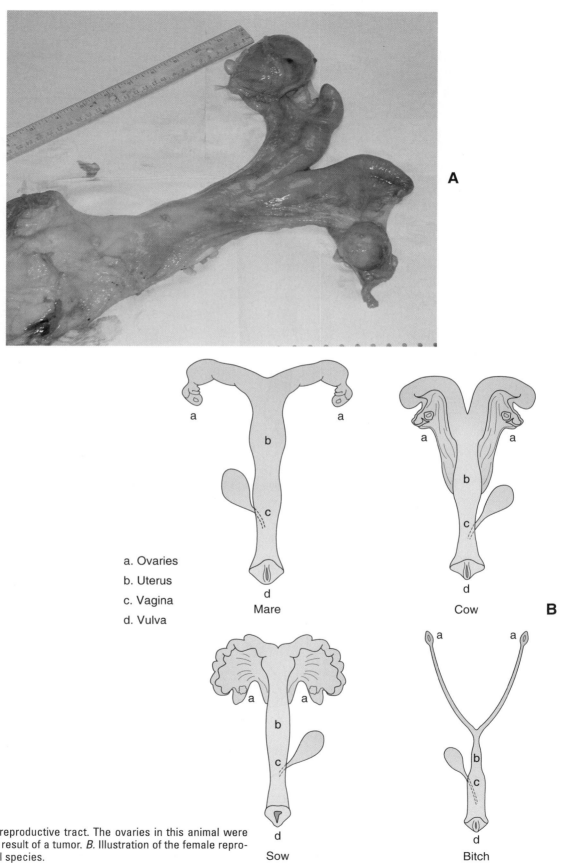

a. Ovaries
b. Uterus
c. Vagina
d. Vulva

Mare

Cow

Sow

Bitch

A

B

FIGURE 8–4 *A.* Bovine reproductive tract. The ovaries in this animal were unusually large, likely a result of a tumor. *B.* Illustration of the female reproductive tracts of several species.

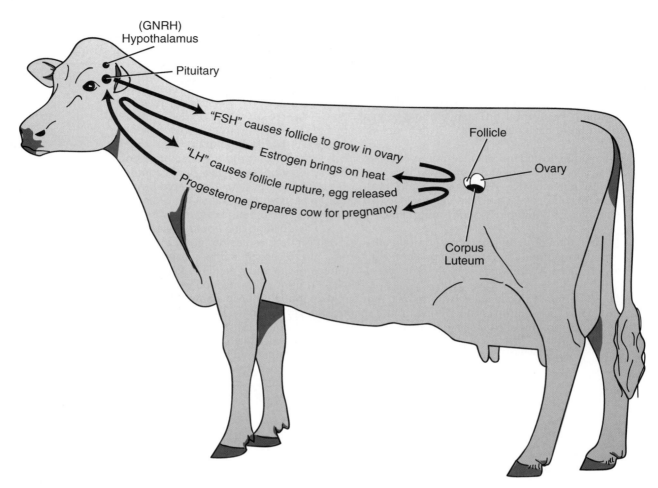

FIGURE 8–5 The hormones of the bovine estrous cycle. GNRH = gonadotropin-releasing hormone.

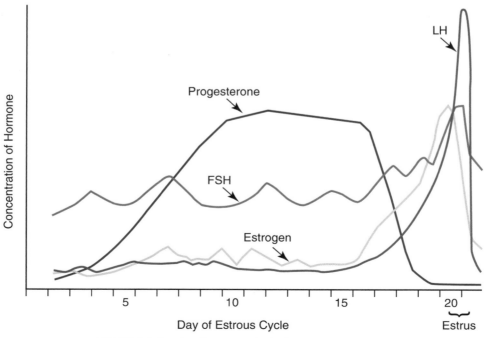

FIGURE 8–6 Levels of hormones in the bovine estrous cycle.

FIGURE 8–7 A cow showing estrus. The lower cow was in standing heat. She stood still, accepting the heat mount by the other cow.

PREGNANCY AND PARTURITION

OBJECTIVE

■ *List the Steps in Establishing Pregnancy and Identify the Stages of Parturition*

The length of daylight affects many seasonal breeders. Horses, for example, are long-day breeders. Horses naturally enter the breeding season in the spring and summer as the days lengthen. For certain breeds, foals born earlier in the year are of more value. Therefore techniques have been developed to use artificial lighting to initiate an earlier start to the breeding season.

Although there is variation among breeds, sheep are classified as short-day breeders. Sheep tend to have a breeding season in the fall as the days shorten. After gestating for 5 months, lambs are born at a time that maximizes their ability to survive, because pasture is usually plentiful in the spring.

Goats are also short-day breeders, with a breeding season that can extend from October to March. Cycling in goats can also be influenced by the introduction of a male. If female goats are housed separately from a buck, his reintroduction can initiate the breeding season several weeks early.

Cats are also seasonally polyestrous; however, the season can be quite long. The breeding season often begins in January as the days begin to lengthen and often continues until September. Cats are induced ovulators. The act of mating stimulates ovulation and a quicker end to estrus. The duration of the cycle varies depending on the timing and occurrence of breeding. Following breeding, a CL is formed, and the cycle may last up to 3 weeks. Without breeding, the cat may return to heat in a much shorter period.

Cats also enter **anestrus** when they are nursing kittens. The body's natural design allows support of an

cycle at proestrus. Following the prostaglandin release the entire cycle begins again, providing the cow another opportunity to become pregnant. If the cow is pregnant, the uterus does not release prostaglandin and the CL does not regress. Progesterone is essential to maintaining pregnancy.

Table 8–1 shows the variation that exists among the estrous cycles of the various species. Cattle and swine are both polyestrus, meaning that in animals that are not pregnant, the cycles continue throughout the year. Sheep, goats, and horses are also polyestrous, but the cycles occur only during certain times of the year (**seasonal polyestrus**). *Anestrus* describes the periods when an animal is not going through estrous cycles.

Species	Age of Puberty	Cycle Description	Cycle Duration	Estrus Duration	Timing of Ovulation
Cattle	12–15 months	Polyestrous	21 days	8–30 hours	10–14 hours after estrus
Sheep	7–12 months	Seasonal polyestrous	17 days	1 day	End of estrus
Goats	6–8 months	Seasonal polyestrous	21 days	12–24 hours	End of estrus
Horse	10–24 months	Seasonal polyestrous	21 days	6 days	Day 5 of estrus
Swine	4–9 months	Polyestrous	21 days	2–3 days	Day 2 or 3 of estrus
Dog	5–24 months*	Monestrous	6–8 months	3–21 days	Day 2 or 3 of estrus
Cat	4–12 months	Seasonal polyestrous	15–21 days	10–14 days†	Induced following breeding

TABLE 8–1 Comparison of Reproductive Cycles

*Larger-breed dogs tend to reach puberty at an older age than smaller breeds.
†Duration of estrus can be shortened, depending on the time of mating.

existing litter before entering into another pregnancy. If the kittens are removed from nursing, the cat enters breeding season again. Male cats may kill a litter of kittens if the female leaves them unprotected. It is presumed that the male does this to stimulate the female back into estrus.

Dogs are unique in being monestrus. Typically a delay of 6 to 8 months exists between heat periods. In dogs, proestrus lasts an average of 9 days but can continue up to 17 days. During this time the female attracts males but is not receptive to them. During proestrus the female typically has a bloody discharge from a swollen vulva. The blood passes through the wall of the uterus and vagina.

Estrus follows, during which the female accepts the male. The duration of estrus in the dog is quite variable, ranging from 3 to 21 days (the average is approximately 9 days). Diestrus, the first day that the female will not accept the male, follows. The corpus luteum is maintained for the same length of time in the pregnant and nonpregnant animal. Pregnancy typically lasts 63 days from the onset of diestrus. The CL regresses at the end of pregnancy and at the same point in time in the nonpregnant female.

The dog then enters anestrus. The next cycle does not occur for another 6 to 8 months. Some dogs may have up to 1 year between cycles. The entire cycle does not vary between pregnant and nonpregnant dogs.

There is a tendency to compare the bloody vaginal discharge in dogs with the menstrual period in human females. In dogs the discharge occurs prior to estrus and ovulation. The human menstrual cycle differs. Following menstruation the uterine lining in women thickens. This thickening continues to increase following ovulation as the uterus prepares to accept a pregnancy. If no pregnancy occurs and the CL regresses, the lining of the uterus is sloughed or shed. Without a pregnancy the uterine lining prepares for the next cycle.

All male domestic animals detect a distinctive chemical emitted by the female in estrus. This chemical is called a **pheromone**. Pheromones serve as a means of chemical communication between the sexes. Pheromones stimulate males to become sexually excited. A nervous reflex constricts the vessels within the penis, resulting in an erection.

For all domestic animals, the female in heat then stands to accept the male, and the male's penis penetrates the vagina of the female. The male's excitement builds until ejaculation occurs. During ejaculation the ducts in the epididymis and ductus deferens contract, propelling the sperm forward. Immediately before ejaculation, the accessory sex glands contribute fluid to create semen. The semen is deposited in the cranial vagina. The sperm then propel themselves forward with the flagella. In addition, the contraction of the walls of the vagina and uterus help to move the sperm into the oviduct.

The sperm meets the egg within the oviduct. The sperm attach to the cell membrane of the egg. The first sperm attaching then penetrates the membrane. Once one sperm enters the egg, the membrane changes to prevent any other sperm from penetrating. The nuclei of the egg and sperm then combine, providing the fertilized egg with a complete number of chromosomes. The fertilized egg then undergoes a series of mitotic divisions, creating an embryo.

The embryo (using the cow as the example; actual details vary with species) stays within the oviduct for 3 to 4 days. Following this the embryo moves into the uterus. Mitosis continues, increasing the number of cells within the embryo. As the cell number increases within the embryo, the major organs and systems develop in a process called differentiation. The presence of the embryo in the uterus inhibits the release of prostaglandin and prevents the cow from entering estrus again. The corpus luteum remains intact, secreting progesterone, which is necessary to maintain the pregnancy.

Beginning at day 28 and completing by day 45, the embryo attaches to the uterus. The points of attachment provide sites for transfer of nutrients and waste between the mother and embryo. In the cow, 80 to 100 of these sites, called placentomes, exist. A placentome has highly folded vessels of the mother closely associated with vessels of the embryo. Separating these layers gives the impression of dividing two layers of Velcro (Figure 8–8). The intertwined vessels provide a large surface area that allows transfer of nutrients, oxygen, and wastes in a mechanism similar to that which occurs

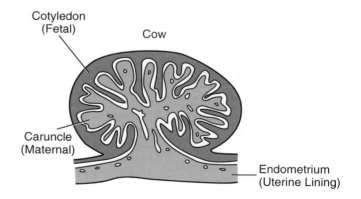

Placentome
1) Fetal attachment to uterus
2) Nutrients for fetus
3) Waste disposal for fetus

FIGURE 8–8 A placentome, the site of attachment between the fetal placenta and the maternal uterus in the cow. The actual blood supply of the two animals does not mix.

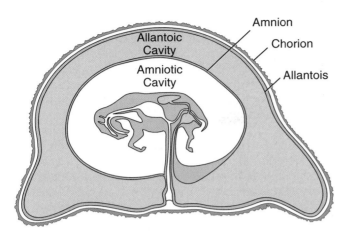

FIGURE 8–9 A developing fetus and the relationship to placenta.

Species	Length of Pregnancy (Days)	Average in Months*
TABLE 8–2 Average Length of Pregnancy		
Cattle	279–292	9
Goats	145–155	5
Sheep	144–151	5
Swine	112–115	3m3w3d†
Horses	330–342	11
Dogs	58–70	2
Cats	58–65	2
Rabbits	30–35	1
Elephants	600–660	20-22

*The length of pregnancy varies significantly with animals and breeds. The pregnancy length is therefore often averaged in months.

†The average length of pregnancy in swine is often considered as 3 months, 3 weeks, and 3 days.

in the alveoli of the lungs. The blood of the embryo and mother do not mix.

The vessels from the embryo's circulation are contained within a set of membranes that surround the embryo (Figure 8–9). The embryo is bathed in a fluid-filled sac called the amnion. In addition, the surrounding membranes form the placenta, which contains the vessels that contact the uterus at the placentome. The vessels of the placenta combine into one large set that enters the embryo (the umbilical cord).

As the embryo undergoes differentiation, it develops the normal appearance and structures of a newborn. At this point the embryo becomes a fetus. The fetus grows relatively slowly early in the pregnancy as differentiation occurs. The most rapid growth of the fetus occurs in the last third, or trimester, of the pregnancy.

To maintain pregnancy, the corpus luteum must continue to produce progesterone. As the fetus develops, the placenta takes over production of progesterone. Eventually the placenta produces enough progesterone that the corpus luteum is no longer necessary to maintain the pregnancy.

As the fetus approaches full term, hormone levels change in preparation for **parturition**. Table 8–2 summarizes the normal **gestation** period for all domestic animals. Parturition, or delivery of the newborn, occurs normally at a point when the fetus is capable of surviving on its own. As the pregnancy ends, progesterone levels begin to decline, and estrogen levels increase. The estrogen prepares the uterus for delivery. The fetus then increases its release of the hormone cortisone, which stimulates parturition to begin.

Parturition develops in three stages. The initial stage prepares for the impending delivery. The animal becomes restless and uncomfortable, and the uterus goes through periodic contractions. The pressure from the fluids and placental membranes stimulates the cervix to dilate, or open. The cervix becomes much softer, and the opening increases.

As the cervix opens and the fetus begins to enter the pelvic canal, the animal enters the second stage of parturition. During this stage of active labor, the uterus undergoes much stronger contractions and abdominal contractions begin. These strong contractions force the cervix to completely open, and the fetus enters the pelvis. The contractions then continue until the newborn animal is delivered.

Following delivery the mother often licks the newborn to stimulate breath. Any fetal membranes that might still be over the nose and mouth are also removed with stimulation from the mother. The sudden stop in delivery of blood through the umbilicus signals the newborn to breath.

In the final stage of parturition the placental membranes are expelled from the uterus. Once the membranes have separated, the uterus begins to shrink back to its normal size. This process is called involution. Uterine size dramatically reduces after delivery. In the Holstein cow an average calf often weighs 40 to 45 kg. The uterus must stretch to accommodate the calf and an equal volume of fluid. Within 3 weeks after delivery, the uterus in a normal cow will have shrunk to only 35 to 40 cm in length and 3 to 4 cm in diameter. Horses also realize a drastic uterine change after birth. Usually within 1 week of delivering a foal, a horse comes into heat and the uterus is ready to carry another pregnancy.

At delivery the fetus has to become independent; it no longer relies on the mother for nutrients and waste elimination. Several important changes occur once the blood supply stops at the umbilical cord. In the uterus the blood supplied through the umbilical

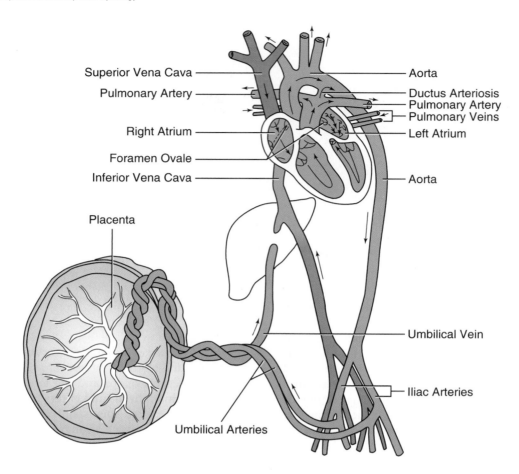

FIGURE 8–10 Fetal circulation. Important differences exist as nutrient- and oxygen-rich blood is supplied from the placenta in the umbilical vein. The fetus does not rely on its own lungs for the supply of oxygen. The foramen ovale and ductus arteriosus lessen the flow of blood to the fetal lungs.

vein is oxygen- and nutrient-rich. The oxygen and nutrients were transferred from the mother to the fetus's circulation in the placenta. Because of this maternal supply, the fetus does not rely on adult-type circulation.

In the fetus an opening called the foramen ovale lies between the left and right atrium (Figure 8–10). In addition, a connection called the ductus arteriosus exists between the pulmonary artery and aorta. Much of the blood delivered to the right atrium bypasses the pulmonary circulation through the foramen ovale and the ductus arteriosus. The fetus does not rely on the lungs to oxygenate the blood; therefore little is delivered there. The highly oxygenated blood delivered by the umbilical vein then directs quickly into the systemic circulation. Blood returns to the placenta through the umbilical arteries to eliminate the wastes and replenish the nutrients and oxygen.

The blood supplied to the fetus is rich in nutrients that have already been processed by the mother's liver. In addition, the liver of the fetus does not fully function. A ductus venosus shunts blood around the fetal liver. About half the blood entering through the umbilical vein is directed around the liver and directly into the caudal vena cava. Note that the blood delivered to the fetus in the umbilical vein is rich in oxygen and nutrients. The vessel delivers blood toward the heart and is therefore a vein, not an artery.

At birth a dramatic transition occurs. The newborn must take a first breath and expand the lungs. This process reduces the pressure in the vessels, allowing blood to flow more freely into them. As the pressure changes within the vessels, the foramen ovale closes with a small valve. With time, this closure becomes permanent. Similarly the ductus arteriosus constricts, thus stopping the flow of blood. This closure becomes permanent with time. As you might expect, the ductus venosus undergoes a similar constriction followed by a permanent closure. These three changes are necessary for the conversion of the fetal circulatory system into the adult-type system.

In mammals the mother supplies nutrition to the infant in the form of milk. The mammary glands,

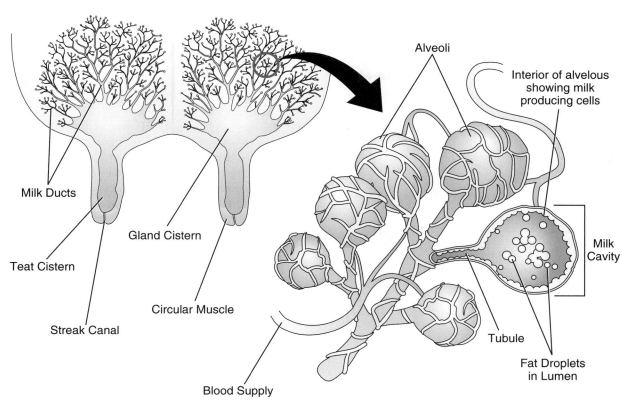

FIGURE 8–11 The udder of a cow.

a collection of epithelial-lined ducts, are housed within connective tissue. The ducts empty into a teat cistern at the base of the teat. The milk can then flow through the teat canal and teat orifice.

Initial development of the mammary system begins at puberty. Development is stimulated by the presence of estrogen and progesterone. The mammary system undergoes dramatic development during pregnancy. The high levels of estrogen stimulate significant growth in the epithelial cells that will eventually secrete milk. The hormone prolactin increases as parturition approaches. Prolactin stimulates the actual production of milk. Oxytocin, a hormone released by the pituitary, actually causes the release of milk from the ducts into the teat cistern (Figure 8–11). This allows the newborn access to the milk as it suckles on the teat.

Following parturition, milk production increases as the newborn grows. Dairy cows have been selected and fed for higher production. Typically these animals produce much more milk than is needed by the calf. Dairy cattle reach peak milk production almost 2 months into their lactation. Production then slowly declines over the next 8 to 12 months.

In most other species the mother continues to produce milk until the nursing animal can consume a normal adult diet. When the infant is **weaned** (stops nursing), the milk production continues initially. How-ever, pressure begins to increase in the mammary gland, which signals the mother to cease milk production.

CLINICAL PRACTICE

OBJECTIVE

■ *Discuss the Clinical Significance of the Academic Material Learned in This Chapter*

Only after the first breath is the newborn no longer dependent on the placental blood supply. In a difficult delivery the umbilical vessels may become obstructed by pressure against the bones of the mother's pelvis. If this restriction takes too long, the fetus will die. The changes in the fetal circulation at birth are not always successful either.

As a fetus prepares to be delivered, the ideal positioning for the animal is to be coming head first, with the two front feet pointing forward (Figure 8–12). Again we will use the cow and calf as our example. As the calf pushes through the pelvic canal, the head works to dilate or open the cervix. As the calf is further delivered, the head becomes exposed by the time the umbilical vessels are pinched over the pelvis. In this presentation the calf is not at risk of inhaling amniotic fluid as it tries to take its first breath.

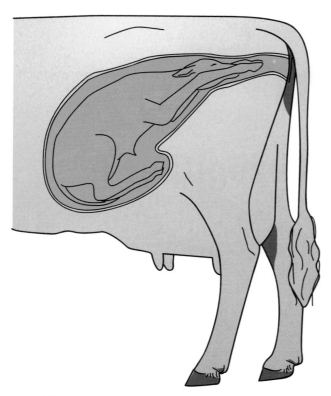

FIGURE 8–12 Normal presentation of a calf at birth.

Calves are often born with both hind legs coming first. Cows usually have more difficulty delivering calves in this position. However, a calf can be removed in this position, even if it takes additional force. In dogs and cats front and rear presentations occur about equally and are both considered normal.

Not all calves are born with a normal presentation. Figure 8–13 shows illustrations of several abnormal calf presentations. In these positions a calf usually cannot be delivered without assistance. In these situations the farmer or veterinarian must correct the positioning of the calf. Cleaning the vulva is the first step. This prevents contamination of the uterus with bacteria. Then, using plastic gloves that reach up to the shoulder, the farmer or vet must reach into the vagina and uterus and assess the positioning of the calf. Knowledge of anatomy is very important in this task. The person delivering must rely on the feel of various structures to identify the positioning of the calf. The head is quite obvious, but it is important to distinguish front legs from back legs.

To reposition the calf, it is pushed deeper into the cow. This provides room to allow for a leg or head to be moved. This can be physically challenging, because the cow's natural reaction is to push. Correcting the position of the calf is usually a multistep process. It is unusual to be able to make the correction on the first

attempt. Usually the calf is forced in deeper and a small correction is made in the position of a leg until the cow forces the calf back. The process is repeated until the calf is moved to a normal forward presentation or with both hind legs coming first.

Once the calf is properly positioned, the cow may still need assistance to deliver. In cattle, **obstetric** chains or straps are placed around the calf's legs and then force is applied to aid the cow (*obstetrics* relates to pregnancy and birthing). This force may come in the form of human power, a come-along, a block and tackle, or a calf jack. A calf jack presses against the cow's hindquarters and has a bar with a winch system to provide tension on the calf (Figure 8–14).

Obviously dogs and cats do not have the size to allow for such intervention. Often just prior to delivery, the pet makes a nest in a secluded location (such as a cat nesting in the sock drawer). A dog's rectal temperature drops below 100°F within 24 hours of **whelping** (normal temperature being 101° to 102°F). As the dog reaches full term, daily rectal temperatures can accurately predict the date for the delivery.

As a dog enters the first stage of labor, she becomes restless and nervous. True labor develops when the dog begins to actively strain. When a dog is in labor more than 5 hours, or more than 2 hours between pups, a problem might exist. It is important for owners to understand that labor may not progress if their dog is upset. Frequent visits and interaction by the owner and family can cause a dog to stop progressing when in early labor. The dog's natural instinct is to find a secluded spot and avoid outside interference.

Dogs often show signs of a false pregnancy. Hormone changes in dogs are the same whether pregnant or not. The life of the CL is the same in both conditions. As the CL regresses and the progesterone declines rapidly, a dog that is not pregnant may show many signs of being pregnant. The dog may also become restless and nervous and begin nesting. Stories of dogs taking stuffed animals to mother are common. This condition is temporary, and the animals return to normal without treatment.

Whenever any animal reaches a point where the veterinarian is unable to complete the delivery, a **cesarean section** (C-section) is required. The C-section is a surgical procedure in which the body wall and uterus are incised and the fetus is extracted. In dogs and cats a ventral midline incision is commonly used. In this surgery the pet lies on its back and an incision is made down the centerline of the abdomen. The uterus is brought outside the body, and an incision is made into it (Figure 8–15). A puppy or kitten is then removed from within the uterus. The umbilical cord is clamped and cut and the puppy or kitten is quickly handed to an assistant. The assistant then stimulates the newborn and suctions any liquid from its mouth and nose. The incision in the uterus is

Abnormal Birth Positions

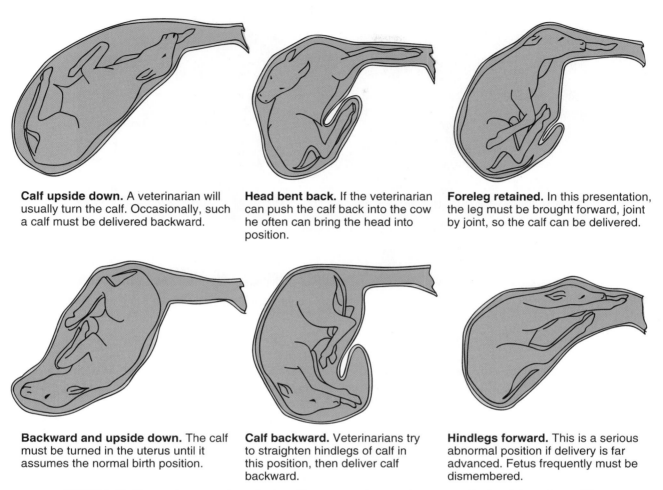

Calf upside down. A veterinarian will usually turn the calf. Occasionally, such a calf must be delivered backward.

Head bent back. If the veterinarian can push the calf back into the cow he often can bring the head into position.

Foreleg retained. In this presentation, the leg must be brought forward, joint by joint, so the calf can be delivered.

Backward and upside down. The calf must be turned in the uterus until it assumes the normal birth position.

Calf backward. Veterinarians try to straighten hindlegs of calf in this position, then deliver calf backward.

Hindlegs forward. This is a serious abnormal position if delivery is far advanced. Fetus frequently must be dismembered.

FIGURE 8–13 Abnormal presentations of calves. The presentation must be corrected before delivery is possible.

then sutured, followed by the body wall. Often the dog or cat is spayed following the C-section. In cattle one method is to do the surgery with the cow standing. The abdomen is entered with an incision in the abdominal wall on the left side.

As mentioned in the beginning of this chapter, spays and neuters or castrations are common surgical procedures in small animal practice. In the spay the female's abdomen must be opened surgically, and the ovaries and uterus are removed (ovariohysterectomy). The ovarian and uterine arteries and veins must be **ligated,** or tied off, before the organs are removed.

Spaying is extremely valuable in controlling pet populations. In addition, the procedure has some distinct health benefits for the pet. Obviously, spaying prevents an unwanted pregnancy and the risks that may occur at delivery. In addition, spayed animals no longer come into heat, which eliminates the behavioral

challenges associated with estrus. For example, cats become very vocal during estrus. Both dogs and cats are much more likely to roam when in heat, increasing the risk of other problems, such as being hit by a car.

In addition, removing the uterus prevents a very serious disease called **pyometra,** an infection within the uterus, which often occurs in elderly dogs and cats. This disease can be life threatening if left untreated. The standard treatment is to spay the animal. Unfortunately, instead of spaying a healthy young animal, the veterinarian must perform the surgery on an elderly sick animal with an infected organ to remove. The risks are much higher in this type of surgery. Spaying pets before their first estrous cycle dramatically decreases the risk of mammary tumors.

Castration in pets is an easier surgical procedure. In dogs a skin incision is made in front of the scrotum on the midline. The testis is pushed into the incision, and the connective tissue is incised until the testis can

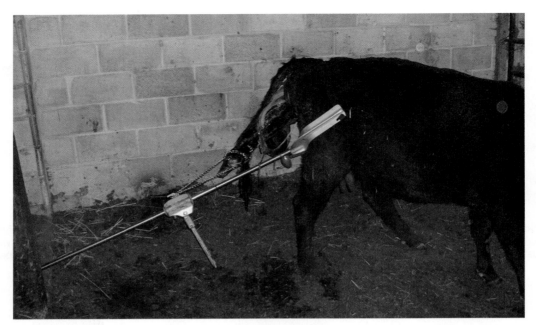

FIGURE 8–14 A fetal extractor (calf jack) in action. The extractor applies pressure to the hind limbs of the cow as traction is applied to the calf.

be removed. The spermatic cord, which includes the testicular artery and vein and the ductus deferens, is ligated and the testis removed. The second testis is then removed through the same incision.

Castration also provides distinct health benefits. Neutering a male at an early age greatly decreases the risk of tumors and enlargement of the prostate gland. Many male dogs develop such problems as they age.

Castration also has behavioral benefits. Neutered males are much less likely to roam and are often less aggressive.

Remember that the testes originate within the abdomen of the fetus and then descend into the scrotum. One or both of the testes may fail to enter the scrotum, a condition called **cryptorchidism** (today's gravy splasher!). The retained testis may be in the abdomen,

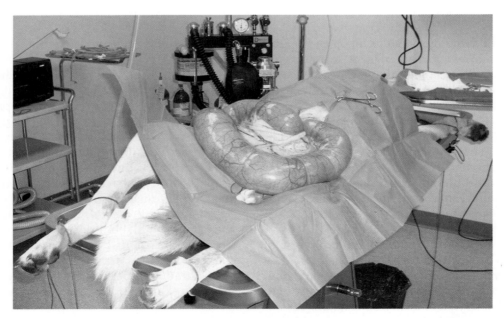

FIGURE 8–15 Intraoperative photograph of a C-section in a dog. The distended uterus has been exteriorized. The next step is to incise the uterus to remove the pups.

in the inguinal canal, or under the skin in front of the scrotum. This testis is much smaller than a descended testis. The higher temperature surrounding this testis prevents its full development. Cryptorchid dogs have a higher rate of tumors in the retained testicle. Surgery to castrate a cryptorchid dog is a little more difficult. The retained testis must be located, which

may mean opening the abdomen. The small size of the testis can make it difficult to find.

Cows that have a very large calf are more likely to have a prolapsed uterus. Following a difficult delivery, all the tissues in the vagina and vulva are stretched and irritated. The cow may continue to strain, causing the uterus to begin to turn inside out (Figure 8–16).

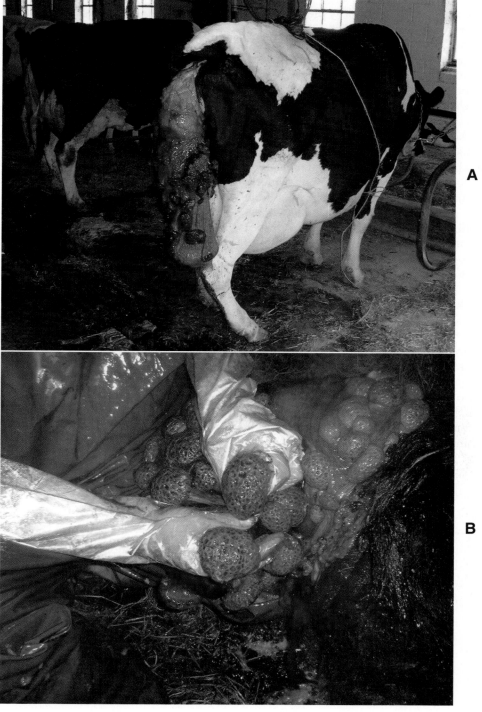

A

B

FIGURE 8–16 *A.* Cow with prolapsed uterus. *B.* Close-up view of prolapsed uterus showing placentomes.

When the tip of the uterus enters the pelvic canal, even more straining occurs. Eventually the entire uterus is exposed outside the body.

This can be a life-threatening situation. The uterine arteries must stretch to travel that far. These vessels can tear, and the cow can bleed to death. In extreme temperatures the exposed uterus, rich in blood vessels, causes the animal to lose or gain tremendous amounts of heat, and the animal can die. One of our clients had a beef cow that began running as he tried to catch her. The cow's entire uterus tore. She rapidly bled to death.

A cow with a prolapsed uterus may first be given epidural anesthesia. A local anesthetic is injected in the space surrounding the spinal cord. This decreases the sensation in the entire pelvic region, minimizing straining as the uterus is replaced.

After being cleaned, the uterus is held close to the body at the level of the vulva, so that the weight of the uterus does not keep pulling it outward. Then, using the back of the fist or the flat portion of the hand, the uterus is pushed back into the cow. This can be a difficult task, because the uterus can be very fragile and the points of the fingers can puncture it. The longer the uterus has been prolapsed, the more swollen it becomes. The swelling makes it more difficult to push it through the vulva. Once the entire uterus has been replaced, it is important to completely correct the positioning. If a portion of the horn is left turned inside out, the condition could recur.

At herd checks, rectal examinations are performed when palpating cows' reproductive tracts. The cervix, uterus, and ovaries are felt through the thin rectal wall. The size of the uterus is judged to evaluate if the cow is ready to be bred, and the ovaries are checked for structures. Ovarian structures include the corpus luteum, a follicle, or a cyst. A follicle feels much like a blister and can be up to 1 inch in diameter. The CL is a firm structure that often has a bump at one margin where the follicle had released its egg. A follicular cyst is a follicle more than an inch in diameter. This abnormal condition occurs when a follicle does not release its egg normally. While an animal is cystic, it will not become pregnant. Some follicular cysts produce excess estrogen. Cows with this type of cyst show estrus very frequently and irregularly.

The changes in the circulation after birth are quite dramatic, although there are times where the process is not completed. For example, if the ductus arteriosus does not close, a heart murmur is present. The pressure in the aorta is higher than in the pulmonary circulation, so blood is shunted from the aorta. This murmur exists throughout the entire cardiac cycle and is called a machinery murmur. When the ductus partially closes, the leakage is less severe. These animals do not have normal circulation, and as they grow, the signs worsen. Many of these animals do not grow well and are inactive. Surgery can be performed to ligate the ductus arteriosus, and a complete cure can be expected. The surgery is quite delicate and is usually performed by a surgical specialist. The cost of the surgery can be quite high. The last puppy that I can remember having this type of repair was being raised as a seeing-eye dog. The surgery was a success, and the puppy recovered nicely.

The portocaval shunt can remain open, again to varying degrees. Animals with this condition show signs as they grow in size. (Review the circulation to the liver and the function of the liver.) Normally the blood from the intestinal tract goes through the liver for processing. With a portocaval shunt, much of this blood goes into the systemic circulation. Certain nutrients absorbed from the intestines, such as ammonia, become toxic at high levels. The blood flows through the body, including the brain, instead of into the liver, where the ammonia could be removed. Seizures can occur, usually a short time after a meal. One of the keys to diagnosing this disease is a history of the signs worsening after eating. This condition too can be repaired by surgically ligating the vessel, redirecting blood into the liver.

SUMMARY

The male reproductive system and the associated hormones allow for the production and delivery of sperm cells. Likewise, the female reproductive system and the associated hormones aid in the production of eggs. The joining of the egg and sperm result in pregnancy and hopefully a successful parturition.

REVIEW QUESTIONS

1. Define any 10 of the following terms:
   parturition
   spay
   (ovariohysterectomy)
   castration
   prolapsed uterus
   epidural
   lidocaine
   estrous cycle
   puberty
   estrus
   seasonal polyestrus
   polyestrus
   anestrus
   pheromone
   gestation
   weaned
   obstetric
   whelping
   cesarean section
   ligated
   pyometra
   cryptorchidism

2. True or False: Luteinizing and follicle-stimulating hormones function in both the female and male.

3. The _____ produces sperm.

4. The _____ is the only accessory sex gland present in the dog.

5. Parturition divides into _____ stages.

6. Cystic ovaries can interfere with _____.

7. What organ becomes infected in pyometra?

8. Name one species that exhibits induced ovulation.

9. In which trimester does the fetus experience the most rapid growth?

10. Which hormone released by the fetus stimulates parturition?

11. When does the peak of lactation occur in dairy cattle?

12. Does a dog's temperature rise or fall prior to whelping?

13. Can females continue to produce additional numbers of eggs throughout life?

14. List the four stages of the estrous cycle.

15. Describe a normal presentation in delivery of calves.

ACTIVITIES

Materials needed for completion of activities:
   sample of frozen bull semen
   light microscope
   slides
   cover slips
   beaker with water
   thermometer

1. Contact an artificial inseminator or farmer and obtain a sample of frozen semen.

   Thaw the semen at 95°F, and place a small drop on a warm glass slide. Cover the drop with a cover slip and examine under a microscope. Healthy sperm should progress forward. If the slide cools, this will slow quickly. Examine the sperm for normal structure. The sperm cell should have a head with a straight and moving tail, or flagella. Examine the slide for defective sperm cells (such as missing or coiled tails).

2. Visit <http://128.192.20.19> to examine ovarian structures in various stages, as well as ovarian abnormalities such as cysts. Key words are theriogenology, bovine theriogenology, and reproductive structures.

3. You are called by Farmer Nitecaller at 2 A.M. to help deliver a calf. Mr. Nitecaller had reached in and found two hooves but no head. He figures it must be coming backward but could not pull the calf. You arrive on the farm and need to determine the calf's position. When reaching into the cow, the calf's entire leg can be felt. What structures or joints must be felt to distinguish a front leg from a back leg? Examine a photograph of a cow, calf, or skeleton to answer this question.

9

The Nervous System

OBJECTIVES

Upon completion of this chapter, you should be able to:

- *Describe the neuron, the nerve impulse, and the synapse, and explain the components of a reflex arc.*
- *Identify the major structures of the brain and name associated functions.*
- *Discuss the anatomy and function of the spinal cord.*
- *Compare and contrast the function of the sensory somatic system to the autonomic nervous system and differentiate between the two branches of the autonomic system.*
- *Discuss the clinical significance of the academic material learned in this chapter.*

KEY TERMS

circling disease
listeriosis
epilepsy
cervical disk disease
equine protozoal
  myeloencephalitis
neuron

volt
polarization
myelinated nerves
coma
myelogram
sensory somatic
  system

autonomic system
plexus
sympathetic system
parasympathetic
  system
dilate
constrict

nystagmus
ataxia
atrophy

INTRODUCTION

The discussion of tissues introduced basic information about nervous cells and tissues. This chapter reviews these facts and goes into further detail on the structure and function of the nervous system, which allows animals to interact with and react to their environments.

A DAY IN THE LIFE

Circling, Seizing, and Stumbling...

Sunday morning the phone started ringing early. It was 6:20 A.M. and I was in the middle of breakfast when I got a call from an upset goat owner. One of his goats had begun to hold its head funny yesterday and was walking in circles. This morning she was down on her side, unable to rise. The signs were classic—this goat likely had **circling disease,** or **listeriosis.** How could this infectious disease in the brain begin?

Two weeks ago I went to the office and was told that an emergency was coming. A dog had begun to have seizures overnight and the owners were quite concerned. Sandy was a 3-year-old golden retriever that had been quite healthy until the seizures occurred. Upon arrival, the dog began to have a seizure in the office. The dog was lying on her side as her legs began to thrash wildly and her jaws chomped quickly (Figure 9–1). The seizure lasted only about 30 seconds. Even so, seizures are always frightening events to watch. Seizures seem to last a long time but usually are only a couple minutes in length. This dog's seizure stopped as I was administering medication, but I cannot take credit for stopping it. At least the medication would be available to help prevent any further seizures. I submitted a blood chemistry profile and did not find any underlying problems. I had to conclude that this dog had **epilepsy.**

I just returned from the office after seeing Jack, a 9-year-old yellow Labrador retriever. Jack developed a limp yesterday, which was much worse this morning. He was in obvious pain. I felt Jack's leg. He did not seem to mind all the bending and poking. It was only when I bent his neck that he reacted. Jack would hold his neck to the right but yelped in pain if I brought it to the left. I concluded that Jack had **cervical disk disease.** One of his intervertebral disks pressed on the

FIGURE 9–1 A seizing dog. During a seizure, dogs typically are unresponsive, laying on their sides, paddling their legs, and chomping their jaws.

nerves or spinal cord in that region. As a result, he felt intense pain in his neck and leg. I started Jack on high levels of a medication that would help relieve the swelling and pain associated with his condition.

In preparing to write this chapter, I remembered a case I had seen in veterinary school. It is one thing to help a staggering dog, but it was quite intimidating to lead a 1000-lb horse that stumbled. This particular horse had lost its coordination and would often trip as it walked. The clinicians on the case suspected **equine protozoal myeloencephalitis** (EPM). In this disease, protozoa invade the brain and cause a variety of neurologic signs. The outcome was not favorable, and this horse was eventually euthanized.

Diseases affecting the nervous system often produce dramatic symptoms. The clinical signs are usually explainable with an understanding of the structure, anatomy, and function of the nervous system.

NEURON FUNCTION

OBJECTIVE

■ *Describe the Neuron, the Nerve Impulse, and the Synapse, and Explain the Components of a Reflex Arc*

The **neuron** is a specialized cell that lies within the nervous system. Neurons conduct electrochemical signals along their length. The neuron has a larger region called

the body and a long, thin extension called the axon. The axon, a membrane-bound extension of the cytoplasm, transmits the signal long distances to other structures. Groups of axons run together in visible structures called nerves. Neurons may also have shorter extensions, dendrites, which receive signals from other neurons.

Other accessory cells may be closely associated with the neuron. Schwann cells have extensions of the cytoplasm and cell membrane that wrap around

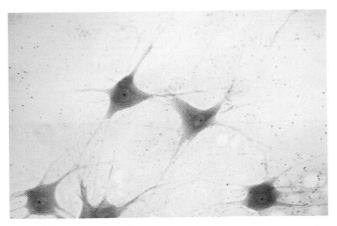

FIGURE 9–2 Photomicrograph of multiple motor neurons from the spinal cord of a cow. *(Photograph courtesy William J. Bacha, PhD, and Linda M. Bacha, MS, VMD.)*

the axons of neurons. The multiple wraps from the Schwann cell increase the diameter of that region (Figure 9–2). Schwann cells create the myelin sheath, which increases the speed of the nerve signal. The points where the Schwann cells meet are called nodes of Ranvier. Not all nerves have a myelin sheath.

Neurons divide into three classifications. The first are sensory neurons, which deliver a signal from a specialized receptor to the central nervous system (CNS). The receptor may detect some mechanical force, light, sound, or chemical. Interneurons are responsible for delivering a signal from one neuron to another. Interneurons provide the complex pathways present in the brain and spinal cord. Motor neurons then deliver the signal from the CNS to the muscle or gland being stimulated for a response (Figure 9–3).

The nerve impulse is the electrochemical signal that transmits along the length of the neuron. This very complex process requires an input of energy by the cell. In the resting neuron the interior of the cell is more negatively charged than the extracellular region. This resting potential is normally in the range of –70 millivolts (mV). A **volt** is the unit of measure in electricity describing the force associated with a difference in charge across a region. **Polarization** is also used to describe the condition in which one region of a cell has a different charge than an adjacent region.

The cell remains polarized by actively pumping sodium (Na^+) ions to the extracellular fluid. The concentration of Na^+ is approximately 10 times higher in the extracellular fluid than in the cytoplasm. Potassium (K^+) ions are distributed in the opposite manner, with the higher concentration within the cytoplasm. The total concentration of K^+, however, is much less than the Na^+, resulting in the interior of the cell being more negatively charged.

The natural tendency is for the Na^+ to diffuse inward and the K^+ to diffuse outward. The neuron must continuously pump the ions to maintain the concentrations. Mitochondria provide the energy for this active transport. Stimuli to the nerve cell increase the permeability of the cell membrane to Na^+. With enough stimulation, the resting potential will approach –50 mV, which is called the threshold potential.

When this threshold is reached, there is a dramatic increase in the permeability of the membrane and Na^+ rushes into the cell. The interior of the cell actually takes on a positive charge for a short time (Figure 9–4). This sudden change in electrical charge is called depolarization. The changes occurring at one point on the membrane stimulates adjacent membranes to change as well. In this manner the signal is sent along the length of the axon. The depolarization

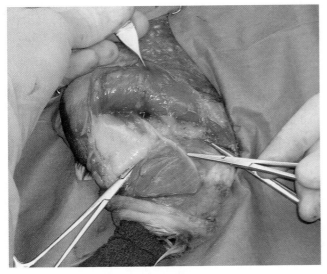

FIGURE 9–3 Intraoperative photograph of a sciatic nerve exposed during an amputation procedure. The cat's leg had been severely damaged by trauma from being struck by a car.

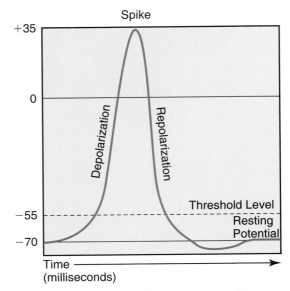

FIGURE 9–4 Voltages in an action potential.

sweeps down the neuron making up the nerve impulse or action potential.

It is tempting to compare the axon to an electrical wire, but this is not correct. Electricity flows at the speed of light. Nerve impulses are much slower. Nerve impulses flow in the range of 10 to 500 m/sec. (The speed of light is approximately 300,000 km/sec.) The nerve impulse is more accurately compared to lighting a fuse or a line of gunpowder.

Myelinated nerves (those with a myelin sheath) transport a signal much faster than nonmyelinated nerves. In myelinated axons the action potential only occurs at the nodes of Ranvier. The impulse occurs at one node and then jumps across the Schwann cell's membrane to depolarize at the next node of Ranvier. Myelinated nerves may transmit an impulse up to 50 times faster than in those axons lacking a myelin sheath.

The instant the nerve impulse has left a region, the process of repolarization begins. Potassium ions rapidly diffuse out of the cell, reestablishing the normal polarity of the cell. During this time, called the refractory period, the nerve cell cannot transmit another signal. For most neurons the refractory period is in the range of 1 to 2 milliseconds. The neuron must reach its normal resting potential before it can transmit another impulse. When nerve cells fire, it is all or none. For a greater response to occur, a greater number of nerve cells must be recruited.

The nerve impulse must then be transmitted to another neuron or cell, such as a muscle or gland. The junction where this occurs is called the synapse (Figure 9–5). The cell carrying the nerve impulse is the presynaptic (*pre–* is a prefix meaning "before") neu-

ron. This axon widens as it ends in a synaptic knob. Once a nerve impulse reaches the synaptic knob, a neurotransmitter, such as acetylcholine (ACH), is released at the synapse. Many more neurotransmitters exist in the CNS.

A neuron may have hundreds of synapses on its dendrites and cell body. ACH, released at the synapse, causes the membrane to become more permeable to Na$^+$. When enough ACH is released from multiple synapses, the neuron reaches the threshold potential and another nerve impulse begins. ACH is a chemical transmitter, allowing the signal to flow from one cell to the next. ACH is then broken down by an enzyme, cholinesterase. The ACH must be removed to prevent repeated stimulation from one incoming signal.

Some synapses may actually be inhibitory. These synapses release a transmitter that allows K$^+$ to diffuse out of the cell. This results in the neuron becoming more polarized, further from the threshold potential. The neuron may still fire, but it would require a higher number of excitatory synapses to reach its threshold potential. These inhibitory synapses are very important in allowing coordinated movement. The discussion of muscles showed that when one muscle group contracts, another antagonist group must relax. These inhibitory synapses allow for the smooth motion with muscle contractions, as one group contracts while another relaxes.

The reflex arc is the simplest unit of function within the nervous system. In a reflex the body reacts without requiring conscious thought. A withdrawal reflex is a commonly encountered example. As an illustration, picture a cat that has jumped onto a kitchen counter and walks toward the stove. The cat takes a step and places its foot onto a hot burner. The cat quickly pulls back its paw to prevent further injury. How did this occur?

The reflex arc begins with a stimulus, a hot stove in our example. Sensors in the foot detect heat, which then triggers a nerve impulse in the sensory nerve (Figure 9–6). The sensory nerve synapses onto an interneuron within the spinal cord. The interneuron is stimulated and, depending on the particular reflex, stimulates other interneurons or goes directly to a motor neuron. The motor neuron then stimulates a muscle to pull back the leg, preventing further damage.

This describes the pathway in a very simplistic reflex arc. The order of events is correct, but in the animal the number of pathways is actually much larger. There are numerous sensory inputs, large numbers of interconnections between interneurons and multiple motor neurons that must be stimulated. The significant features in a reflex arc, though, are illustrated. The large number of connections allows for the response in a reflex to have variation, depending on the severity of the event.

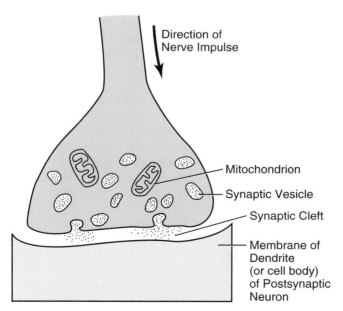

Direction of Nerve Impulse

Mitochondrion
Synaptic Vesicle
Synaptic Cleft
Membrane of Dendrite (or cell body) of Postsynaptic Neuron

FIGURE 9–5 A synapse. Nerve impulse from the axon stimulates the release of a neurotransmitter.

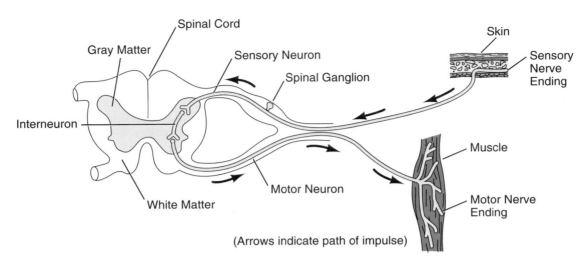

FIGURE 9–6 The pathway for a reflex. Painful stimuli are detected by sensory nerve endings in the skin. The signal is transmitted by a sensory neuron to an interneuron. The interneuron then stimulates a motor neuron. The result is a muscle contraction, pulling the leg away from the painful stimuli.

Notice that no signal is sent to the brain to cause the motion. The entire reflex occurs at the level of the spinal cord. The brain, however, can influence a reflex arc. To some degree, conscious thought can be used to overcome the events in a reflex. Many people are able to overcome a natural reflex of blinking when placing contacts in their eyes. It is a reflex to blink when an object is coming toward the eye. The contact wearer consciously keeps the eye open, knowing that no damage will occur. During a reflex the brain does receive signals of the events. The cat in our example required no input from the brain to pull its foot away. However, the brain was sent signals telling the animal that the stove was hot. As a result, the cat changed its course and the reflex prevented a severe burn.

BRAIN STRUCTURE AND FUNCTION

OBJECTIVE

■ *Identify the Major Structures of the Brain and Name Associated Functions*

The nervous system divides into the central nervous system and the peripheral nervous system (PNS). The peripheral nervous system detects stimuli and informs the CNS. The PNS also carries the signal to cause a response at the level of the muscles and glands. The CNS receives all the signals from the peripheral system and coordinates all the activity. The spinal cord and brain comprise the central nervous system.

The skull and vertebrae provide protection for the CNS. Within the bony casing, the meninges, a group of three membranes wrap around the brain and spinal cord. In addition, the CNS is bathed in cerebrospinal fluid (CSF), which provides additional protection. The CSF gives cushioning when trauma occurs to the head or backbone. There are openings within the brain itself, called ventricles, that also contain the CSF.

Cerebrospinal fluid is produced within the ventricles of the brain from a complex group of capillaries called the choroid plexus. The CSF flows from within the ventricles in the brain to the space between the brain and meninges. With activity from the animal, the CSF flows around the brain and spinal cord. Another complex of capillaries absorbs the fluid and returns it to the bloodstream. CSF is constantly being produced and then absorbed. In the average dog this amounts to about 3 ml every hour.

Visually the brain divides into three regions: the cerebrum, the cerebellum, and the brain stem (Figure 9–7). The brain stem controls most of the functions necessary to maintain life. The cerebellum provides coordination in the animal's movement. The cerebrum, the largest and most prominent region of the brain, controls the remainder of voluntary movement and thought.

Within the brain stem lies the medulla oblongata and the pons, which appear as a swollen end to the spinal cord. This region of the brain stem controls respiration and circulation. All the nervous control of functions such as heart rate, blood pressure, and respiration rate originates from this region of the brain. Damage to this region of the brain instantly causes death.

In front of the medulla and pons sits a region called the midbrain. The midbrain provides communication between regions of the brain and also from sensory organs, such as the eyes and ears, to the brain. This region of the brain controls balance and many reflexes of the eye. Also within this region lies the reticular activating system, which is involved in the animal's

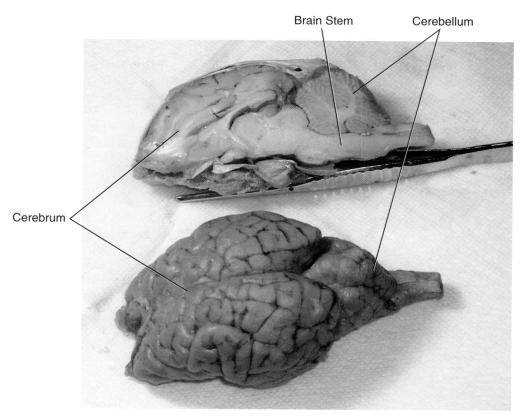

Brain Stem   Cerebellum

Cerebrum

FIGURE 9–7 The external structure and midline section of the brain.

awareness. When the system is actively firing, the animal is quite aware of its surroundings and is focused. This region basically stimulates the cerebrum. Damage to this region can result in **coma,** a prolonged state of unconsciousness.

At the top of the brain stem sits the thalamus. The thalamus works to process all the incoming sensory signals (except smell) before transferring the signal to the appropriate regions of the cerebrum. The thalamus works to coordinate the signals between the spinal cord and the cerebrum.

Immediately below the thalamus are the hypothalamus and the pituitary gland. This region helps to control water balance, thirst, hunger, and temperature regulation. The hypothalamus also acts as an endocrine gland. Hormones such as oxytocin and antidiuretic hormone are produced by the hypothalamus. They are then stored in and released by the pituitary gland. Many of the hormones of the pituitary gland have been discussed in previous chapters and are further detailed in Chapter 10.

Toward the back of the brain is the cerebellum. The cerebellum's main role is to control the coordination of movement. The cerebrum initiates most movement, but the signal passes through the cerebellum to provide controlled, coordinated motion. The cerebel-

lum also monitors signals from the eyes and the balance center to aid in coordination.

The largest and most cranial structure of the brain is the cerebrum. The cerebrum divides into two halves, or hemispheres. The hemispheres are joined by white matter in the center of the brain. The white matter is made of the myelinated axons (Figure 9–8). The outer region, or cerebral cortex, consists of cell bodies, or

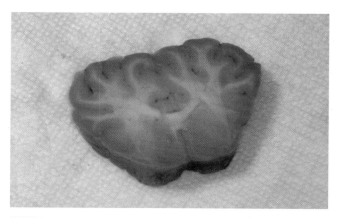

FIGURE 9–8 The brain in cross-section. Note the arrangement of gray matter (cell bodies) in the outer regions and the internal white matter (myelinated axons).

gray matter, which have elevations and grooves that increase the surface area within the brain.

The cerebrum is the sight of most voluntary and conscious processing. Voluntary movement begins in this region, and much of the sensory input is processed there as well. The cerebrum controls all thought, learning, memory, judgement, language, and personality. The cerebrum is quite large to allow for all the pathways and connections necessary to process this information.

The cerebrum divides into several lobes. Different regions within the cerebrum control different functions. There are regions that process the different sensory information, movement, and thought. These regions are well mapped in humans. With this knowledge it is possible to predict the effects of damage to a certain region. For example, a patient requiring surgery on a brain tumor can be told what defects might exist following the procedure.

ANATOMY AND FUNCTION OF THE SPINAL CORD

OBJECTIVE

■ *Discuss the Anatomy and Function of the Spinal Cord*

The spinal cord is the second component of the central nervous system and functions as the link between the peripheral nervous system and the brain. In addition, the cord functions in many aspects of coordination through the reflex arcs. It extends from the base of the brain through the canal formed by the vertebrae. The actual cord reaches the level of the sixth or seventh lumbar vertebra in the dog. From that point, nerves pass through the spinal canal until they exit between the lower vertebrae.

The organization of the spinal cord is opposite of the brain. In the spinal cord the white matter is on the outer region (Figure 9–9). As in the brain, the white matter is composed of the myelinated nerves. The gray matter, the cell bodies, is housed in the inner portion of the cord. The meninges and cerebrospinal fluid surround the spinal cord, just as in the brain.

At each vertebral segment, two nerve branches exit the spinal cord. The dorsal root carries sensory nerves. The ventral root has motor function. In each root a collection of nerve cell bodies creates an enlargement in that region. A collection of nerve cell bodies outside the central nervous system is called a ganglion. The two branches join to create a mixed nerve, and then mixed branches divide in the periphery.

Nerves of similar function run together within the spinal cord. These collections are called tracts. Many nerve tracts cross somewhere within the brain or spinal cord. This is important in localizing the site of a

disease process. Damage to a region in the right side of the brain may result in a weakness in the left side of the body.

In Chapter 3, intervertebral disk disease was discussed. In review, a portion of the intervertebral disk protrudes and puts pressure on the spinal cord or nerves. This information can now be interpreted with the knowledge of the structure of the spinal cord and nerves. Depending on the direction of the disk protrusion, the clinical signs vary. The disk may bulge to the side and put pressure on the spinal nerves exiting the cord. This can result in pain and motor deficits at the sight of the defect.

The disk may also bulge dorsally into the spinal cord. The signs are dependent on the degree of damage that occurs. The disk may just put pressure on the cord or may cause significant swelling and bleeding. The extent of the damage defines how severe the signs are. Complete paralysis can occur in the legs behind the damage.

One means of testing for damage is to evaluate the reflexes in the legs. If damage has occurred at the level where the spinal nerve for that reflex joins the cord, the reflex is depressed. If the damage occurs above the level of the reflex, the signs are often exaggerated. This shows the effect of inhibitory neurons. The brain is not necessary for the reflex, but it does help to control it and keep it coordinated. As a result, if the cord is unable to transmit the inhibition, the signs of the reflex can be greater than normal.

One diagnostic test that is extremely useful in identifying the site of the problem is a **myelogram**. In the myelogram a dye is injected into the epidural space (that is, into the CSF). Because the CSF moves, it carries the dye along with it. The dye is unable to flow around an area where the disk has compressed the

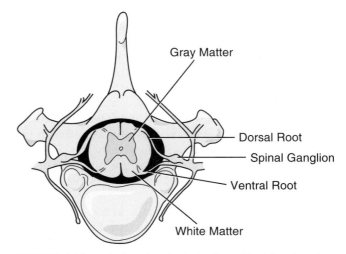

FIGURE 9–9 The spinal cord and spinal column. Note that the orientation of the gray and white matter is the opposite of that in the brain.

Gray Matter

Dorsal Root

Spinal Ganglion

Ventral Root

White Matter

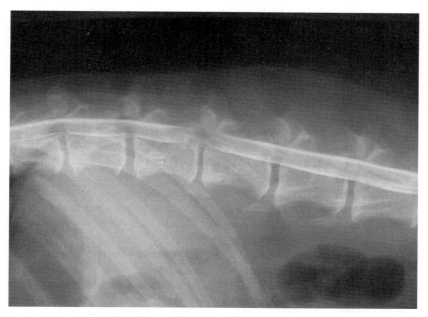

FIGURE 9–10 Myelogram showing intervertebral disk disease. Dye injected into the spinal column shows the compression of the spinal cord. *(Photograph courtesy David Sweet, VMD.)*

cord (Figure 9–10). A radiograph is taken to trace the flow of the dye. The dye is visible on a radiograph and helps to show the location of the problem. In severe cases, surgery is necessary to remove the pressure from the spinal cord. The myelogram identifies the location where surgery is required.

SENSORY SOMATIC AND AUTONOMIC NERVOUS SYSTEMS

OBJECTIVE

■ *Compare and Contrast the Function of the Sensory Somatic System to the Autonomic Nervous System and Differentiate Between the Two Branches of the Autonomic System*

The peripheral nervous system consists of all the nerves and neurons outside the brain and spinal cord. The PNS divides into two systems: the **sensory somatic system** and the **autonomic system**. The sensory somatic system operates all the motor activity of the body and includes all the receptors and neurons associated with detecting changes in the environment.

The sensory somatic system includes twelve pairs of cranial nerves. These nerves enter the brain directly, not the spinal cord (Figure 9–11). Several of these nerves are sensory only and therefore have no motor function. These include the nerves for sight, smell, and sound. Table 9–1 lists the 12 cranial nerves and describes their function. These nerves are responsible for

controlling the functions of structures within the head. In addition, some of the nerves control functions in regions outside the head. For example, the vagus nerve controls many functions in the organs in the chest and abdomen.

The spinal nerves enter the spinal cord at each intervertebral opening. As mentioned earlier, the nerves enter the cord in dorsal (sensory) and ventral (motor) branches. In the dog, typically 36 pairs of spinal nerves exist. The dorsal and ventral branches join to create a mixed nerve. In many regions the spinal nerves combine and then further branch into smaller divisions. This extensive network of nerves is called a **plexus**. An example is the brachial plexus, which is the origin of the nerves to the front leg (Figure 9–12). The nerves in this region are derived from the last three cervical and first two thoracic vertebrae regions. The nerves join within the brachial plexus and then reorganize as nerves to the muscles of the front leg.

The other component of the peripheral nervous system is the autonomic system. The autonomic nervous system has two divisions: the **sympathetic system** and the **parasympathetic system**. The autonomic system is an involuntary system that controls the internal environment of the animal. Special sensors within the internal organs and the central nervous system (especially the hypothalamus) monitor the internal conditions. The system then adapts to any changes and influences any corrections. Table 9–2 compares the two systems, describing many of the functions under its control.

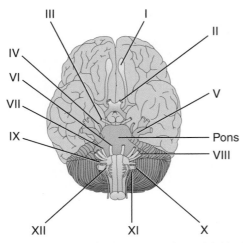

I. Olfactory
II. Optic Nerve
III. Oculomotor Nerve
IV. Trochlear Nerve
V. Trigeminal Nerve
VI. Abducent Nerve
VII. Facial Nerve
VIII. Vestibulocochlear Nerve
IX. Glossopharyngeal Nerve
X. Vagus Nerve
XI. Accessory Nerve
XII. Hypoglossal Nerve

FIGURE 9–11 Cranial nerves and ventral surface of the brain. Table 9–1 lists the cranial nerves and their function.

The autonomic system has two groups of motor neurons. The first, or preganglionic, neuron is in the CNS. Its axon exits through the ventral root of the spinal nerve and synapses in a ganglion. In the sympathetic system this ganglion lays on either side of the spinal column. In the parasympathetic system the ganglia are much closer to the organ being stimulated. The preganglionic neurons use acetylcholine as the neurotransmitter.

The second neuron in the system is called the postganglionic neuron, which sends an axon to the affected organ. In the sympathetic system the axon releases the neurotransmitter norepinephrine at the synapse. The parasympathetic system uses acetylcholine in the second neuron, as well as the first.

In general terms the sympathetic system stimulates organs in preparation for fight or flight. Consider the reaction of your body when you have been suddenly startled. With sympathetic stimulation, heart rate and blood pressure increase. Blood shifts away from the skin and abdominal organs to the muscles, brain, and heart. The bronchi are stimulated to open, allowing more air to enter the lungs. The pupils also dilate (open), allowing more light into the eyes. All these changes prepare the animal for an increase in physical activity, such as running from a predator.

Certain drugs and medications stimulate the sympathetic system. Caffeine, nicotine, and amphetamines are all example of chemicals that stimulate this system. The clinical effects of this are all related to the signs associated with the sympathetic system.

The parasympathetic system basically has the opposite effect of the sympathetic system. The main nerve of this system is the vagus, or tenth cranial nerve. Other nerves exit from the lower spinal cord. Activation of the parasympathetic system slows the heart rate and lowers the blood pressure. The pupils constrict and blood is shifted back to the skin and abdominal organs. Peristalsis is also increased under the stimulation of the parasympathetic nervous system.

TABLE 9–1 The Cranial Nerves

Number	Nerve	Sensory Function	Motor Function
I	Olfactory	Smell	None
II	Optic	Vision	None
III	Oculomotor	Position of eye	Move eye, constrict pupil, focus lens
IV	Trochlear	Position of eye	Move eye
V	Trigeminal	Sense in face and teeth	Muscles of chewing
VI	Abducens	Position of eye	Move eye
VII	Facial	Taste buds	Blinking, facial expression
VIII	Auditory	Hearing, Balance	None
IX	Glossopharyngeal	Taste buds	Muscles in swallowing
X	Vagus	Sensory in internal organs	Parasympathetic to internal organs
XI	Spinal accessory	Muscles of shoulder	Muscles of neck and shoulders
XII	Hypoglossal	Muscles of tongue	Muscles of tongue

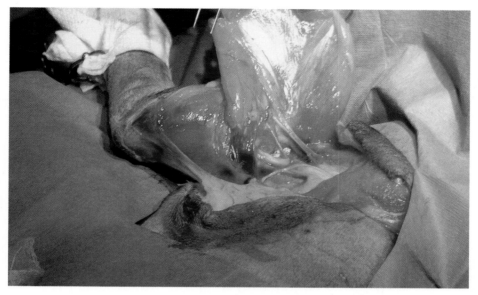

FIGURE 9–12 Intraoperative photograph of the brachial plexus showing nerves entering the leg and the underside of the scapula. This cat was undergoing an amputation because of severe injuries suffered from a gun shot.

Receptors are the means by which the nervous system is able to detect changes in the environment. The environment can be outside the animal's body or within it. Heat receptors are present in the skin and are sensitive to the gain or loss of heat. These receptors are most sensitive to changes in temperature. There are also temperature receptors within the hypothalamus that detect changes in the body's internal temperature. As temperature rises, vessels dilate in the skin and sweating begins. If the body's core temperature declines, blood vessels in the skin constrict and shivering occurs.

The body also has mechanoreceptors, including those of pain and pressure. The skin has receptors sensitive to light touch. These receptors are close to the surface of the skin and are closely associated with hair follicles. These receptors are stimulated by light touch to the skin or even movement of the hair.

The skin also has deeper receptors sensitive to pressure. These receptors, pacinian corpuscles, are deformed with pressure on the skin (Figure 9–13). Each pacinian corpuscle is attached to a sensory neuron.

Once adequate pressure is reached, the neuron sends a nerve impulse to the CNS. Differences in the amount of pressure are signaled by the frequency at which the nerve impulse is sent. In addition, increased pressure activates more pacinian corpuscles, sending more signals to the brain.

The body has no particular receptors to detect pain. Pain results when the neurons receive such a deep stimulus that a nerve impulse is created. Lacking a receptor, these neurons require a stronger stimulus to signal the brain, and this is interpreted as pain.

Mammals also have important internal mechanoreceptors within the tendons and skeletal muscles. These proprioceptors detect stretching or contraction of the muscles and tendons. Proprioception is the body's ability to recognize the position of the body and limbs without having to visually observe it. These receptors are critical in maintaining the body's posture and allowing movement.

Several chemoreceptors have already been discussed in previous chapters. For example, the hypothalamus is able to detect changes in water or salt concentrations. Chemoreceptors are also used in the taste buds of the tongue. Four types of taste buds are distributed in certain regions on the tongue. These receptors detect the tastes of sweet, sour, salty, and bitter. Special chemoreceptors within the nasal cavity detect different smells. There is a close relationship between smell and taste.

The ear is obviously adapted to detect sound. The waves created in sound are converted to the nerve signal sent to the brain in the eighth cranial nerve. In addition, the inner ear contains the structures called

TABLE 9–2 Summary of Autonomic Nervous System	
Sympathetic	**Parasympathetic**
Dilates pupil	Constricts pupil
Dilates bronchi	Constricts bronchi
Increases heart rate	Decreases heart rate
Slows gut activity	Increases gut activity
Decreases saliva	Increases saliva

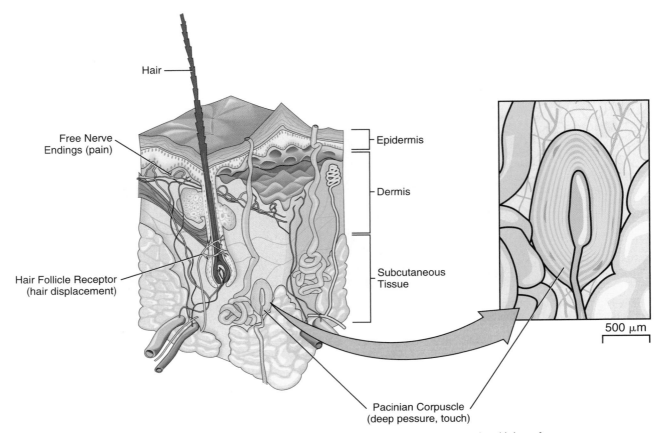

FIGURE 9–13 A pacinian corpuscle. This receptor is sensitive to pressure on the skin's surface.

the vestibular apparatus. The vestibular apparatus detects the position of the body relative to gravity, as well as the motion of the body. This structure gives animals the ability to maintain balance.

The eye houses specialized receptors sensitive to light. The eye works on the same principle as a camera. In the eye a single lens focuses the image onto a layer of light-sensitive cells. The eyeball, a roughly spherical structure, consists of three layers (Figure 9–14). The outermost layer is the sclera, the white portion of the outer eye. The very front of the sclera is specially adapted to be transparent or clear. This clear portion of the eye is called the cornea.

The middle layer, the choroids, is pigmented and contains the blood vessels of the eye. The iris, or choroid, the portion of the eye that gives color, is located at the front of the eye. In the center of the iris is an opening, called the pupil. The pupil **dilates** (opens) and **constricts** (closes) in response to changes in the amount of light entering the eye. The lens sits immediately behind the iris and focuses the incoming light onto the back of the eye. The lens is suspended by ligaments and has muscles that can alter its shape. The shape of the lens changes to maintain focus on the back of the eye.

The iris and lens divide the eye into two chambers. The anterior chamber, in front of the iris, is filled with a liquidlike aqueous humor. This aqueous humor is constantly produced within the eye and then drains into the bloodstream. If the ability to drain the aqueous humor decreases, the pressure within the eyeball increases. This increase in pressure is called glaucoma. Initially glaucoma can cause discomfort and redness of the eye. If left untreated, the animal will develop blindness as the light-sensitive cells are damaged.

The vitreous chamber is behind the iris and lens. This chamber is filled with a thick jellylike material called vitreous humor. Both the aqueous and vitreous humors are transparent to allow light to be transmitted to the back of the eye.

The innermost layer, the retina, contains specialized light receptors. Two cell types, rods and cones, adapt to detect changes in light. The rods are very receptive to light. The rods detect objects in very dim light. Many rods attach to one neuron. The rods contain a special pigment that absorbs light, triggering the signal to be sent through the neuron. The rods provide a very coarse, colorless image. Rods make up about 95% of the receptors in dogs.

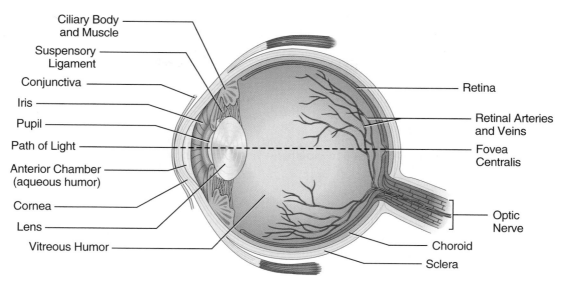

Ciliary Body
and Muscle

Suspensory
Ligament

Conjunctiva

Iris

Pupil

Path of Light

Anterior Chamber
(aqueous humor)

Cornea

Lens

Vitreous Humor

Retina

Retinal Arteries
and Veins

Fovea
Centralis

Optic
Nerve

Choroid

Sclera

FIGURE 9–14 The structures of the eye.

The other cell types, the cones, work in bright light and are receptive to colors. Cones make up a small percentage of the receptor cells in animals. As a result, animals are not as perceptive of colors as humans. The human eye has three different cones, each sensitive to a different wavelength of light or color (blue, red, and green). The sensitivity results from the type of pigment present within the cell. It is the relative response of the three cell types that allows humans to interpret color.

CLINICAL PRACTICE

OBJECTIVE

■ *Discuss the Clinical Significance of the Academic Material Learned in This Chapter*

A thorough physical examination aids in determining the location of a defect in the nervous system. A complete neurologic examination can be very extensive, testing the function of both the cranial and spinal nerves. A description of all the possible steps in a neurologic examination is beyond the scope of this text. A few of the common tests performed are discussed here.

The neurologic examination begins with general observation of the patient. The animal is evaluated for its mental status, determining alertness or depression. In addition, posture, head position, standing, and walking are all checked. Any defects in these signs can help to localize a problem. Table 9–3 summarizes these signs and the region of the nervous system being evaluated.

Many specific reflexes can also be evaluated. Table 9–4 summarizes several of the reflex tests available to check the cranial nerves. The menace response is an easy reflex to test. To do so, the hand is brought rapidly toward the eye. Care is taken to not touch the skin or hairs around the eye. In a normal animal the hand is seen and recognized as a threat, and then a signal is sent to close the eyelids. If the reflex is intact, it shows that the optic nerve (cranial nerve II), the pathways within the brain, and the facial nerve (cranial nerve VII) are all functioning. The animal can therefore see the threat, react to it, and signal the muscles to blink.

Another test is called the pupillary light reflex. In this test a light is shined into an eye. Normally the pupil will constrict or shrink, preventing too much

TABLE 9–3 General Observations of the Patient and the Region Involved

Region Evaluated	Normal	Abnormal
Cerebrum	Alert	Depression or coma
Reticular activating system	Alert	Depression or coma
Cerebellar	Balanced	Loss of balance
Vestibular/Cerebellar	Normal head position	Head tilt
Proprioceptive system	Normal posture/gait	Ataxia, limbs knuckling

TABLE 9–4 Cranial Nerve Testing

Test	Nerve Tested
Menace (rapid movement toward eye)	Optic Nerve (II) and Facial (VII)
Pupillary light	Optic (II) and Oculomotor (III)
Blink when skin near eye is touched	Trigeminal (V) and Facial (VII)
Normal jaw tone	Trigeminal (V)
Hearing (response to loud noise)	Vestibulocochlear (VIII)
Head in normal position	Vestibulocochlear (VIII)
Gag reflex (response to finger deep in mouth)	Glossopharyngeal (IX) and Vagus (X)
Normal tongue strength	Hypoglossal (XII)

light from damaging the retina. This test requires that the optic nerve is intact but also that the oculomotor nerve (cranial nerve III) is functioning to constrict the pupil. Due to the communication between the optic nerves, the opposite pupil will constrict equally. When both pupils react, the interconnection is intact.

Observing the eyeballs for normal positioning and movement helps to evaluate cranial nerves III, IV, and VI. In addition, a normal animal with its head held still, should have no rhythmic jerking to the eye. In **nystagmus** the eyes jerk back and forth in a rhythmic manner. The eye jerks one direction in a quick fashion and then returns more slowly. The presence of nystagmus can result from damage in the inner ear, cranial nerve VIII, brain stem, or cerebellum. This condition can be created in people riding a spinning-type ride at the amusement park. When the ride stops (and the rider feels dizzy), the eyes will show nystagmus.

Reflexes are also used to evaluate the spinal nerves. Proprioceptive reflexes occur when tendons or muscles are stretched. One simple test is done with the pet standing. The foot is turned so that the top of the foot is touching the ground. In a normal animal, the foot is placed into the correct position almost immediately. Even without observing the foot, the animal recognizes the incorrect position and corrects it. In animals with nerve damage, the foot's position may not be corrected or only corrected after a prolonged period.

A classic reflex is the knee jerk. In this test the patellar ligament is struck with a reflex hammer. This ligament stretches from the patella to the top of the tibia. The signal is sent to the spinal cord as if the knee is flexing. In response, the reflex signal is sent through the motor nerve to quickly extend the knee joint. A lack of this response signals nerve damage in the lower lumbar region.

An animal having a seizure often begins seizing by falling and losing consciousness. The animal shows violent uncontrolled motor activity. In small animals the pet paddles the legs and opens and closes the jaws (chewing gum motion). Many muscle groups also seem to twitch. During the seizure the animal may drool, urinate, and/or defecate. Seizures result from excessive firing of neurons in the gray matter of the cerebrum.

Following a seizure the animal may show abnormal behavior, such as disorientation. Some pets even appear blind for a time following a seizure. The severity of these signs usually depends on the severity and duration of the seizure. Many of these animals have an elevated body temperature due to the heat given off during the extensive muscle activity.

Clients with pets that have seizures are advised to be very careful around the animal's mouth. The seizing pet is unaware of its action during the seizure, and the owner could potentially be bitten. The pet should be protected from falling off furniture or down stairs during the seizure.

The majority of seizures last only a few minutes or less. The duration of seizures often seems much longer, because seizures are so stressful to watch. Short seizures are, in general, not life threatening. Occasionally an animal's seizure lasts for more than 30 minutes. This can become life threatening.

There are many causes of seizures, including trauma, tumors, toxins, and certain infectious diseases. The most common seizure diagnosis in private practice is epilepsy. In epilepsy, no underlying cause can be determined. The seizures usually begin in pets from 6 months to 5 years of age. Often the seizures are infrequent, occurring several months apart. As the pet ages, the seizures often increase in frequency.

There are several medications used to control epilepsy, including phenobarbital. Usually the pet requires treatment for its entire life. The medication does not cure the disease but merely controls the frequency and severity of the seizures. With long-term treatment, the dosage often requires adjustment when the seizures increase in frequency. The liver adapts to medication exposure. With time, the liver increases the amount of enzyme present that breaks down the drug. As this occurs, the same dosage to the pet results in lower levels in the bloodstream. To compensate, the dosage must be increased.

Certain insecticides (called organophosphates or carbamates) are toxic to the animal's nervous system. As a review, certain neurons (motor neurons and parasympathetic neurons) release acetylcholine at the synapse. The ACH is rapidly broken down by an enzyme, cholinesterase. The ACH is broken down to prevent repeated overexcitation at the synapse. These insecticides block the action of the cholinesterase.

The signs are therefore predictable, knowing where the ACH will accumulate. At the muscle level, the increased ACH results in excessive stimulation with muscle twitching, tremors, and stiffness. Due to the effect on the parasympathetic system, the animal will drool, have constricted pupils, experience diarrhea, and show a slow heart rate. The brain can also be involved, resulting in signs of nervousness and even seizures. Medications are available that help to block the effects of the ACH or to reverse the effect of the insecticide on the cholinesterase.

Listeriosis (circling disease), an infection that affects the brain, is most commonly seen in cattle, sheep, and goats. Humans are susceptible as well, with cases usually resulting from consumption of unpasteurized dairy products. The organism, which is often found in soil and can therefore contaminate feed, enters through the mucosa in the roof of the mouth. It then tracts along the branches of the trigeminal nerve (cranial nerve V) to enter the brain. The clinical signs then result from the location where the infection becomes established.

The signs may include difficulty chewing and swallowing. The muscles of the face can become paralyzed, producing a drooped ear or eyelid (Figure 9–15). The vestibular system can be infected, resulting in the circling, incoordination, and head tilt. As the disease progresses, the animal is unable to rise. Death occurs as a result of the infection in the brain or from dehydration, because these animals are often unable to drink.

If caught early, animals are treated with high levels of antibiotics. The medication can stop the infection and allow the damaged tissue to heal. Unfortunately, many animals do not survive, even with medication. I am pessimistic about the chances of success for the goat that I introduced at the beginning of this chapter. It has been my experience that treating down animals is usually unsuccessful.

Equine protozoal myeloencephalitis (EPM) is caused by a protozoa (*Sarcocystis neurona*) that invades the brain and spinal cord of the horse. Protozoa are a class of single-celled microscopic organisms. The clinical signs of the disease vary tremendously, because the organism can attack any region of the central nervous system. The disease is usually progressive, starting with mild signs and then worsening. The signs include **ataxia,** an uncoordinated movement. These horses may stagger and stumble. In some instances, certain muscles may begin to shrink, or **atrophy.** Muscles that lose nerve stimulation shrink because they are not being used.

Regions of the brain can also be involved, resulting in signs similar to those seen in listeriosis of cattle, sheep, and goats. The signs can vary among animals

FIGURE 9–15 A cow suffering from listeriosis. The cow is unable to rise. Note the obvious drooped ear.

but include head tilt, circling, paralysis of facial muscles, and difficulty chewing and swallowing.

Treatment of horses with EPM has had relatively poor and inconsistent results. A positive diagnosis is possible only with an autopsy. However, blood tests are available to show exposure to the organism but not disease confirmation. The clinical signs and the positive blood test do help support a diagnosis. Recently a new medication has gained approval and offers hope in treating these horses.

SUMMARY

Understanding of the nervous system begins with the ability to describe the neuron, nerve impulse, synapse, and components of a reflex arc. An ability to explain brain structures and their associated functions is also important.

The brain and the spinal column comprise the central nervous system, whereas the nerves in the limbs constitute the peripheral nervous system. Conversely, the nervous system can also be divided into the sensory somatic and autonomic nervous systems and between the two branches of the autonomic system. The nervous system in its entirety controls many body functions.

REVIEW QUESTIONS

1. Define any 10 of the following terms:

   listeriosis (circling disease)
   epilepsy
   cervical disk disease
   equine protozoal myeloencephalitis
   neuron
   volt
   polarization
   myelinated nerves
   coma
   myelogram
   sensory somatic system
   autonomic system
   plexus
   sympathetic system
   parasympathetic system
   dilate
   constrict
   nystagmus
   ataxia
   atrophy

2. True or False: All nerves have a myelin sheath.

3. True or False: Damage to the right side of the brain causes weakness in the left side of the body.

4. The long, thin extension of a neuron is called the _____.

5. Gray matter is housed in the _____ region of the spinal cord.

6. The dorsal root, which exits the spinal cord, carries _____ nerves.

7. Sympathetic stimulation causes the heart rate to _____.

8. Does a reflex occur with conscious thought?

9. Are the rod or cone cells of the eyes receptive to colors?

10. Can an underlying cause of epilepsy seizures be found?

11. How can humans contract listeriosis?

12. Name the two systems of the peripheral nervous system.

13. Name the junctions where nerve impulses are transmitted.

14. List the three regions of the brain.

15. List the four types of taste buds found on the tongue.

ACTIVITIES

Materials needed for completion of activities:
   flashlight
   reflex hammer
   six small paper bags
   six small scoops of mini jelly beans in three different flavors (lemon, grape, cherry)
   six marking pens
   yardstick
   graph paper

1. To evaluate the reflexes of the eye, have a subject seated in a dimly lit room. The dim light allows the pupil to dilate, but it should be light enough to observe the size of the pupil.

a. Gently and carefully touch the skin at the corner of the eye closest to the nose. What happened? Did the subject blink? It is possible for the subject to consciously overcome this reflex. In a normal person this should result in blinking.

b. Observe the size of the pupil in the eye. Describe the size, estimating in millimeters the diameter of the opening.

c. Take a small flashlight and shine it toward the eye from the side. Do not use an extremely bright flashlight. Shine the flashlight into the eye from the side for a short period. What happened to the size of the pupil? Repeat on the opposite side. Were the results the same?

d. Once again, shine the flashlight into one eye. This time watch the opposite eye. Does this pupil show the same response? This pupil should react the same as the pupil with the light. This is called a consensual response and proves that the nervous connection between the eyes is functional.

2. Evaluate the knee jerk reflex. The subject should be seated with legs dangling and should try to relax. Identify the subject's patella and tibia. The patellar ligament runs from the patella to the front and top of the tibia. Physicians use a small rubber hammer to strike this tendon. Use the tips of the fingers held together or the side of the hand (such as in a karate chop). Strike the ligament firmly and quickly. This causes a stretching of the ligament that is detected by the spinal cord. The cord then signals the muscles to contract, resulting in the knee being quickly extended. There is only a small amount of movement as the reflex compensates for the stretching. It is very possible for the subject to overcome this reflex if the muscles are not relaxed. It is also possible for the reflex to be exaggerated when the subject anticipates the result.

3. The senses of taste and smell are closely associated. Anyone who has had a severely congested nose can relate to the fact that food just does not taste as good. To test this relationship, the subject should be blindfolded and have his or her nose tightly held closed. Test the ability of the subject to identify distinct tastes. Items such as oranges and onions are excellent test foods.

4. Test the speed of nerve conduction and reaction time by having a subject sit with hand ready to grasp the end of a yardstick being held by the evaluator. Signal the subject to grab the yardstick at the same moment that the evaluator drops it. Perform each step of the following steps three times and average the results.

a. The evaluator simply drops the yardstick. The subject will see the drop and try to quickly catch it. Record the point on the yardstick where it was caught.

b. Have the subject keeps his or her eyes closed. Tap the shoulder of the subject at the moment the stick is dropped.

c. Tap the foot of the subject as the stick is dropped. Again the subject should not observe the drop.

Average the three trials in each step. Estimate the length of the nerve conduction in each step (for example, foot to brain to hand in the third test). Plot the results with length of nerve conduction on the X-axis and distance on the yardstick on the Y-axis. The Y-axis is basically a measure of time required for the signal to be conducted through the nerves and synapses necessary to process each step. A stick is available that converts the distance to an estimate of time. Using the graph, is it possible to show that the longer the pathway, the longer the time required?

10

The Endocrine System

OBJECTIVES

Upon completion of this chapter, you should be able to:

- *Describe the endocrine system.*
- *Name the major endocrine glands, list the hormones secreted by each gland, and describe the functions of these hormones.*
- *Discuss the clinical significance of excesses or deficiencies of endocrine-related hormones.*

KEY TERMS

Addison's disease	hypoglycemia	rickets	Cushing's disease
diabetes insipidus	shunting	alopecia	iatrogenic
diabetes mellitus			

INTRODUCTION

You have now seen how the nervous system provides electrochemical communication among regions of the body. The endocrine system provides a chemical means of controlling distant regions of the body. Numerous ductless endocrine glands are present to help control many aspects of the body's metabolism and regulation.

Much of this chapter involves review of the many hormones already covered in the discussion of other organ systems. This chapter helps to summarize much of this information and provides further detail on the control and regulation of the endocrine system.

A DAY IN THE LIFE

Between a Rock and a Hard Place...

There are times when medicine and physiology just hit a little too close to home. I was examining a heifer for pregnancy today. With the farmer's help we had run her behind a gate to examine her. This usually works quite well, but occasionally the animal decides that it doesn't want to stay. This heifer elected to turn around. I waved my hand in front of her eyes, a technique that often works to discourage the movement. This heifer was not about to change her mind, and she lunged toward me. Unfortunately, my finger was between her head and the metal gate latch.

A reflex was immediately called into action. My finger sensed the pain, sent the signal to my spinal cord, and the motor signal quickly pulled back my hand. Unfortunately the reflex was not quick enough to prevent some damage. My sympathetic nervous system then went into action as my heart rate increased and my pupils dilated. In addition, the stress of the events stimulated my endocrine system (adrenal glands) to release hormones to adapt to the stress of the event.

There are cases that just stick in your mind for years. It was many years ago, in veterinary school, when

I saw a pony that was almost 30 years old. It was June, but this pony had a very thick, long hair coat. It was warm, but this horse was sweating quite abnormally. The veterinarians on the case finally diagnosed the problem as a tumor in the pituitary gland. The pituitary gland was producing too much hormone, which was stimulating the adrenal gland to produce too much of its hormone.

It was only 3 months ago when Sebastian, a 9-year-old rottweiler mix, presented for severe weakness, poor appetite, and vomiting. Sebastian and his owners had just recently moved to the area. The owner informed me that Sebastian had been under treatment for low production by the adrenal gland for several years. This condition is called **Addison's disease.** Table 10–1 shows the results for several blood tests. Sebastian had several electrolytes in the abnormal range. His medication was no longer keeping his condition under control. The subsequent test results show how these values changed after his dosage was increased.

TABLE 10–1 Sebastian's Blood Results

Test	Units	Reference Range	30-Jul	2-Aug	15-Aug	7-Sep
Glucose	mg/dl	65–120	86	101	78	81
Urea Nitrogen	mg/dl	6–24	79	53	23	23
Creatinine	mg/dl	0.4–1.4	3.5	1.9	1	1.1
Sodium	mEq/dl	140–151	135	138	147	145
Potassium	mEq/dl	3.4–5.4	8.7	7.2	5.6	5.3
Calcium	mg/dl	7.9–12.0	13.3	11.4	11.2	11.4
Phosphorus	mg/dl	2.1–6.8	8.1	5.8	4.3	5.8
Amylase	units/liter	400–1400	1185	538	433	709

Red values are above the reference range.
Blue values are below the reference range.
30-Jul: Chemistry profile shows azotemia (high urea nitrogen, creatinine), hyponatremia, hyperkalemia, and hyperphosphatemia. These values are consistent with Addison's disease and a renal insufficiency.
2-Aug: Three days into treatment the values begin to improve. Dosage of medication had been increased.
15-Aug: Treatment at the higher level has brought the blood levels almost to normal. Potassium still slightly above the reference range.
7-Sep: Current treatment dosage has brought blood levels into normal range.

ENDOCRINE SYSTEM

OBJECTIVE

■ *Describe the Endocrine System*

Endocrine glands do not contain ducts and release the hormones into the extracellular fluid or the bloodstream. The hormones are then transported throughout the body. The hormone may have effects throughout the body or may only target specific cells. Once delivered, the hormones influence the activities of these target cells.

Hormones have a wide variety of roles in the body. They help to regulate growth, sexual development, and the metabolism of the cells. In addition, the endocrine system is essential in maintaining homeostasis in the internal environment.

Hormones are divided into four chemical groups. One group of compounds are derived from fatty acids (Figure 10–1A). This group of hormones includes the prostaglandins, one of which has been discussed in the estrous cycle. The second group is called steroids, which are derived from cholesterol (Figure 10–1B). A great deal of attention is given to the role of cholesterol

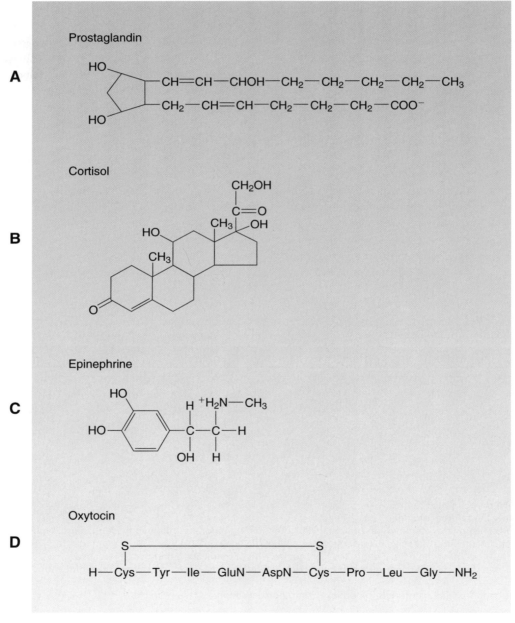

FIGURE 10–1 Four chemical groups of hormones. *A.* Fatty acid derivatives (prostaglandin). *B.* Cholesterol derivatives (cortisol). *C.* Amino acid derivatives (epinephrine). *D.* Peptide hormones (oxytocin).

in heart disease of humans. Cholesterol is an essential compound in the physiology of mammals. Estrogen is one example of a steroid hormone.

A third group of hormones are derived from amino acids (Figure 10–1C). For example, the thyroid hormones are a combination of the amino acid tyrosine and iodine. Chemically these are the simplest of the hormones. The final class is the peptide hormones (Figure 10–1D). These are the largest of the hormones and can be a short peptide chain (such as oxytocin, with nine amino acids) or a large protein.

Once the hormone is delivered, it must signal the target cell to alter its activity. Hormone receptors are present on the cell membrane or within the cytoplasm or nucleus. The analogy of a lock and key is used to describe the receptor and the hormone (Figure 10–2). The molecule making up the hormone has a distinct shape (like the key). The receptor therefore has to accept this specific shape (like the lock).

The steroid and thyroid hormones are relatively small molecules that enter the cells. The receptors for these hormones are within the cytoplasm or in the nucleus. The larger peptide hormones have receptors on the surface of the cell membrane. A mechanism must then exist to convert this extracellular signal to one that influences the interior of the cell. When the hormone attaches to the receptor, an enzyme is activated that creates a second messenger. This second messenger then increases the activity of the target cell. The final result of the hormone depends on the type of specialized cell that is activated. The second messenger must also be destroyed to prevent excess stimulation by the hormone.

The endocrine system is in general regulated by a system of negative feedback. In this type of system the hormone is secreted in response to a change within the internal environment. Once significant correction is accomplished, the secretion is stopped. An example of this is the release of insulin when the blood sugar ele-

vates. As the blood sugar drops, insulin secretion is stopped. In this system the effect of insulin is relatively slow. To prevent the blood sugar from falling excessively, glucagon is released to elevate the blood sugar. It is with this mechanism that the sugar within the blood is held in a tight range.

ENDOCRINE GLANDS

OBJECTIVE

■ *Name the Major Endocrine Glands, List the Hormones Secreted by Each Gland, and Describe the Functions of These Hormones*

The hypothalamus and pituitary gland closely associate in location and function. The hypothalamus not only functions as an endocrine gland but helps to control much of the body's entire endocrine system. The hypothalamus provides a link between the nervous and endocrine systems. The pituitary gland, or hypophysis, lies at the base of the brain, contacting the hypothalamus (Figure 10–3). The pituitary divides into two regions, the anterior and posterior lobes.

There is a nervous connection between the hypothalamus and the posterior lobe of the pituitary gland (Figure 10–4). This region of the gland actually develops from brain tissue in the embryo. The posterior lobe acts as a reservoir for hormones produced in the hypothalamus. The posterior lobe then releases the hormone when appropriate nervous stimulation occurs.

Two peptide hormones are released from the posterior lobe. One hormone, oxytocin, is responsible for contraction of smooth muscles and stimulates the muscles in the wall of the uterus and ducts of the mammary gland. This hormone plays an important role in the process of parturition, or birthing. The release of oxytocin results in contraction of the uterus to aid in the delivery of the newborn.

Once born, the newborn suckling the mother causes a sensory stimulation to be sent to the pituitary. The pituitary then releases oxytocin to stimulate the smooth muscles within the mammary gland. This forces milk into the teat cistern and canal, providing an ample supply of milk to the newborn. Oxytocin is available in an injectable form. This drug is often used to stimulate uterine contraction during delivery of a newborn. It is also used to stimulate release of milk from the mammary gland. Release of milk is also called letdown.

Mastitis is an infection within the mammary gland. The gland can become so swollen that the animal's natural oxytocin release may be ineffectual in allowing the release of milk. Supplementing the cow with an oxytocin injection can allow for a more complete release of the milk from the infected gland.

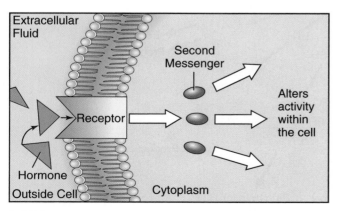

FIGURE 10–2 A hormone receptor in a cell membrane joins with the hormone in a lock and key manner. This causes the release of a second messenger that is active within the cell.

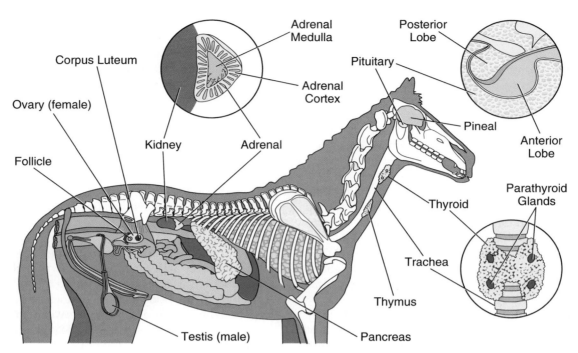

FIGURE 10–3 General location of endocrine glands in a horse.

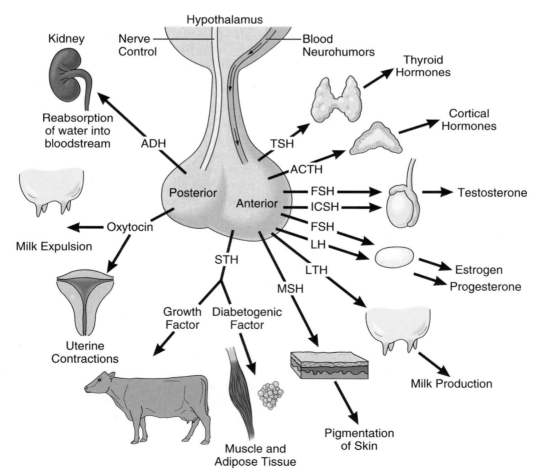

FIGURE 10–4 Hypothalamus—pituitary structure. The posterior lobe has neural secretion. The anterior lobe is stimulated by releasing factors carried in the blood from the hypothalamus.

This can help to remove the bacteria and toxins trapped within the udder and speed the recovery of the animal.

Antidiuretic hormone (ADH) is also released from the posterior lobe of the pituitary gland. A review of Chapter 6 will emphasize the importance of ADH in the regulation of urine formation and water reabsorption. ADH, also a peptide hormone, has two major functions. One role causes constriction of arterioles, which increases blood pressure. The other role causes the distal tubules within the kidneys to become more permeable to water. As a result, more water is retained and the urine becomes more concentrated. The release of ADH is controlled by receptors within the hypothalamus, which detect changes in the water content of the blood. It is predictable that a lack of ADH results in very dilute urine. This condition is called **diabetes insipidus**, and affected animals have an uncontrolled thirst and urinate excessively.

The anterior lobe of the pituitary gland is a classic endocrine gland. It receives signals through the blood and releases its hormones into the bloodstream (Figure 10–4). The hypothalamus also helps to regulate the function of the anterior lobe. Rather than the neural control of the posterior lobe, the hypothalamus produces several releasing and inhibiting factors to control the anterior lobe. For example, the hypothalamus produces a growth hormone releasing factor and a growth hormone inhibiting factor. These two compounds respectively increase and decrease the pituitary's production of growth hormone. The anterior lobe produces and releases at least six polypeptide hormones under the control of the hypothalamus.

The anterior pituitary secretes growth hormone, or somatotropin. As the name implies, somatotropin is responsible for growth of the animal. Growth hormone stimulates all the cells in the body to increase the synthesis of protein. In addition, it signals the cells to utilize fat reserves and conserve carbohydrates. The increase in size results from the increased protein synthesis.

Growth hormone does not directly stimulate the growth of bones and cartilage necessary for the increase in size of young animals. This hormone does stimulate the liver to produce another molecule, somatomedin. Somatomedin is the substance that provides the direct stimulation of bone and cartilage.

Prolactin is another peptide hormone of the anterior pituitary. Prolactin stimulates the development of the mammary glands in preparation for milk production. In addition, prolactin signals the epithelial cells of the mammary gland to produce milk as the pregnancy comes to an end. Prolactin works in coordination with other hormones, such as growth hormone, to control this milk production.

The pituitary gland is called the master gland because of the control it has over the body and other endocrine glands (Table 10–2). One example of this master control is the means by which the pituitary controls the thyroid gland. The pituitary secretes thyroid-stimulating hormone (TSH), which in turn stimulates the thyroid gland to produce thyroxine. The role of the thyroid gland is discussed later in this chapter. Increased levels of thyroxine provide a negative feedback, causing the release of TSH to diminish. The increasing level of hormone, therefore, provides the signal to slow its release.

Luteinizing hormone (LH) and follicle-stimulating hormone (FSH) were discussed extensively in Chapter 8. In males, LH stimulates the interstitial cells of the testes to produce testosterone. FSH helps to control sperm production within the testes. In females the same hormones are essential in regulating the estrous cycle. FSH stimulates the formation of the follicle, which then produces estrogen. A surge in LH signals ovulation and the formation of the corpus luteum. The corpus luteum produces progesterone, essential in maintaining a pregnancy.

The release of LH and FSH is controlled by the hypothalamus. The hypothalamus produces and releases

TABLE 10–2 Summary of Pituitary Gland: Hormones and Actions

Location	Hormone	Target	Action
Posterior lobe	Oxytocin	Uterus, Mammary gland	Stimulates smooth muscle contraction
	Antidiuretic hormone (ADH)	Kidneys	Stimulates water reabsorption
Anterior lobe	Growth hormone	Non specific	Stimulates growth
	Prolactin	Mammary gland	Stimulates milk production
	Thyroid stimulating hormone	Thyroid gland	Stimulates thyroxine production
	Luteinizing hormone	Ovaries/testes	Stimulates ovulation and CL formation Stimulates testosterone production
	Follicle stimulating hormone	Ovaries/testes	Stimulates follicle formation Stimulates sperm production
	Adrenocorticotropic hormone	Adrenal glands (cortex)	Stimulates aldosterone and cortisol production

gonadotropin-releasing hormone (GnRH). GnRH is released in surges in response to the level of other reproductive hormones. GnRH is necessary to maintain a normal estrus cycle.

Adrenocorticotropic hormone (ACTH) is a peptide hormone that helps to regulate the function of another endocrine gland. In this case, ACTH regulates the function of the adrenal glands. The adrenal glands are also discussed in more detail later in this chapter. ACTH stimulates the adrenal cortex (outer region) and its production of aldosterone and cortisol. Aldosterone was mentioned earlier in the text in relation to control of blood pressure and urine production. When the animal is under stress, the increase in ACTH can stimulate a very rapid rise in the blood levels of cortisol. As the stress on the animal declines, the elevated cortisol signals the pituitary to decrease its production of ACTH. This is another example of a negative feedback loop.

The pancreas was discussed in Chapter 7. The pancreas plays the dual role of endocrine and exocrine gland (see Figure 7–11). Clusters of cells, called islets of Langerhans, are scattered throughout the gland. These cells are responsible for the endocrine functions, producing insulin and glucagon. There are two cell types within the islets that secrete these hormones. The cells are labeled as beta cells, producing insulin, and alpha cells, producing glucagon.

The two hormones, insulin and glucagon, work together to maintain a tight control over the level of blood sugar. Following a meal rich in carbohydrates, the blood sugar levels increase rapidly. In response the pancreas releases insulin, which is transported through the bloodstream. Insulin transported through the hepatic portal vein stimulates the liver to convert glucose to glycogen and fat. Insulin stimulates cells throughout the body to become more permeable to glucose. Once in the cell, the sugar can be metabolized. The net effect is that blood sugar level declines.

If this were the only associated regulating mechanism, blood sugar would fall too low. Without another meal to supply more carbohydrates, the use by the liver and the body's cells would make the blood sugar decline to dangerously low levels. As the blood sugar

declines, the pancreas releases glucagon (Figure 10–5). The net effect of glucagon is to increase the level of blood sugar. Glucagon stimulates the liver to break down glycogen to produce more glucose. In addition, glucagon stimulates the liver to convert amino acids and fats into new glucose molecules.

Diabetes is a disease in which an animal is consistently hyperglycemic (has elevated blood sugar levels). The complete term for this disease is **diabetes mellitus**. This distinguishes it from the ADH deficiency in diabetes insipidus. In lay terms, *diabetes* is generally used to refer to diabetes mellitus. The classic signs of diabetes in animals are similar to those in humans. The common history for pets with diabetes includes an animal that is drinking and urinating excessively. In addition, these animals often lose weight in spite of having an aggressive appetite.

In dogs, one common form of the disease occurs when the beta cells of the pancreas deteriorate. This destruction may actually occur as a result of the body's own immune system attacking these cells. More than 75% of the beta cells must be destroyed before signs of diabetes become evident. The standard treatment for diabetes is to supplement insulin. Insulin, a peptide hormone, is destroyed in the stomach if given orally. Therefore insulin must be given by injection. Fortunately, the hormone is given with a very small needle, depositing the medication under the skin. In general, animals do not react painfully to such a small injection.

In controlling diabetes it is imperative to maintain a consistent diet on a regular schedule. The dosage of insulin is determined by monitoring blood sugar levels periodically throughout the day after the injection. If the diet or the level of activity changes, the insulin requirement also changes. The goal is to keep the blood sugar levels close to normal throughout the day.

Over the lifetime of an animal, diabetes can result in many side effects. Poorly regulated diabetes increases the risk of such side effects. Vision is often damaged by diabetes. Elevated sugar levels often cause the development of cataracts (an opacity in the lens of the eye) or damage to the retina itself. In addition, the long-term effects of high sugar levels include

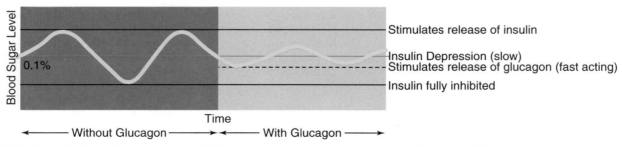

FIGURE 10–5 Regulation of blood sugar levels. With only insulin, blood sugar undergoes wide swings. With the combination of insulin and glucagon, the fluctuations are much less.

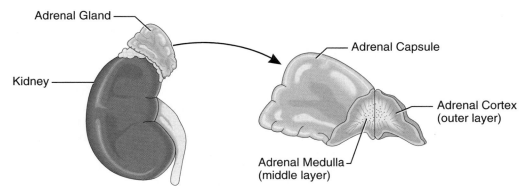

FIGURE 10–6 Location and structure of the adrenal glands.

damage to the nerves and the kidneys. The damage can eventually result in kidney failure. The signs of nerve damage are quite varied, depending on what nerves are involved.

Controlling diabetes requires careful attention. The needs of the animal can change with time and also with differences in appetite and exercise. Other illness, such as vomiting, which limits the intake of the animal greatly, lowers the need for insulin. An overdose of insulin results in **hypoglycemia**. Initially the animal becomes weak, lethargic, and ataxic, or unable to coordinate muscle movement. If the blood sugar level falls further, the animal can enter a coma or develop seizures. Death can result if hypoglycemia is not treated. When caught in the earliest stages, owners can rub syrup on the gums of the animal to provide a rapid source of glucose. More severe cases may require the intravenous delivery of glucose by a veterinarian.

If inadequate insulin is delivered, blood sugar can reach such high levels that the animal also becomes very ill. These animals also become weak and lethargic. The signs can be difficult to distinguish from hypoglycemia. These pets often begin to vomit as well. Treatment in these cases requires additional insulin and often intravenous fluids to keep the animal hydrated. It is extremely important to distinguish these two conditions because treatments are so different.

The pancreas is an endocrine gland closely associated with the digestive system. The stomach also has endocrine function, with the release of gastrin. Gastrin, a short peptide, is released into the bloodstream. When the gastrin is returned to the stomach, it stimulates the release of hydrochloric acid (HCl) by the parietal cells. The duodenum also has endocrine functions. The cells lining the duodenum secrete three hormones, cholecystokinin (CCK), secretin, and gastric inhibitory peptide (GIP). CCK stimulates the pancreas and gall bladder to release their digestive enzymes. Secretin stimulates the release of sodium bicarbonate from the pancreas and the liver to produce bile. GIP is released to slow the activity of the stomach. A more thorough discussion of these hormones and the digestive system can be found in Chapter 7.

The adrenal glands lie just cranial to each kidney (Figure 10–6) and are divided structurally into a central medulla and a larger outer region, the cortex. These two regions have distinctly different functions.

The adrenal medulla is closely associated with the sympathetic nervous system. This region secretes epinephrine (adrenaline) and norepinephrine (noradrenaline) (Figure 10–7). Norepinephrine is the same chemical used as the neurotransmitter in the sympathetic nervous system. The release of these hormones is under the

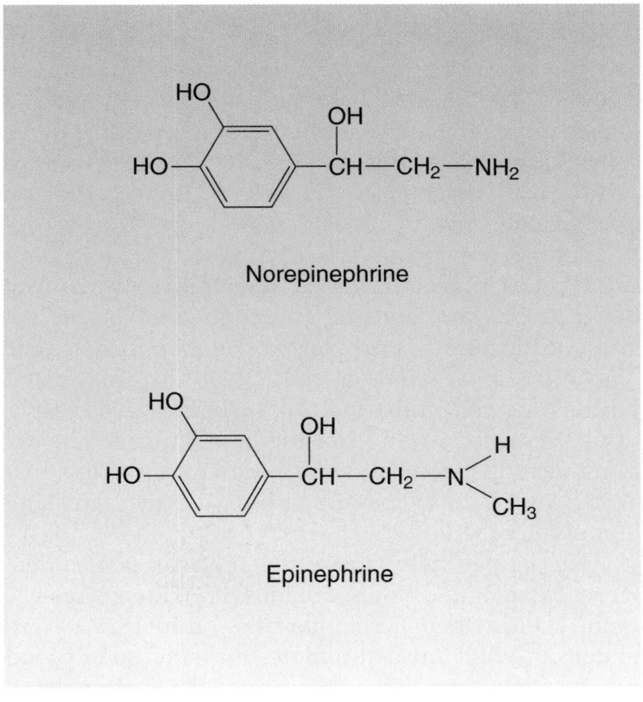

FIGURE 10–7 Chemical structure of norepinephrine and epinephrine.

control of the sympathetic nervous system. The cells within this region are actually modified nerve cells.

The adrenal medulla releases these hormones in periods of stress, such as injury or fright. The effects of epinephrine and norepinephrine are the same as those of the sympathetic nervous system (see Chapter 9). Most notably the heart rate increases rapidly, along with an increase in blood pressure, as arterioles are constricted. Flow of blood is shifted from the skin and intestinal organs to the skeletal muscles, coronary arteries, liver, and brain. The metabolic rate and blood sugar increase. The bronchi dilate to provide more oxygen to the tissues. The pupils also dilate. These changes prepare the animal for sudden increases in physical activity, such as fleeing a predator.

The stress that stimulates the adrenal medulla to release its hormones also causes a release of ACTH from the pituitary gland. The ACTH stimulates the adrenal cortex. All the hormones secreted by the cortex are derivatives of the molecule cholesterol. The adrenal cortex produces two major types of hormones, mineralocorticoids and glucocorticoids.

Cortisol, or hydrocortisone, is the primary glucocorticoid produced by the adrenal cortex. The action of cortisol is primarily at the level of the liver. Cortisol stimulates the liver to convert fat and protein into glucose. This action increases the supply of glucose and glycogen at the level of the liver and also increases the blood sugar level. In addition, cortisol has the potent effect of minimizing inflammation.

In times of stress the cellular demand for energy is often increased. Initially epinephrine is released by the medulla, **shunting** blood supply to areas needing it the most. Epinephrine has only a short duration of action. The elevated stress signals a release of ACTH from the pituitary. The increased cortisol level then provides a longer-term effect, maintaining adequate fuel to the cells. As the stress subsides, the elevated cortisol suppresses the release of ACTH and cortisol returns to a normal level.

Aldosterone is the primary mineralocorticoid. Aldosterone has been discussed in relation to the control of blood pressure and mineral control in Chapters 4 and 6. Aldosterone's primary action is to increase the reabsorption of sodium ions (Na^+) in the distal tubules and collecting ducts of the kidney. As a result, more water is also held in the bloodstream. The actions of aldosterone therefore maintain sodium balance but also influence blood pressure by increasing blood volume.

The kidney helps to control the release of aldosterone. When blood pressure falls, the kidneys release renin. Renin then accelerates the formation of angiotensin, which also stimulates an increase of blood pressure. The angiotensin then signals the adrenal cortex to produce more aldosterone.

The thyroid gland is located in the neck, with a lobe on each side of the trachea. The actual position is variable between animals. The thyroid is unique, being the only endocrine gland that can be palpated. In most animals, however, this is only possible if the gland becomes enlarged. The thyroid gland produces two hormones, thyroxine and calcitonin.

Thyroid hormones, or thyroxine, are produced from the amino acid tyrosine combined with iodine. Two forms of thyroxine, T_3 and T_4, are labeled based on the number of iodine atoms within the molecule (Figure 10–8). T_3 is the more active and potent form of the hormone. T_4 is often converted into the more active form at the tissue level. Thyroxine enters the cell and binds a receptor on the nucleus.

Thyroxine increases the metabolic rate in almost all tissues. As a result, oxygen consumption is also increased. In the heart, thyroxine increases the speed and strength of contraction. In addition, it makes the heart more sensitive to the effects of epinephrine from either the sympathetic nervous system or the adrenal glands. The thyroid hormone stimulates the breakdown of adipose or fat tissue and stimulates erythropoiesis. Thyroid hormone is also necessary for normal growth and development.

Control of thyroxine production is influenced by thyroid-stimulating hormone, already discussed with the pituitary gland. As mentioned, this control is a classic negative feedback loop. Thyroid-stimulating hormone increases the production and release of thyroxine. As a result, thyroxine levels increase in the bloodstream. It is this increase in levels that then signals the pituitary to release less TSH. This feedback loop maintains homeostasis within the animal.

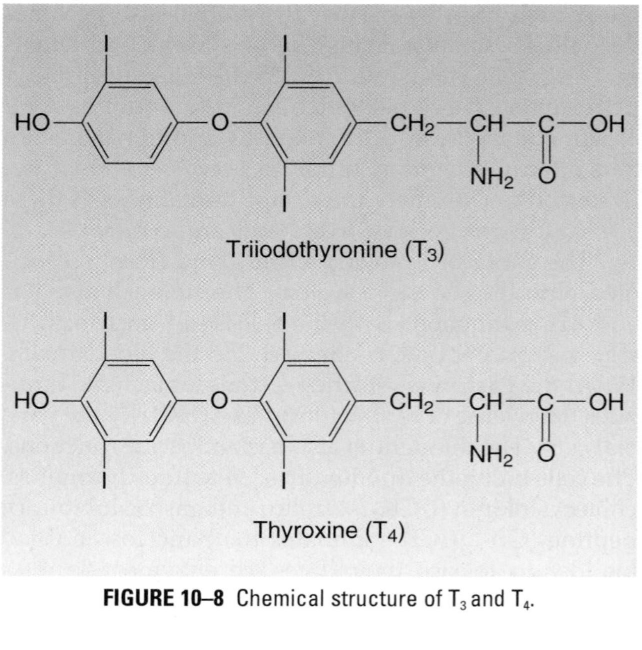

FIGURE 10–8 Chemical structure of T_3 and T_4.

As mentioned, the thyroid also produces calcitonin. Calcitonin is involved in the regulation of calcium levels, along with the parathyroid glands. The parathyroid glands lie adjacent to the thyroid gland. The parathyroid glands may even be embedded within the thyroid gland (Figure 10–9). Parathyroid hormone is a protein hormone also involved in calcium metabolism.

The general effect of parathyroid hormone is to increase the blood level of calcium. This action occurs at three levels. One action is to increase the mobilization of calcium from bone. Parathyroid stimulates the actions of osteoclasts, releasing calcium. It also stimulates an increased absorption from the gastrointestinal tract. Finally, parathyroid hormone stimulates the kidney to increase the reabsorption of calcium while decreasing the absorption of phosphate (PO_4^{---}).

Vitamin D is essential for the absorption of calcium from the intestinal tract. Vitamin D can be ingested in feeds but is also produced in skin exposed to ultraviolet light (sunlight) (Figure 10–10). Vitamin D is a derivative of cholesterol. To become activated, the vitamin D is transported to the liver, where it undergoes its first modification. The kidneys then pick up this intermediate molecule, and it is converted to the active form. This final modification is accelerated by the presence of parathyroid hormone. Lack of vitamin D prevents normal absorption of calcium and prevents adequate deposition in the bone. If this deficiency occurs during childhood, it results in a condition called **rickets**. Animals with rickets have deformed and weakened bones.

As the level of calcium increases, the parathyroid glands are suppressed. In addition, calcitonin is re-

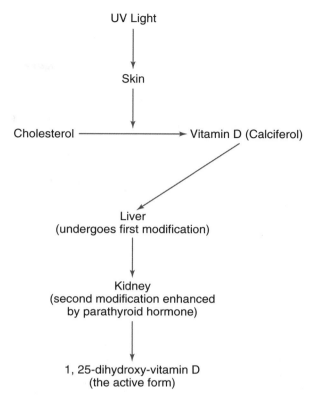

FIGURE 10–10 Summary of vitamin D metabolism.

leased from the thyroid gland. Calcitonin is an antagonist to the actions of parathyroid hormone. Calcitonin decreases the activity of the osteoclasts and decreases the reabsorption of calcium from the kidney and the gut. The action of the two antagonistic hormones helps to maintain homeostasis.

The kidney has also been mentioned in several sites as an endocrine gland. The kidney releases renin, which helps to regulate blood pressure and the release of aldosterone. The kidney converts vitamin D into the active form necessary for calcium metabolism. Erythropoietin also originates from the kidney and is essential to stimulate bone marrow in the production of blood cells.

CLINICAL PRACTICE

OBJECTIVE

■ *Discuss the Clinical Significance of Excesses or Deficiencies of Endocrine-Related Hormones*

Many disease conditions can result if there is either an excessive or deficient production of a hormone. In Chapter 4 the prefixes *hypo–* ("below") and *hyper–* ("above") were introduced to describe the level of compounds within the bloodstream. The same prefixes are

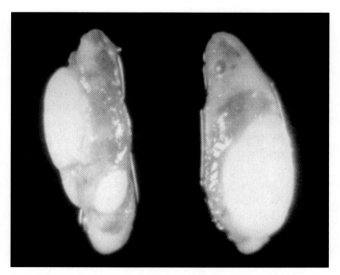

FIGURE 10–9 Thyroid and parathyroid glands from a cat. The white and tan areas are very enlarged parathyroid glands. Normal parathyroid glands would not be visible.

used to describe disease conditions when hormone levels lie outside the normal range. Many diagnostic tests are available to measure the level of hormones within the bloodstream. For example, hypothyroidism describes a deficiency in thyroxine. Hyperthyroidism describes an excessive production.

With an understanding of the function of the hormone, it is possible to predict many of the signs in these disease conditions. Thyroxine plays an essential role in the rate of cell metabolism. In hypothyroidism the body's cells lack stimulation by thyroxine. As a result the body's metabolism slows. The common signs of lethargy, weight gain, weakness, and intolerance to cold are all predictable. In addition, many of these animals first present because of hair coat or skin problems. These animals often have areas of thin hair coat or baldness (**alopecia**). Infection of the skin is also a common side effect of hypothyroidism. Treatment of these animals involves oral dosing of a manufactured thyroxine. The treatment replaces the deficiency of thyroxine and in general is required for the lifetime of the animal. The treatment is not designed to cure the thyroid production, but instead replaces it.

Hyperthyroidism is most commonly seen in elderly cats (Figure 10–11). Again, many of the signs are predictable based on the effect of thyroxine increasing cellular metabolism. The most consistent feature of the disease is weight loss, often in spite of an above-normal appetite. Many of these animals drink and urinate excessively and have increased activity. On physical examination, these cats often have an extremely rapid heart rate, and the thyroid gland is enlarged enough to palpate. As the disease progresses, many of these animals develop diarrhea and vomiting.

Several treatment options exist for hyperthyroidism. Surgical removal of the thyroid gland is extremely effec-

FIGURE 10–11 A cat showing signs consistent with hyperthyroidism or diabetes mellitus. *(Courtesy Mark Jackson, North Carolina State University.)*

tive. Often these animals develop hypothyroidism and require daily thyroxine supplementation. A more serious complication occurs when the parathyroid glands are also removed or damaged during surgery. The parathyroid and thyroid glands lie in such close association that the two may not be distinguished during surgery. Removal of the parathyroid glands results in a severe hypocalcemia, which must then be treated.

Another treatment option is radioactive iodine. Iodine is only included in the molecules of the thyroid hormones. When radioactive iodine is administered, the thyroid gland rapidly picks up the iodine from the bloodstream. The regions of the thyroid that are extremely active (overproducing) pick up the majority of the iodine. Being radioactive, this iodine damages the thyroid tissue. The less active thyroid regions are then left undamaged. It is possible to underdose or overdose animals, leaving the cat hyperthyroid or creating a hypothyroid condition. This procedure is generally very effective. Its usage is limited due to the availability of centers able to handle the radioactive material.

The third option is to administer an oral medication (such as methimazole) that blocks the synthesis of thyroxine. The dosage of this drug must be adjusted to bring the thyroxine levels back into the normal range. This dosage often requires adjustment over the lifetime of the animal. If the medication is stopped, the cat will revert to the hyperthyroid condition. The advantages are that it requires no advanced surgical training, avoids the risk of hypoparathyroidism, and does not require handling of the radioactive iodine. The disadvantages are that treatment is generally required for the lifetime of the cat, failure to treat allows the condition to recur, and side effects to the medication are possible.

The adrenal cortex can also be involved in conditions with excessive or deficient hormone production. Hyperadrenocorticism is also called **Cushing's disease** after the doctor who discovered the condition. In this condition the blood levels of cortisol are excessive. This can be a result of a pituitary tumor producing excessive ACTH or an adrenal tumor producing excessive cortisol. This condition can also be caused by excessive use of cortisone medications by a veterinarian. **Iatrogenic** is a term that describes a condition caused by treatment (for example, iatrogenic hyperadrenocorticism).

The clinical signs and history are all consistent with the variety of effects that cortisol has within the body. The most common clinical signs are excessive thirst and urination, excessive appetite, thin skin and hair coat, panting, enlarged abdomen, weakness, and lethargy (Figure 10–12). Most animals do not show all these signs.

Once Cushing's disease is suspected, diagnosis requires an understanding of the adrenal glands and

FIGURE 10-12 A dog with hyperadrenocorticism. *(Courtesy Mark Jackson, North Carolina State University.)*

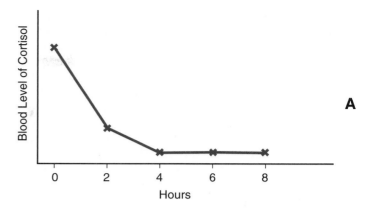

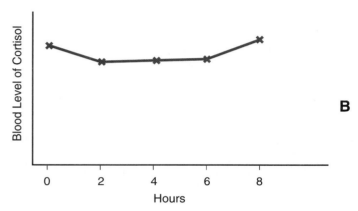

FIGURE 10-13 Dexamethasone suppression test. *A.* Response of blood cortisol levels in a normal animal. *B.* The same test in an animal with hyperadrenocorticism. The levels do not suppress to the low levels and begin to rebound by 8 hours after administration of the dexamethasone.

stimulation by the pituitary gland. A single measure of the cortisol in the bloodstream is helpful but does not prove the diagnosis. Cortisol levels rise and decline during the day, plus the stress of being at the veterinarian's office may result in an elevated level. Several tests are available to confirm the diagnosis.

One such test is called the dexamethasone suppression test. In this test, blood is taken to establish a baseline blood level for cortisol. Then dexamethasone (an injectable corticosteroid) is administered. The cortisol levels are sampled 4 and 8 hours later. In normal animals the dexamethasone is detected by the animal, which causes a decrease in the secretion of ACTH. As a result, cortisol levels decline (Figure 10–13A). In animals with an adrenal tumor, the gland produces cortisol without the need for ACTH. In animals with a pituitary tumor, ACTH is produced without the normal negative feedback from cortisol in the blood. As a result, animals with Cushing's disease do not show the normal decline in cortisol following the dexamethasone injection (Figure 10–13B).

Treatment of animals with hyperadrenocorticism can be surgical or medical. Surgical treatment involves removing the adrenal tumor if that is the site of the problem. Medical treatment involves using a medication that damages the adrenal gland, lowering the cortisol production. The medication is generally required for the life of the animal.

Either treatment can cause such damage that the animal can develop hypoadrenocorticism. This disease can develop naturally as well, if the adrenal glands fail to produce normal levels. Hypoadrenocorticism is also called Addison's disease, after the clinician who discovered the condition. In hypoadrenocorticism there is a deficiency in both the glucocorticoids and mineralocorticoids. Although Sebastian was being treated, he showed all the typical symptoms of hypoadrenocorticism.

The low level of cortisol shows as lethargy, weakness, and weight loss. Often many gastrointestinal signs are evident, including poor appetite, vomiting, and diarrhea. The lack of aldosterone leads to a loss of sodium, chloride, and water. In addition, potassium levels increase. The loss of water can lead to a low blood volume and low blood pressure, which adds to the weakness. As the potassium increases, it can affect the contraction of muscles, including the heart. This too contributes to weakness and fatigue. Often the pet shows no signs until the condition is quite severe. These pets often present as medical emergencies, with severe weakness, vomiting, diarrhea, and low blood pressure.

Hypoadrenocorticism is often suspected when a chemistry profile shows hyponatremia (low sodium) and hyperkalemia (high potassium). While waiting on the blood results, animals in such critical condition must be treated. Based only on the clinical signs, these animals are often treated with intravenous fluids and given glucocorticoids. Fortunately, this symptomatic treatment often yields very positive results while the blood results are pending.

The sodium and potassium changes are very suggestive of Addison's disease. To confirm the diagnosis, an ACTH stimulation test is performed. In this test a resting cortisol level is taken. Then ACTH is administered, and a second cortisol level is taken 1 to 2 hours later. In a normal animal the cortisol increases significantly between the two tests. In an animal with hypoadrenocorticism, ACTH is already quite high. The low cortisol levels stimulate the pituitary gland to release ACTH, trying to increase the levels naturally. In these animals, administering additional ACTH results in very little increase in cortisol levels.

To treat these animals, a mineralocorticoid must be administered. This may come in a long-lasting injectable form or in a pill form. Many of these pets do not require routine treatment with a glucocorticoid. However, when additional stress or trauma occurs, administering more cortisone is helpful. This mimics the body's normal reaction of increasing the cortisone level during times of stress. The story of Sebastian shows that animals under treatment must be evaluated and dosages adapted if the need arises.

In medicine the terms *glucocorticoid, cortisone, cortisol,* and *steroids* are all used interchangeably to describe medications mimicking the natural effect of the glucocorticoids. The multiple terms can at times lead to confusion, but all are referring to the same class of medication. Steroids are used in a variety of cases. The primary usage is for their antiinflammatory and immunosuppressive effects. There is a wide dosage range that varies based on the desired effect.

One of the most common uses for the antiinflammatory effect is in the skin diseases of dogs and cats. Many allergic reactions in pets (such as allergic reactions to flea bites) result in severe itching and skin inflammation. Even when the underlying cause is removed (for example, flea treatment with insecticides), the animal can remain very itchy. Students may relate this to having a mosquito bite. The insect is gone, but the itchiness continues for some time. The biting and scratching that occur can perpetuate the signs (that is, the itching actually makes the skin worse). Glucocorticoids can be administered to relieve this inflammation in the skin and allow it to recover.

Other causes of inflammation, such as trauma, are also treated with glucocorticoids. Trauma to the spinal cord or brain is often treated with this class of drugs. The trauma can even occur from within the animal itself. Intervertebral disk disease traumatizes the spinal cord, and steroids are a standard treatment for this problem.

There are conditions in which an animal's immune system attacks its own cells. (More discussion of this topic is presented in the following chapter.) In this type of disease the dosage level is much higher to provide the immunosuppressive effect. In this type of treatment the function of the immune system is diminished. It is important to realize that this is a general affect, making the animal more susceptible to other infectious diseases.

Other hormones are available to treat animals as well. We have discussed the role of the veterinarian in the reproductive cycle of cattle. I often use hormones to treat cattle, thus influencing their reproductive pattern. Because a goal in dairy cattle production is to have the cow calve on a regular basis, it is helpful to be able to predict when a cow will show estrus. When a cow has a functional corpus luteum, an injection of prostaglandin can be administered. Although this naturally occurs on about day 17 of the cycle, the injection can be given much earlier. Just as with the natural release of prostaglandin, the cow comes into heat following the injection (usually within 3 to 5 days). The advantages of such treatment include fewer days to breeding, better prediction of when a cow will be in heat, and being able to group cows, which improves how well the heat is shown.

Another condition that occurs in cattle is a follicular cyst. A follicular cyst appears as a very large follicle (greater than 1 inch in diameter). Functionally a follicular cyst does not ovulate normally. If left untreated, the cow does not progress through the estrus cycle. Treatment of a follicular cyst is with an injection of GnRH. The medication causes an increase in the release of LH and FSH from the pituitary gland. The follicular cyst may ovulate in response to the GnRH or may form a corpus luteum. When successful, both results start the cow into another estrus cycle.

SUMMARY

As mentioned, the endocrine system via ductless glands allows remote communication with distant areas of the body. Specific hormones were studied in previous chapters, but you can now relate these hormones to total body function. The endocrine system via hormones controls many body processes.

REVIEW QUESTIONS

1. Define the following terms:
   Addison's disease   rickets
   diabetes insipidus   alopecia
   diabetes mellitus   Cushing's disease
   hypoglycemia   iatrogenic
   shunting

2. True or False: Endocrine glands contain ducts.

3. True or False: Luteinizing hormone has no effect in the male body.

4. True or False: Poor vision can result from diabetes.

5. The pituitary gland is often called the _____ gland.

6. The release of follicle-stimulating hormone is controlled by the _____.

7. Vitamin _____ is produced by the skin when exposed to sunlight.

8. In general, is the endocrine system is regulated by positive or negative feedback?

9. What hormone causes milk letdown?

10. Which hormone stimulates the development of the mammary gland?

11. What is the name for low blood sugar?

12. Which gland is the only one in the endocrine system that can be palpated?

13. Renin helps to regulate which vital sign?

14. What is somatotropin?

15. Which two hormones work in tandem to regulate blood sugar?

ACTIVITIES

Materials needed for completion of activities:
   test kit that simulates the detection of pregnancy
   milk samples from a pregnant cow and one in estrus
   milk progesterone test kit

1. Pregnant women secrete a hormone called human chorionic gonadotropin. This hormone is present in the urine of pregnant women. Obtain a kit simulating the detection of pregnancy. Perform the test to learn one method of hormone level evaluation.

2. Visit <http://www.felinediabetes.com> to read an article titled "Diabetes for Beginners." The article gives a thorough explanation on diabetes in cats, including signs, treatment, and monitoring.

3. Obtain milk from a pregnant cow and another cow close to estrus. Run milk progesterone tests on each sample. The pregnant cow will have high levels of progesterone in her body, which will also be present in the milk. The cow in heat will have very low levels of progesterone, which also will be quite low in the milk. The milk progesterone test kit evaluates the level of progesterone and shows the results as a color reaction. (How this test works is discussed in Chapter 11.) Identify which sample is from the pregnant cow.

11

The Immune System

OBJECTIVES

Upon completion of this chapter, you should be able to:

- *Define the term* antigen *and explain its significance in immunity.*
- *Distinguish between passive and active immunity, differentiate between humoral and cellular immunity and their relationship in immunity, and explain primary and secondary immune response.*
- *Discuss the clinical significance of the academic material learned in this chapter.*

KEY TERMS

abscess	edema	active immunity	anaphylaxis
banded	humoral immunity	passive immunity	titer
tetanus	primary response	colostrum	seroconversion
antigen	secondary response	intranasally	ELISA test
lymph	pus	kennel cough	
stocking up	modified live vaccines	pruritus	
phagocytized	killed vaccines	atopy	

INTRODUCTION

The immune system is responsible for protecting the animal from potentially harmful organisms attempting to invade. For many diseases the animal will only become sick from an organism once. The immune system remembers the organism and if exposed again will mount a very quick response, protecting the animal from disease. The immune system is very complicated and detailed. We will try to learn the basics of the immune system to understand how it protects animals from disease.

A DAY IN THE LIFE

Saving Lives and Stomping Out Disease and Pestilence...

In the description of my day I often share the dramatic and interesting cases with you. However, much of my day can be classified as routine. When I am working in the small animal clinic, a large percentage of my time is spent performing routine examinations and giving vaccines to healthy animals (Figure 11–1). Although less exciting to describe, these are the most valuable benefits of my job. By maintaining up-to-date vaccinations, we try to keep animals healthy and prevent disease. For many diseases, such as rabies, there is no cure once the disease is contracted. Prevention is the key. Not quite as exciting as saving lives...but equally important.

A common complaint in small animal work is an itching pet. Dogs and cats often develop allergies that cause them to scratch and bite themselves. The pets may be allergic to fleas, pollens, or even their food. In this situation the immune system, instead of preventing disease, causes problems.

Cows continue to impress me. One poor cow I treated chewed her cud contently despite having a huge swelling on the back of her leg (Figure 11–2). Her owner was concerned about the **abscess**. Infection was intro-

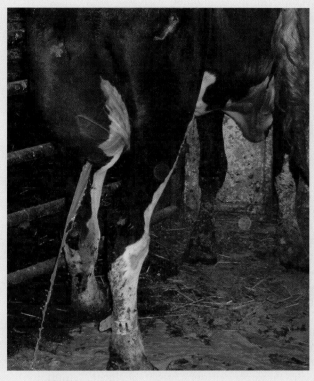

FIGURE 11–2 A draining abscess on a cow's leg.

FIGURE 11–1 Dr. Lawhead talks with his young friend Libby about keeping her cat healthy.

duced under the skin (possibly from an injection), and the cow developed a large abscess (an accumulation of pus). Treatment was gross and dramatic. I lanced the abscess by cutting a large hole into the skin. The very fluid pus squirted from the hole. There were likely several gallons of pus in this abscess, and the cow did not even seem to mind. The cow's immune system was trying to clean an infection and had accumulated a large amount of fluid and debris in this pocket.

There are times when the immune system produces an emergency situation. It has been many years, but I remember one case quite well. A horse had developed an infection, and the attending veterinarian administered an injection of penicillin, an antibiotic. Within a couple minutes the animal began to breath heavily, started sweating, and then became severely ataxic. I scrambled to the truck to get a bottle of epinephrine and a syringe. By the time I returned to the stall, the horse was on its side and experiencing extreme difficulty in

continued

breathing. We immediately administered the epinephrine, but it was too late. In the matter of a few minutes the horse had developed a severe reaction to the medication and died! This case impacted me greatly. How could a treatment meant to cure this animal result in such a horrible outcome?

I went on another call to see a young sheep that was not doing well. The sheep had recently been **banded.** In this procedure a special rubber band is placed at the base of the scrotum, which consequently cuts the circu-

lation to the testes. This is a nonsurgical means of castrating a ram. This sheep was standing, but his stance was very wide and his movement was stiff. On examination I was unable to open his mouth. Unfortunately for this animal, he had developed **tetanus,** or lockjaw. Bacteria had developed in the region of the testes due to the poor circulation. The bacteria then released a toxin that caused all the clinical signs the sheep was experiencing. Treatment was not successful, but could this have been prevented?

ANTIGENS AND IMMUNITY

OBJECTIVE

■ *Define the Term* Antigen *and Explain its Significance in Immunity*

The immune system must be able to recognize substances that are foreign to the body. To accomplish this it must also recognize what is natural to the body. An **antigen** is any foreign molecule that is capable of stimulating an immune response. The term *antigen* does refer to molecules that are within the body itself. Large molecules within the intestinal tract are not considered antigens. These molecules are broken down into smaller fragments before being absorbed into the body.

In general, antigens must be large. Small molecules, such as glucose, do not elicit an immune response. Most antigens are very large proteins or polysaccharides. Smaller molecules may bind to protein and then become antigenic. In addition to being large, the antigens are also complex. This complexity gives the molecule a specific shape that can be recognized by the immune system.

An antigen must not be a naturally occurring molecule within the body (that is, it must be foreign). This feature keeps an animal's immune system from attacking itself. This is also the reason that the immune system will attack a transplanted organ. Large molecules on the surface of cells give them a distinct characteristic that can be recognized by the immune system. Closely related individuals have similar-appearing antigens on the surface of the cells, which makes them better organ donors.

Each antigen is a complex molecule; however, the antigen is recognized by a only a small part of the molecule. This site of recognition is called the antigenic determinant. These regions have a specific shape that can be recognized by the body's immune system. The recognition is very specific for each antigen. Large and complex molecules may have several antigenic determinants and are considered quite antigenic because they greatly stimulate the immune system.

The body has collections of lymph tissue distributed throughout the body to detect antigens quickly. For example, tonsils are collections of lymph tissue within a connective tissue framework. The tonsils are strategically located at the back of the pharynx. In this location they are able to trap many of the invading organisms that might enter through the nose or mouth.

The body has a network of small vessels in addition to the arteries and veins. These lymphatic vessels begin as small capillaries in the tissues and form larger veins that drain fluid back to lymph nodes scattered throughout the body and eventually back to the bloodstream. The fluid that they carry is called **lymph.** Lymph begins as the interstitial fluid that forms between the cells.

As blood enters the capillaries, a portion of the water and small molecules are squeezed from the vessels. This fluid nourishes the cells. The concentration of protein in the blood helps to retain much of the fluid in the bloodstream. The osmotic pressure created by the protein keeps the water drawn into the blood. Animals that develop conditions where the protein declines to very low levels will accumulate more fluid in the tissues. All of this fluid is not reabsorbed into the bloodstream. The lymphatic vessels pick up this fluid, which is then called lymph.

The fluid eventually drains back into the bloodstream. The lymph travels passively; there is no pump to force it through the lymphatics. Activity by the animal helps to keep the fluid flowing. Horses tied in a stall for extended periods may develop a fluid-caused swelling in their legs. The lack of activity allows for fluid to build up because it has difficulty fighting gravity. This condition is called **stocking up.** When the horse returns to activity, the contraction of the muscles surrounding the lymphatics helps to force the lymph upward and back into the circulation.

The lacteals are also lymphatics that are responsible for absorbing lipids from the intestinal tract. (Review Chapter 7.) Each lacteal begins within the villi of the small intestine. These lacteals join into a larger thoracic duct, which empties into the circulatory system in the thorax.

As mentioned, the lymphatic vessels drain into regional lymph nodes (Figure 11–3). These lymph nodes are also lymph tissue encased in connective tissue. The lymph nodes protect the body from invading organisms in that region of the body. The lymph nodes filter the lymph for disease-causing organisms.

The spleen, a large, reddish brown organ located within the abdomen, is also involved in the immune system. Just as the lymph nodes filter the lymph, the spleen filters the blood. The spleen contains a large number of immune cells scanning the blood for antigens. The spleen also houses cells that respond to antigens. Many of the cells produce a large protein called an antibody. Antibodies are an important step in the immune system and are discussed more thoroughly later in the chapter. The spleen also removes aged red blood cells from the circulation.

Although the spleen conducts many important tasks, it is not essential for the animal to live. Fortunately, the lymph nodes can function to detect antigens and house the cells producing antibodies. The spleen is a common site for tumors to develop (Figure 11–4). In addition, the spleen can be damaged by trauma. The spleen has a very rich blood supply, and if it is torn by injury (such as in a dog hit by a car), there can be extensive bleeding. In either situation, tumor or injury, the spleen may be surgically removed.

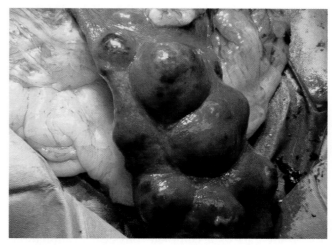

FIGURE 11–4 Intraoperative photograph of a spleen with tumors.

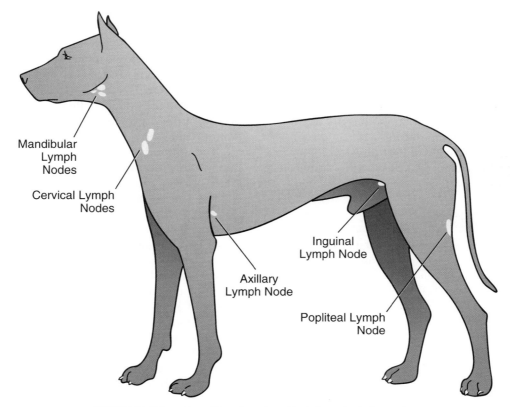

FIGURE 11–3 Location of lymph nodes able to be palpated on a dog.

Mandibular Lymph Nodes

Cervical Lymph Nodes

Axillary Lymph Node

Inguinal Lymph Node

Popliteal Lymph Node

The bone marrow is an essential part of the immune system as well. The bone marrow produces the white blood cells, which are the functional cells of the immune system. When the body first detects an antigen, it is **phagocytized**. Neutrophils and macrophages are the primary cells responsible for this. The macrophages arise from bloodborne monocytes that enter tissues.

Once the antigen (such as a bacterium) is phagocytized, it is broken down within the cell. Fragments of the antigen are then moved to the surface of the cell. These antigen-presenting cells are responsible for stimulating the activity of many other immune cells. The macrophages release substances that that stimulate the general immune response.

These substances have a variety of functions. They may kill viruses or slow their replication. Release of these substances attracts other immune cells to the area and makes antigens easier to phagocytize. Some of the factors also kill damaged cells. One of the factors affects the hypothalamus, which controls the body's temperature. This results in a fever or elevated body temperature, which is a common occurrence in infectious diseases. Monitoring body temperatures is an important part of a complete physical examination. Table 11–1 lists the normal resting body temperature for many animals. It is critical to recognize that excitement and high external temperature and humidity can alter the animal's temperature without an underlying illness.

Fever makes an animal feel poorly but is specifically designed to aid in the fight against pathogens. The higher temperature may hinder the replication of an infectious organism or actually kill it. The fever helps to destroy virally infected cells. The activity of lymphocytes, phagocytes, and antibodies are all stimulated by higher body temperatures.

Inflammation can be a result of physical injury or a reaction to the invasion of a pathogen. Numerous factors released by the cells of the immune system are involved in the control of regulation. These products cause local blood vessels to dilate and the capillaries to become more permeable. This is specifically designed to deliver more white blood cells and antibodies to the area.

The increase in blood flow often results in the area becoming warm and red. The increased capillary permeability results in more fluid entering the tissues, causing **edema,** or swelling. The excess fluid in the tissues results in a higher pressure, which in turn compresses nerve endings. As a result, pain is often associated with these signs. The four classic signs of inflammation are warmth, redness, edema, and pain.

With inflammation a large number of white blood cells enter the infected area. Lymphocytes come into contact with the antigen presenting cells. Each lymphocyte is programmed to respond to a specific antigen. The exposure to the antigen stimulates the cells to undergo repeated mitosis. The result is a large number of lymphocytes specifically designed to react to this particular antigen. Because all the resulting cells are identical, this rapid growth in cell number is called clonal expansion (Figure 11–5).

One type of lymphocyte is called a B cell or B lymphocyte. The B is added because they develop and mature in the bone marrow. Again, each B cell responds to only one antigen. During this clonal expansion, many of the B cells develop into plasma cells. Plasma cells develop a large amount of rough endoplasmic reticulum and Golgi apparatus for protein production and secretion. Plasma cells secrete a specific type of protein, called antibody. The production of antibody in response to an antigen is called **humoral immunity.**

The antibody binds to the same antigenic determinant that stimulated the B cell. Antibodies are large protein molecules made of four polypeptide chains. The resulting molecule takes on a Y shape (Figure 11–6). The arms of the Y have regions that are responsible for detecting the antigen, the receptor region. This region varies among different antibody molecules.

As with all receptors, this binding occurs as the shape of the antigen fits into the antibody (as in a lock and key) (Figure 11–7). The protein found in blood divides into albumin and globulin. Antibodies are large proteins found in the globulin fraction. The antibody protein is also called immunoglobulin (Ig). A given plasma cell produces large amounts of identical antibody molecules, all designed to attack the same antigen.

A pathogen has numerous antigenic determinant sites on its surface. Many antibody molecules bind to

TABLE 11–1 Normal Body Temperatures*

Species	Temperature Taken Rectally (degrees F)
Cat	101.5
Cow	101.5
Dog	102
Goat	102
Horse	100
Swine	102.5
Sheep	103

*It is important to recognize that the temperatures are in a range around the number in the table. Body temperatures vary during the course of a day and are altered by external temperatures, activity level, and excitement.

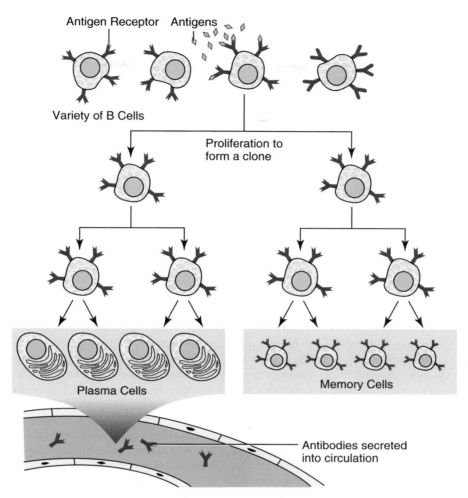

FIGURE 11–5 Summary of the clonal expansion of lymphocytes.

the surface of the pathogen. This antibody-antigen complex stimulates phagocytic cells and is more readily phagocytized and destroyed. The antibodies binding to the surface may themselves inactivate the pathogen. One type of pathogen, a virus, must attach to the surface of a cell before it can invade it. An antibody-coated virus is incapable of attaching to the cell surface.

Antibody production occurs primarily in lymph nodes but also in the spleen and bone marrow. The lymph nodes are strategically located so that a rapid immune response may prevent the pathogen from entering the bloodstream and rapidly spreading throughout the body. For example, there are lymph nodes in the chest and abdominal cavities, closely associated with the lungs and intestines.

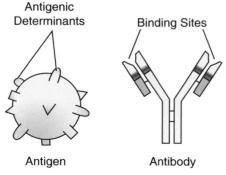

FIGURE 11–6 Structure of an antibody and antigen.

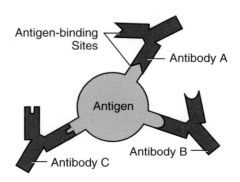

FIGURE 11–7 Multiple antibody molecules bound to an antigen.

IMMUNITY AND IMMUNE RESPONSE

OBJECTIVE

■ *Distinguish Between Passive and Active Immunity, Differentiate Between Humoral and Cellular Immunity and Their Relationship in Immunity, and Explain Primary and Secondary Immune Response*

When an animal is first exposed to an antigen, it generally requires 3 to 14 days for a significant amount of antibody to be produced. This **primary response** triggers the recognition by the lymphocytes, the clonal expansion, and the production of antibodies. The plasma cells can produce millions of antibody molecules every hour once they are activated. With this tremendous demand, plasma cells usually survive only for 4 or 5 days. Not all the cells of the clonal expansion develop into plasma cells.

Some of the clonal cells develop into memory cells. These cells do not actively produce huge amounts of antibody like plasma cells. These memory cells, however, survive for long periods. The role of the memory cell is to mount a much quicker response the next time the same antigen is encountered. With the presence of memory cells, much less antigen is required to stimulate the reaction by the immune system. Not only is the response quicker, but also even higher amounts of antibody are produced (Figure 11–8).

It is this **secondary response** that prevents an animal from developing an infectious disease a second time. One may wonder why humans have repeated episodes of the cold or flu. The secondary response is highly effective at preventing the same disease, but the response is very specific. Diseases such as the human common cold are caused by a virus, which has many strains different enough to not stimulate the secondary response.

The immune response involves more than just humoral immunity provided by antibody production.

The immune system also mounts a cell-mediated response. In addition to the B lymphocytes, there are also T lymphocytes. T cells also originate from the bone marrow, but they then mature in the thymus gland. At birth the thymus gland is a relatively large lymphoid organ that is found in the thorax (cranial and ventral region). T lymphocytes enter the thymus and mature into a cell that is capable of responding to an antigen (the cell becomes immunocompetent). Each T cell is receptive to a specific antigenic determinant. Any immune cells that might respond to the animal's own antigens are destroyed.

Once mature, the T cells leave the thymus and enter other lymph tissues. The thymus is very active in the fetus and early in life. As the animal matures, the thymus shrinks into a small residual structure. In the dog the thymus begins to lose lymphoid tissue after 5 to 6 months of age. Fat is often deposited in the connective tissue framework of the organ. Although the thymus might be quite small, remnants are often detectable even in an elderly animal.

In general terms, cell-mediated immunity works in combination with humoral immunity. Antibody production enhances cell-mediated immunity. T cells and macrophages are essential in destroying infected cells, altered cells, pathogens, and any foreign antigen. Cell-mediated immunity plays critical role in the reaction that destroys organ or tissue grafts from other animals.

When the T cells are exposed to a foreign antigen, they undergo a clonal expansion, just as the B cells do. Most of these T cells leave the lymph node to attack damaged cells. T cells can detect some cancer cells because they develop different surface antigens. They can also release enzymes that damage cells and stimulate the remainder of the immune system.

A small fraction of T lymphocytes also develop into memory cells. Just like the memory B cells, these lymphocytes are long-lived cells that provide a quick response with subsequent exposure to the same antigen.

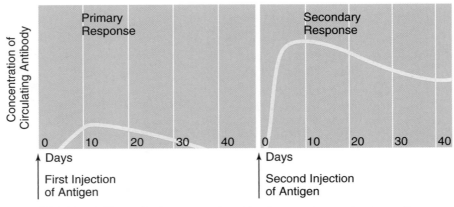

FIGURE 11–8 The antibody response found in primary and secondary immunity.

Active immunity also relies on the cell-mediated response to quickly destroy the invading pathogen. Whereas antibody levels are able to be measured in the blood, cell-mediated immunity is not easily measured.

A tremendous number of cells migrate to the site of infection. Neutrophils, lymphocytes, and macrophages all are present to help fight against the invading pathogen. In addition, there is the invading pathogen, destroyed and dying tissue cells, and tissue fluid. The accumulation of all the cells and cellular debris is called **pus.** (The correct adjective when discussing pus is *purulent.* It is incorrect to add –*y* or –*sy* to the end of pus. For example, the correct usage is, "The wound has a purulent discharge.") When the pus becomes trapped in a pocket, it is called an abscess. The cow discussed in the introduction showed a dramatic example of the accumulation of pus in an abscess.

The ability of an animal to defend itself from a second infection was detected by investigators. The role of memory cells to respond quickly and prevent infection is the underlying mechanism for vaccines. Edward Jenner, an English physician, is credited for discovering the first vaccination in 1796. Cowpox is a viral disease causing sores on the teats of cows. Dr. Jenner recognized that farm laborers who had been exposed to cowpox were not susceptible to the much more serious human disease smallpox.

Dr. Jenner performed an experiment in which he inoculated a human volunteer with material from the sores of a cow infected with cowpox. The volunteer only developed a localized sore at the site of the inoculation. Several months later, Dr. Jenner exposed the volunteer to smallpox. The subject did not develop any signs from the exposure. What Dr. Jenner had discovered was that the two viruses causing cowpox and smallpox were very closely related. The patient had developed antibodies to cowpox that were protective against the smallpox virus. The two viruses shared similar antigens that allowed the antibodies to prevent a new infection.

Protection against disease can come from a natural infection or from vaccination. Vaccination or immunization relies on the development of memory cells, preparing the animal to respond quickly upon exposure to the natural disease. Vaccines can be made from modified strains of the pathogen. The strain is weakened, allowing it to divide after injection but not create clinical disease. These vaccines are called **modified live vaccines.** Other vaccines are made from a killed pathogen. The pathogen will no longer divide but still has all its antigens to stimulate an immune response (Table 11–2).

Some diseases are not caused by the invading bacteria, but by the toxin that they produce. In this type of disease a modified toxin is used as the vaccine. This toxoid is still antigenic but is modified enough that it does not cause clinical signs.

In recent years it has been discovered that some pathogens share common antigens. For example, a related group of bacteria, called gram negative, share similar antigens. Immunity to these core antigens provides protection against all the bacteria in this class. Development of these core antigen vaccines allows for one vaccine to protect against several diseases.

For **killed vaccines** the first shot is given to stimulate the primary immune response. Two to four weeks later a booster shot is given, which activates the secondary response. High levels of antibody are produced, and memory of the antigen is established. For many modified live vaccines, only a single vaccination is required to establish a strong immunity. Because the pathogen in the vaccine replicates within the animal, it presents a high level of antigen. In effect it is simulating a natural infection without causing clinical signs.

The use of vaccines or exposure to natural infections provides the animal with active immunity. Exposure to the antigen allows the animal to mount an immune response and to develop a memory for the disease. With any subsequent exposure the animal's memory response is quickly activated and the disease-causing organisms are killed. **Active immunity** can last for very long periods. The immunity generally lasts at

TABLE 11–2 Vaccine Comparison

Type of Vaccine	Benefits	Disadvantages
Passive vaccines	Protection is immediate	Provides only temporary protection
Modified live vaccine (MLV)	High level of protection	Easily inactivated with sunlight and
	Longer duration of protection	disinfectants
	Booster not always needed	Must be handled carefully
		Some vaccines cannot be used in pregnant animals, may harm the fetus
Killed vaccine	Very safe vaccines, can be used in pregnant animals	Must be boostered
	Good protection	Shorter duration of protection compared with MLV

least 6 months but often provides protection for years or even a lifetime.

For many diseases, annual revaccination is recommended. This annual booster activates the immune response and elevates the level of antibody present. Once again memory cells are produced, and the animal is prepared for exposure to the natural disease.

Passive immunity, on the other hand, develops when antibodies are transferred from one animal to another. This routinely occurs from the mother to her offspring through the first milk produced. During pregnancy the first milk that the mother produces is very high in antibodies. This antibody-rich milk is called **colostrum**. Very early in life, the gastrointestinal tract of the newborn is able to absorb these antibodies without destroying them. (Remember that antibodies are protein molecules and would be digested and inactivated in normal digestion.) The antibodies are absorbed into the bloodstream, providing the newborn with immediate immunity. The ability to absorb antibodies declines rapidly (Figure 11–9). It is extremely important for the newborn to nurse soon after birth.

Passive immunity can also be transferred through commercial products. Antibodies can be harvested from the plasma of highly vaccinated animals. The globulin portion of the plasma is isolated and packaged to allow other animals to be treated. These products can be in an injectable form. They can also be provided as an oral product that is given shortly after birth to supplement the colostrum.

Passive immunity provides only temporary immunity. The duration of this type of immunity is generally several weeks to months. The higher the level of antibody transfers, the longer the duration of the immunity. Passive immunity must be considered when vaccinating young animals. This is evaluated in developing a vaccination schedule for puppies and kittens. When a modified live vaccine is administered to a puppy, the antibodies derived from the mother may immediately inactivate it. Because the pathogen in the vaccine does not replicate, the puppy develops very little immunity from that vaccine. As the maternal antibody declines, the puppy eventually is able to respond to the vaccine and develop an active immunity.

The problem arises because it is impractical to determine when a young animal will respond to the vaccine. Within a litter of offspring there will be significant variation in when the response will occur. An aggressive healthy kitten that nurses quickly after birth and drinks a large quantity of colostrum will have a high level of passive immunity (Figure 11–10). A kitten that is born weak and is slower to nurse and drinks less has a lower level of immunity. This second kitten will respond to the vaccine at an earlier age due to the shorter duration of the passive immunity.

Therefore vaccination schedules are developed to give periodic boosters in attempt to provide active immunity as early as possible. There is not one set vaccination schedule that is to be followed for all animals. Vaccination schedules need to be adapted to individual settings and the risk that the animal faces. For puppies a typical vaccination schedule would be for the first vaccine to be given at 6 to 8 weeks of age. Boosters are then given every 3 to 4 weeks, until the pup is 14 to 16 weeks of age. Animals at higher risk may require a modified vaccine schedule. The vaccines may need to be given at an earlier age, given more frequently, or continued longer.

Even with such a vaccination schedule there can be a window of opportunity for a pathogen to invade. Imagine that the animal is vaccinated at a time when the maternal antibody is just high enough to inactivate the vaccine. Over the next weeks the antibody levels continue to decline (Figure 11–11). If the animal is

FIGURE 11–9 The absorption of colostral antibodies as it relates to age of the newborn.

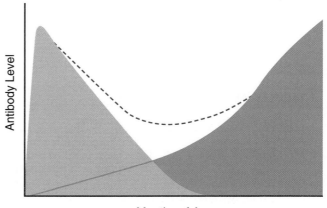

FIGURE 11–10 Decline of passive immunity and immune stimulation.

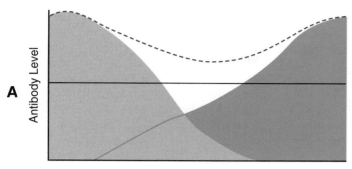

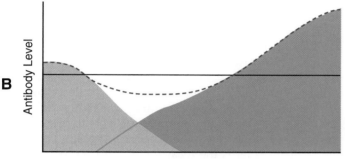

■ Maternal Antibody Level
■ Active Immunity=Antibody Level
---- Total Antibody Level
—— Level of Antibody Needed for Protection

FIGURE 11–11 Maternal antibodies begin at a high level and then decline. During that time, natural immunity develops. The ideal situation occurs when the total antibody level stays high enough to prevent disease. a. The animal starts with high levels of maternal antibodies. The total antibodies stay at a level that is protective against disease. b. The animal starts with a low level of maternal antibodies. The total antibodies fall below the protective level, leaving the animal susceptible to disease.

exposed before the next vaccine, there may be inadequate levels of antibodies to protect it. That is why animals in higher risk environments may require more frequent vaccinations. (For example, a puppy in a large breeding kennel may have a higher exposure risk than a single puppy in an individual's home.)

CLINICAL PRACTICE

OBJECTIVE

■ *Discuss the Clinical Significance of the Academic Material Learned in This Chapter*

Much of the discussion of the immune system has revolved around prevention of disease with the use of vaccines. The sheep suffering from tetanus could have

been spared the dreaded disease had the farmer planned ahead. A toxin released from the organism, *Clostridium tetani*, causes tetanus. This bacterium survives in wounds where there is little to no oxygen. Banding as a means of castration provides an environment where this kind of infection is possible. This disease is common in farm animals, especially following castration or dehorning.

The farmer could have prevented this disease with proper vaccination. One option would have been for the animal to be given passive immunity with an injection of antitoxin. Antitoxin is antibody against the toxin that is harvested from animals that are highly vaccinated. Administering the antitoxin would have provided passive immunity for the time in which the animal was at risk. The passive immunity is temporary, but the benefit starts immediately after the injection.

This animal could also have been vaccinated with a toxoid, or inactivated toxin. The toxoid has no harmful effects on the animal but does allow the animal to mount the immune response against it. Ideally this type of vaccination should have been done at least a week or two prior to the period of risk. This would allow the animal to develop an active immunity before it was at risk.

Most vaccines come in an injectable form. Several vaccines are administered **intranasally** (in the nose). The intranasal vaccines are generally modified live vaccines that allow the animal to develop a very strong local immunity. These vaccines are for respiratory diseases that enter through the nasal passages. By increasing the levels of antibodies in the nasal passages, the virus is unable to attach and the disease is prevented. **Kennel cough** in dogs is one disease that can be prevented this way. Kennel cough, which may be caused by several different organisms, is spread between dogs through the nose. As the name implies, dogs suffering from this disease develop a severe cough. The disease is much more common in situations where dogs are kept in close quarters, such as a kennel. The vaccine is very effective in preventing this disease.

All immune reactions do not work to prevent disease in animals. Some immune response occurs against antigens that are not harmful. An example is when a dog mounts an immune response to pollens common in the air (such as ragweed, trees, and grasses). Only a low percentage of animals exposed to these antigens develop an immune response (that is, they are weak antigens). *Sensitization* describes the animal's immune response to such allergens. IgE is the specific type of immunoglobulin that is produced in high amounts in these animals. The IgE is bound to the surface of basophils and mast cells. When the IgE comes in contact with the antigen, it causes the cell to release histamine and serotonin. These products cause smooth muscle constriction and increase permeability in the capillaries. In pets the reaction is often severe itchiness (**pruritus**).

Atopy describes this kind of sensitization to foreign antigens. In dogs and cats, atopy can be to a wide array of pollens, dusts, and dust mites. These animals typically present because of hair loss and pruritus. The animal may irritate its skin so severely that a secondary infection develops. Humans who develop the same type of sensitization show the signs of hay fever. The release of histamine in the upper airways and eyes causes the signs of runny nose, sneezing, and watery itchy eyes.

Antihistamines are used to treat this type of allergic condition. An antihistamine binds the same receptor that histamine uses to cause its effect. Antihistamines are much more effective if they are administered before the exposure to the antigen. Once the histamine has

bound the receptor, antihistamines will not reverse the effect. The benefit of antihistamines is that they prevent the attachment of histamine to the receptor.

When the allergic reaction is generalized and life threatening, it is termed **anaphylaxis**. The horse that died from exposure to penicillin succumbed to anaphylaxis. The horse had been treated with penicillin before without any bad effects. Unfortunately, this horse had become sensitized to the molecule. Penicillin is a relatively small molecule, but it is complex enough that certain animals mount an immune response to it. The same problem can develop in humans as well. The generalized release of histamine and serotonin causes bronchoconstriction and leaky vessels. The animal developed respiratory distress and shock. Epinephrine counteracts the physiologic effects of histamine and is the treatment of choice. Unfortunately, in this case our treatment was unsuccessful.

The immune system can also be used to help make a diagnosis in many infectious diseases. In cattle a group of infectious diseases cause abortion (premature delivery of a nonviable fetus). Tests are available to measure the level of antibodies against these diseases in the bloodstream. **Titer** is the measure of antibody levels in the bloodstream. The serum is diluted repeatedly, and the last tube that is positive defines the titer. Titers are reported as a ratio describing this dilution. For example, a titer of 1:256 has more antibody than a titer of 1:64.

If a single sample of blood is tested and comes back with a high titer, it proves that the animal had been exposed to the pathogen. Unfortunately, it does not prove when the exposure occurred and may even be the result of a vaccination. To prove that the infection was recent and the cause of the abortion, paired tests are run. For this a sample is taken immediately (the acute sample) following the abortion and then another is taken 2 to 4 weeks later (the convalescent sample). A titer is evaluated on each sample. To prove the cause of the abortion, the titer must change by four times. This is called **seroconversion**.

Two possibilities exist in seroconversion (Figure 11–12). The first occurs when the abortion results immediately after exposure to the pathogen. In this situation the titer is very low on the first sample. Over the following weeks the animal mounts an immune response and the second sample has a much higher titer (four times or more). In the other situation the actual infection occurred prior to the abortion and the titer on the first sample comes back high. Because the infection is cleared, the antibody levels decline with time and the titer of the second sample is much lower (decreases by four times). This too is considered a seroconversion. This paired testing is used to diagnose many infectious diseases.

In the exercises at the end of Chapter 10 an **ELISA test** was used to measure the level of progesterone in milk. ELISA is the abbreviation for enzyme-linked

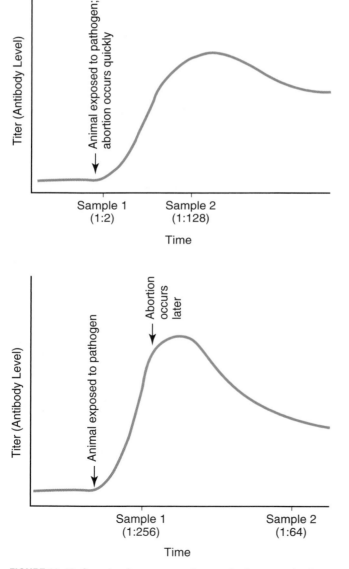

FIGURE 11–12 Sample of seroconversion—paired sera evaluation.

immunosorbent assay. Many of these tests are available to measure either antigen or antibody in the blood. Many different tests are commercially prepared, so that they may be run without expensive laboratory equipment and without extensive knowledge. One parasitic disease, heartworm, is commonly diagnosed with an ELISA test (Figure 11–13).

In the milk progesterone test, antibody against progesterone is bound to the bottom of the test cup. Milk is added to the cup, and any progesterone in the milk is bound to the antibodies. Then progesterone bound to an enzyme is added to the cup. Any unbound antibody is then filled with this enzyme-bound progesterone. A substrate is added to the test kit that causes a color reaction with the enzyme.

If the milk is high in progesterone, all the antibodies will be filled and no color reaction occurs. If the milk is low in progesterone, much of the enzyme-bound progesterone is attached in the test cup. When the final substrate is added, a color reaction occurs. This same principle is used in many diagnostic tests. Many variations exist in which radioactive or fluorescent materials are bound. Special equipment is then required to evaluate the results of these tests.

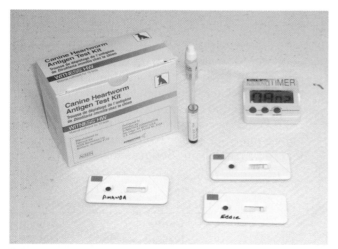

FIGURE 11–13 An in-office ELISA test. This test evaluates for the presence of heartworm disease. Three tests are shown. The first test is in process (left). A positive result, showing two red lines (upper right). A negative result, showing only the red line of the control (lower right).

In Chapter 3 the condition of joint ill was discussed. In this condition, bacteria enter the umbilical vessels, spread through the bloodstream of the newborn animal, and deposit in the joints. Newborns have a capable immune system; they are able to mount an immune response to foreign antigens. However, they have basically no circulating immunoglobulin of their own to fight off invading organisms. Their initial protection depends on the transfer of antibodies from the mother through ingested colostrum. (It is interesting to note that antibodies are able to cross the placenta from the mother's blood and into the baby in humans. Human babies therefore are born with circulating immunoglobulin.) As already discussed, this passive immunity is temporary. The higher the initial antibody level, the longer the protection lasts.

If the newborn animal fails to gain adequate amounts of antibody from the colostrum, it is much more susceptible to diseases. These animals suffer a much higher incidence of joint ill, as well as diarrhea, pneumonia, and brain infections. This lack of colostral antibodies is called failure of passive transfer. Newborns have the greatest ability to absorb antibodies in the first 6 to 12 hours. The absorption declines rapidly over the first day. Absorption has basically ceased by the end of the first 24 hours. Continuing to feed colostrum does not increase the circulating antibody level but does provide some local protection by producing higher levels in the gastrointestinal tract. Having these antibodies in the gut helps to keep ingested pathogens from attaching to the lining of the gastrointestinal tract.

Several factors may contribute to failure of passive transfer. One factor is when colostrum quality is poor. This more often occurs in younger animals, those that deliver early, or those that leak colostrum before the birth. Even when colostrum quality is high, the animal must consume an adequate volume. This may not occur if the newborn is weak, such as after a difficult delivery. These weak animals often have a delay before nursing. The longer the delay, the lower the number of antibodies absorbed. Several tests are available to measure the level of antibody in the blood, evaluating the passive transfer. Measuring these levels may help the veterinarian care for the individual, but it is also quite helpful in evaluating the management on a farm.

SUMMARY

Foreign molecules that stimulate an immune response are called antigens. Immunity to antigens can be passive (that is, obtained from drinking colostrum or injected as a vaccine) or active (that is, established after infection). The primary immune response occurs in the first days after invasion of the antigen, whereas the secondary immune response prevents a second episode of the same disease. The immune system response is strengthened by overall good animal health.

REVIEW QUESTIONS

1. Define any 10 of the following terms:

   abscess
   banded
   tetanus
   antigen
   lymph
   stocking up
   phagocytized
   edema
   humoral immunity
   primary response
   secondary response
   pus
   modified live vaccines

   killed vaccines
   active immunity
   passive immunity
   colostrum
   intranasally
   kennel cough
   pruritus
   atopy
   anaphylaxis
   titer
   seroconversion
   ELISA test

2. True or False: Horses may experience leg swelling when tied for an extended period.

3. True or False: All pathogens only have one site of determinant.

4. Lymph begins as _____ fluid that forms between the cells.

5. What is the typical life span of plasma cells?

6. Do larger or smaller molecules typically elicit the immune response?

7. What is the site of recognition on an antigen called?

8. What removes aged red blood cells from circulation?

9. What gland controls body temperature?

10. What type of molecule are antigens?

11. What human disease is closely associated with cowpox?

12. Who is credited for the first vaccination?

13. What two cells have the primary responsibility for phagocytizing antigens when first detected by the body?

14. Do naturally occurring molecules cause an immune reaction?

15. Do blood vessels dilate at the site of infection, causing reddening or inflammation?

16. Why does fever aid in the fight against pathogens?

ACTIVITIES

Materials needed for completion of activities:
   colostrum samples
   colostrometer

1. The discussion of passive transfer emphasized that the quality of the colostrum influences the amount of antibody absorbed by the newborn. A colostrometer is a tool that measures the antibody level in colostrum. This exercise requires collection of colostrum samples from local farms. Colostrum can be frozen without damaging the antibody level. This allows samples to be collected over time and accumulated. Each sample should be recorded with the animal's identification and her age or lactation number. In addition, any comments about the animal should be recorded. Important facts could include any illness, leaking of colostrum, parturition in relationship to due date, and the number of milkings when the sample was taken. Measure each colostrum sample with the colostrometer. Can any comparison be made with the other factors recorded about the animal using the samples obtained? (For example, did the younger animals have poorer quality colostrum? Did the quality decline with later milkings?)

2. Farmer Watkins calls because his cows have been aborting. You elect to take blood samples on the cow that most recently aborted and another herd-mate that was also pregnant but did not lose her calf. You follow up with another blood sample three weeks later. Titers are evaluated on several of the diseases that commonly cause abortions in cattle. (Do not be concerned about the actual disease name; many of these are discussed later in the text.) The results are shown as follow:

Disease	Cow #416 (aborted 9/16)		Cow #184	
	Acute	Convalescent	Acute	Convalescent
IBR	1:8	1:8	Negative	Negative
BVD	1:16	1:512	1:8	1:4
Lepto	Negative	Negative	1:100	1:100

Farmer Watkins calls your office and asks if you found anything on those blood samples. Can you make a diagnosis?

a. Did either cow seroconvert to any pathogen? If yes, what organism?

b. If there were any changes in titers, what degree of change was there?

c. If there was a seroconversion, was the infection very recent or had it occurred much before the time of the abortion?

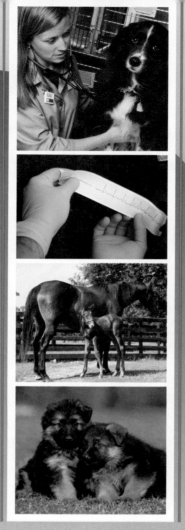

UNIT II

Nutrition

12

Basic Nutrients

OBJECTIVES

Upon completion of this chapter, you should be able to:

- *List the six major components of animal diets, and discuss their structure and significance in nutrition.*
- *Discuss the clinical significance of the academic material learned in the chapter.*

KEY TERMS

autopsy	flatulence	rodenticide	dry matter
postmortem	calorie	free radicals	hemolysis
constipation	hydrolysis		

INTRODUCTION

In Chapter 1 we learned about the basic structure of lipids, carbohydrates, and proteins and their importance in cell structure. Later we learned how these different molecules are digested. In this chapter we review many of these nutrients and discuss their importance in an animal's diet.

A DAY IN THE LIFE

I Love My Job...

One of the greatest rewards of my profession is the people that I get to meet. This is especially true of the farm clients that I get to see every few weeks. The farmers and I typically work together for hours and have plenty of time to talk and get to know each other. Over the years friendships develop. I get to see farm children grow up, and it is quite rewarding to see their ensuing successes.

I really enjoy my job. Most of my farmer clients like to joke a bit. Many a time I have been moving down a line of cattle, performing pregnancy checks, when I inevitably start to struggle and experience difficulty in finding the uterus or cervix. I look up and see the farmer smiling and realize that I have been trying to perform a pregnancy check on a bull!

I visited a farm just yesterday that made me think of these kinds of pranks. I was only on the job a few months when I went to Mel's farm for the first time. As a new veterinarian, I was trying hard to make a good impression. This was especially true when I first visited area farms. At this particular farm, I needed to work on a cow's sore hind foot. For this procedure I used a rope to hold the affected leg. In many barns I can loop a rope over an exposed beam. In Mel's barn I needed to use a beam hook. This is a hard metal hook that is driven into each side of a beam. When the rope is threaded through the loop in the bottom, it pulls the hooks tighter into the beam, making it very secure.

I retrieved my beam hook and asked Mel for a hammer. Mel came back with a rubber mallet. I looked at the rubber hammer, realizing that I could not drive the metal hook with that. My mind raced with questions: Doesn't he know you can't hammer metal with a rubber hammer? What do I do next? Should I tell him this is stupid? I would love to have seen my expression of bewilderment as I looked up and saw him smiling. He pulled the real hammer from behind his back as we both started laughing. He was just testing me and keeping the first meeting on a light note. I knew at that moment I was going to have a lot of fun in my chosen vocation.

Most farm animals being sold or shown in competition require health charts from a veterinarian. These health charts show that the animal has been examined and is free of visible signs of infectious or contagious diseases. These animals may require some testing for contagious diseases or even vaccination to prevent diseases. The requirements vary considerably among locations.

Two years ago I had the opportunity to examine two rams heading to a show. The animals appeared quite healthy, and I issued the paperwork. I received a call from the owner the day these animals were trucked to the fair. He told me he took them early in the morning, and they appeared healthy then. Now, late in the afternoon, both were quite sick. Officials requested that he pick them up, so he brought them by the office for examination.

As I waited for the animals to arrive, I wondered what I missed. The two rams were well grown, strong, and in good body condition. When these animals arrived, it was quite obvious that they were seriously ill. They were showing weakness and had pale gums and very dark urine. How could they have become so sick so quickly? All the signs were suggestive of an excess of copper in their diet. A small mistake in their diet over a long period had initiated a major illness.

I was on an emergency call one late November evening. I had just treated a milk fever case and hoped that my day was finally finished. I called into the office to discover that another farmer had a weak and shaking cow. I was disappointed but of course drove to the next emergency. By the time I arrived, the cow could not stand. She seemed perfectly normal earlier in the day and had just come inside after spending some exercise time in the pasture.

The producer had just changed the herd ration, and I figured the cow had low calcium levels. I began to treat her with intravenous calcium and dextrose. As I treated her, another cow in the barn collapsed. Worse yet, the first cow was not responding to my treatment. I was totally confused. Then I grew concerned that the cattle had ingested some toxin in the pasture. I took a blood sample from the second cow and submitted it to the diagnostic laboratory. Neither cow was responding to my treatment, so I needed more information. Table 12–1 shows the results of the blood test. Find what changes are present.

Last summer I visited a farm with two down heifers. One heifer had already died the day before, and now two more were showing symptoms. Both heifers were laying flat on their sides with their heads stretched upward. These animals appeared to be blind, and their muscles quivered. The signs were suggestive of a deficiency of vitamin B_1, thiamine. Moreover, I could not rule out a brain infection. I started treatment with high levels of antibiotics and thiamine. Unfortunately, another animal died and was submitted for an **autopsy,** or **postmortem.**

TABLE 12–1 Blood Chemistry Results: Star—3-Year-Old Female Jersey

Test	Units	Result	Reference Range
Glucose	mg/dl	90	40–74
Urea nitrogen	mg/dl	14	10–35
Creatinine	mg/dl	0.8	0.8–1.5
Sodium	mEq/l	178	135–157
Potassium	mEq/l	3.7	3.7–5.8
Chloride	mEq/l	135	90–110
Bilirubin	mg/dl	0.1	0.0–0.5
Total protein	g/dl	8.8	6.0–9.0
Calcium	mg/dl	8.9	9.0–11.5
Phosphorus	mg/dl	6.1	4.3–8.6
Magnesium	mEq/l	2.3	1.1–2.3

This cow had mild hyperglycemia (elevated glucose) and mild hypocalcemia (low calcium)
This cow had dramatic hypernatremia (elevated sodium) and hyperchloremia (elevated chloride)

NUTRIENTS

OBJECTIVES

■ *List the Six Major Components of Animal Diets, and Discuss Their Structure and Significance in Nutrition*

Diets are divided into six classes of nutrients. A nutrient is a component of food necessary to support life. The first three nutrients, carbohydrates, proteins, and fats, supply energy and structure for the cells of the body. Water and minerals are not used for energy but are utilized in many of the functions and structures of the body. The final class, vitamins, plays a key role in metabolic functions of the cell.

Carbohydrates supply energy and provide structure within the cell. Carbohydrates in the diet also provide energy to the animal. In addition, the carbohydrates supply fiber and bulkiness to the diet. The absolute necessity of carbohydrates in the diet remains low, because the animal can convert amino acids and fats into sugars.

Carbohydrates provide glucose to maintain blood sugar levels, which is critical for maintaining cell function. Carbohydrates create the lactose that is found in milk. Carbohydrates are also converted to glycogen as a cellular reserve of energy, and to fat as a body reserve. Carbohydrates are also necessary to complete the metabolism of fat.

The simplest carbohydrates, the monosaccharides, are easily absorbed from the intestines. Glucose and fructose provide two common examples of monosaccharides (Figure 12–1). The monosaccharides form the building blocks for more complex carbohydrates. The disaccharides, such as lactose, are a combination of two monosaccharides. In digestion, these sugars are broken down by enzymes released in the intestines.

The polysaccharides are long chains of simple sugars. Two types of chemical bonds can be found in

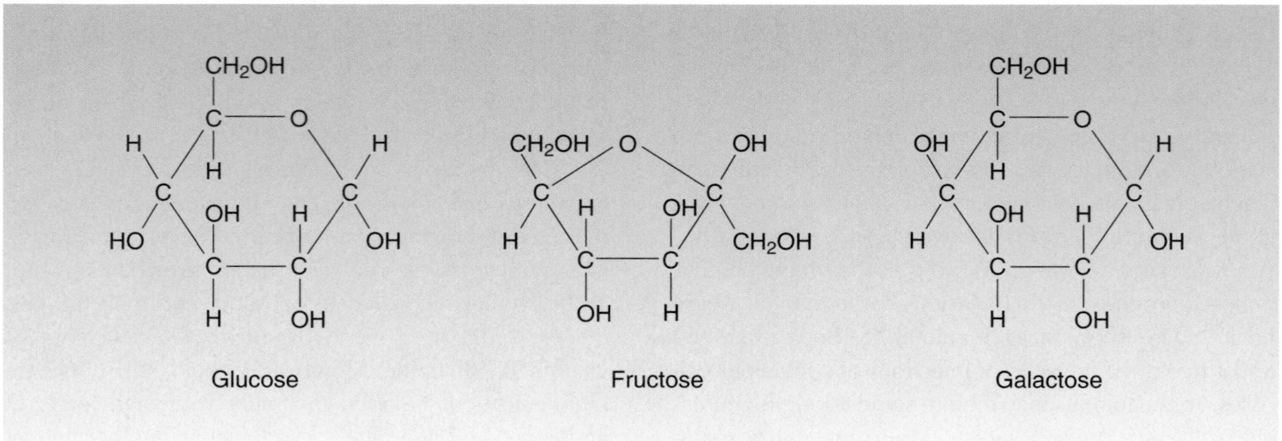

FIGURE 12–1 Chemical structure of simple sugars.

the polysaccharides. Starch is an example of a polysaccharide, a chain of glucose molecules joined by alpha bonds. The starch molecule can have many side branches all identically linked. The alpha bond is significant, because the amylase enzyme secreted during digestion can break it down. Amylase converts the long-chain starch into single molecules of glucose and disaccharides that can be further digested.

Dietary fiber or roughage is also composed of polysaccharides. The fiber derives from plants, where it is used for structural support. Fiber is also composed of chains of monosaccharides, but these are linked with beta bonds. The enzymes secreted in the digestive system of mammals are unable to digest these polysaccharides. The microorganisms found within the gastrointestinal tract of animals may be able to digest this fiber and allow the host animal to use the nutrients. This concept of symbiosis was introduced in the discussion of ruminants and horses in Chapter 7.

Cellulose serves as a classic example of the long-chain polysaccharide found in plant fiber. Just as in starch, glucose forms the building block of the cellulose molecule. These long straight chains of glucose molecules are linked with beta bonds (Figure 12–2).

Plant fiber has many components, including cellulose but also hemicellulose and lignin. These different fibers have different digestibilities by the microorganisms found in the intestinal tract.

Fiber, even though it is not digested in monogastrics, does affect the digestive tract. Fiber in the diet slows the emptying of the stomach and helps to protect the lining cells in the intestines. Fiber within the feces increases the amount of water held. This effect helps to maintain a normal rate of movement through the intestines. Adding fiber to the diet can be helpful in both diarrhea and **constipation**. In diarrhea stools are very liquid and move through the animal too quickly. Additional dietary fiber helps to hold more of the water and slows down the movement of feces through the intestines. Constipation occurs when the feces is too dry and moves too slowly. Again, fiber helps to retain water within the feces and increases the rate of passage to a more normal level.

Added fiber within a diet does increase the volume of feces. Owners will note that pets placed on a high-fiber diet have much larger stools than before the dietary change. In addition, higher fiber levels can increase **flatulence** in pets. Flatulence is an accumulation

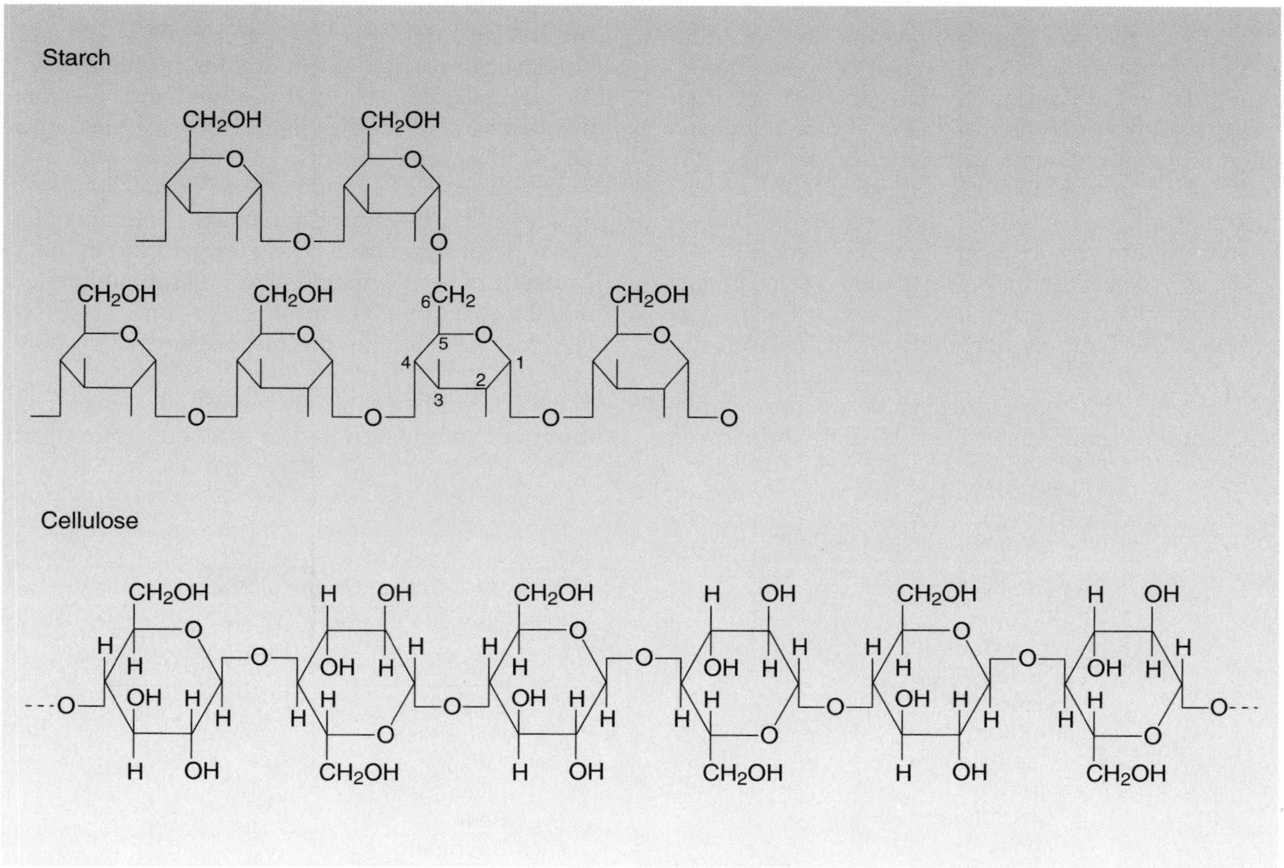

FIGURE 12–2 Chemical structure of starch and cellulose. Notice that the bonds joining the molecules are oriented differently in the two molecules.

of gas in the intestinal tract, which is produced as intestinal bacteria ferment the fiber. *Flatus* describes when the gas is passed through the anus. The majority of the gas produced is odorless; however, a small percentage that contains sulfur contributes to the foul odors produced. If the condition becomes too severe, changing to a highly digestible low-fiber diet can be helpful.

Proteins are composed of chains of amino acids. As we have seen throughout the text, proteins play a key role in the structure and many of the functions of cells. Proteins serve in the structure of cells but also act as enzymes, hormones, and antibodies. Proteins can also be utilized as a source of energy for the animal.

The amino acids are either essential or nonessential. The essential amino acids must be supplied in the diet, whereas the nonessential ones can be synthesized from other amino acids. Each species has its own list of essential amino acids. Table 12–2 lists the 10 essential amino acids of the dog. Cats have the same 10, plus the additional requirement of taurine. Cystine and tyrosine are occasionally listed as essential amino acids. This is the case if the diet does not contain adequate methionine and phenylalanine, which are necessary to produce cystine and tyrosine.

A deficiency of protein in the diet often shows clinically as poor growth or low body weight. These animals many times have a poor appearance to their hair coat and are more susceptible to disease (poor immunity). On the other hand, excessive protein is used as an energy source or converted to fat. This metabolism does produce nitrogenous waste that must be cleared by the kidney. In animals with decreased kidney function, the nitrogenous wastes increase in the bloodstream, worsening the clinical signs of the condition.

Biologic value describes the quality of a protein source. The value is based on how efficiently the protein is utilized. The more closely the protein matches the amino acid needs of the animal, the higher the biologic value. Proteins with high biologic value are more highly digested and result in less waste than those with lower value. Biologic value is expressed as a percentage, comparing the amount retained by the body with the amount excreted. The animal requires less protein if the biologic value closely matches its requirements for essential amino acids. For dogs and cats, animal proteins generally have a higher biologic value than plant proteins. Table 12–3 shows the biologic value of several protein sources found in dog food.

The third class of nutrients is fats and oils. Fats are solid at room temperature, whereas oils are in a liquid state. The simple lipids come in a form with three fatty acid molecules bound to a molecule of glycerol. Lipids can be conjugated with other molecules such as proteins.

Fats can be used as an immediate supply of energy or stored in the fat reserves of the animal. A deposit of fat not only provides a source of energy, but also aids in insulating the animal from the cold and provides some degree of protection to organs as well. Lipids are used in producing certain hormones and structurally within cell membranes. Fats generally increase the palatability (or tastiness) of the food. Fat in the diet is necessary to allow for the absorption of certain fat-soluble vitamins.

Lipids are a source of fatty acids. Certain fatty acids are required to be supplied within the diet and, just as with amino acids, are called essential. The essential fatty acids vary with species as well. In dogs, linoleic acid assists in production of linolenic and arachidonic acids. Cats, on the other hand, also require arachidonic acid in the diet. Fats are found in both plant and animal sources. Arachidonic acid is only found in animal sources.

These first three nutrients are all potential sources of energy for the animal. The energy found in the form of chemical bonds metabolizes within the cell. The animal utilizes much of the energy; some is released as heat, which maintains the normal body temperature. A **calorie** is the unit of measure that defines the energy contained within a food. By definition, a calorie is the amount of energy required to raise one gram of water

TABLE 12–2 Essential Amino Acids of the Dog*

Arginine
Histidine
Isoleucine
Leucine
Lysine
Methionine
Phenylalanine
Threonine
Tryptophan
Valine

*Cats have all of the same essential amino acids plus Taurine.

TABLE 12–3 Biologic Value of Protein Sources Found in Dog Food

Source	Biologic Value (%)
Egg	100
Fish meal	92
Milk	92
Liver	79
Beef	78
Soybean meal	67
Corn	45

one degree Celsius. Many labels report the value in kilocalories (kcal or C), which is 1000 calories.

Feedstuffs are burned in a special device, called a calorimeter, to measure the amount of heat released and therefore determine how many calories it contains. In addition, the efficiency of digestion is considered to establish the total energy that can be obtained from an ingredient. In human physiology it is common to report the values of 4 kcal/g for both carbohydrates and protein and 9 kcal/gram for fat. Using standard values for the efficiency at which average dog and cat foods are digested, the values lower to 3.5 kcal/g for carbohydrate and protein and 8.7 kcal/g for fat.

Although protein, fat, and carbohydrate play an essential role in the diet of animals, water remains the most critical nutrient. An animal can lose almost all its fat and up to half its protein and survive, but dehydration can quickly become life threatening. In Chapter 6 the significance of fluid balance was emphasized. A loss of 10% of the body's water becomes very serious, and losses above 15% are generally life threatening.

Some water is actually produced during metabolism. This is a relative small amount compared with how much is obtained in feed ingredients and is directly consumed. Depending on the type of diet, the amount of water consumed in feed can be quite significant. A cat eating canned food takes in much more water in its food than a cat eating strictly dry food. An adult Holstein dairy cow often consumes more than 100 lb of total feed that has more than half its weight as water. In the feed alone, the cow will have consumed 50 lb of water.

Water plays a tremendous role in many functions. Up to two thirds of the body is composed of water, including that found within and among the cells and in the bloodstream. Within cells, water provides a medium for chemical reactions to occur. In addition, the process of **hydrolysis** adds water to a molecule to cleave it into smaller parts. Water has also been shown to play a role in transporting nutrients, wastes, and hormones in the blood and lymph. A developing fetus is bathed in liquid that is supportive and protective.

During panting and sweating, evaporation of water helps to control body temperature. In addition to these losses, the body also loses water in urine and feces. The tremendous loss of water with diarrhea, and its affect on dehydration has already been discussed. In lactating animals, producing and secreting milk also causes tremendous water loss. It is quite common for the average Holstein cow to be producing 70 to 100 lb of milk a day. A major portion of milk is composed of water, and therefore milk production needs to be considered in evaluating the water requirements of the cow.

Vitamins are necessary in relatively small amounts to maintain the health and function of the animal. Vitamins may act as an enzyme, helping to control a chemical reaction. Other vitamins act as coenzymes by combining with a protein to create an active enzyme. In addition, other vitamins serve as precursors to enzymes. These vitamins are converted into an active enzyme once they are absorbed into the body.

Vitamins are divided into two classes: fat soluble and water soluble. The water-soluble group includes eight B vitamins and vitamin C. Vitamins in this class are not stored in the body, and any excess is excreted in the urine. Because these vitamins are not stored, they basically require daily intake to maintain adequate levels in the body. Table 12–4 summarizes the vitamins and their functions.

The fat-soluble vitamins are A, D, E, and K. These vitamins are absorbed from the intestinal tract with fat. Any excess absorbed is stored in the fat reserves of the body. Daily input is not required, because reserves are available. Because these vitamins are stored, it is possible for toxicity to occur with excessive intake. One class of **rodenticide** (poisons for mice and rats) uses high levels of vitamin D to cause a toxic effect. Excessive intake creates severe hypercalcemia. This causes damage to many organs, including the kidneys and heart. These products are designed to kill mice but are also toxic to pets when consumed in high enough levels.

Vitamins A, C, and E, along with the minerals selenium, copper, iron, manganese, and zinc, are all considered antioxidants. Antioxidants help to protect the lipid membranes by neutralizing **free radicals** released within the body. Free radicals have a single free electron, which attracts another electron from neighboring atoms. Free radicals have function, including the destruction of invading pathogens by the immune systems. Left uncontrolled, these free radicals can then damage the neighboring cells. These vitamins and minerals work to eliminate the free radicals, therefore protecting the body's cells.

Minerals are divided into macrominerals and trace minerals, or microminerals. The categories are based on the relative amounts required in the body. The amount of macrominerals in a diet is generally expressed in the percent of **dry matter**. Dry matter is the percent of the feed ingredient that remains when all the water is removed (100% = Percent dry matter + Percent moisture). The microminerals or trace minerals are generally expressed in parts per million (ppm). One part per million is equivalent to 1 mg in 1 kg of feed. As an example, cattle require macrominerals in grams per day and the microminerals in milligram or microgram quantities.

The common macrominerals are calcium (Ca), potassium (K), sodium (Na), phosphorus (P), magnesium (Mg), sulfur (S), and chlorine (Cl) (Table 12–5). Many of these minerals have already been discussed in other chapters. Calcium and phosphorus play an

TABLE 12–4 Summary of Vitamins

Vitamin	Functions and Required For:
A	Essential in the pigments of the retina, esp. low light vision Necessary for healthy cell division, bone growth, blood cell formation Aids in the health of epithelial tissues; important in reproduction
D	Synthesized in skin with exposure to UV light from the sun Maintenance and formation of bone and teeth Effects calcium and phosphorus metabolism (through kidneys, bone, and intestine)
E	Closely associated with selenium Antioxidant—protects lipids and cell membranes; stabilizes red blood cells Important role in immune and reproductive systems
K	Essential for blood clotting
B1 (Thiamin)	Carbohydrate and energy metabolism; maintenance of the nervous system
B2 (Riboflavin)	Carbohydrate, fat, and protein metabolism; involved in many enzymes Important in growth, maintaining healthy skin, and nervous system
B5 (Pantothenic Acid)	Carbohydrate, fat, and protein metabolism; necessary for antibody production
B3 (Niacin)	Carbohydrate, fat, and protein metabolism; healthy oral tissues
B6 (Pyridoxine)	Amino acid and energy metabolism; necessary for growth Maintenance of healthy nervous and immune system; important in hemoglobin formation
Folic Acid	Necessary for nucleic acid synthesis and cell division Closely associated with B12 in red blood cell formation
Biotin	Synthesis of glucose and energy metabolism; growth Necessary for healthy skin and hooves
B12 (Cobolamin)	Closely associated with folic acid in red blood cell production Nervous system function; fat, carbohydrate, and energy metabolism
C	Important in wound healing and collagen formation Maintain strength of capillaries and mucosa

important role in bone structure. Calcium also serves in the function of muscles. Sodium and potassium allow for nerve conduction and, along with sodium, are involved in the regulation of fluid balance in the blood.

The microminerals are commonly involved in enzyme reactions, helping to speed the chemical reaction. The trace minerals may also be included in hormones, such as iodine in thyroid hormone. Table 12–6 lists many of the trace minerals and summarizes some of their functions.

Deficiencies and excesses of the trace minerals can occur. Many of the trace minerals are absorbed from the intestinal tract by the same mechanism. Feeding an excess of one micromineral can result in a deficiency of another. For example, the absorption of copper is decreased as sulfur and molybdenum increase.

TABLE 12–5 Macrominerals

Mineral	Functions
Calcium	Development of bone and teeth, muscle and nerve function Activates many enzymes, necessary in blood clotting
Chloride	Essential for acid/base, osmotic pressure, and fluid control in the blood
Magnesium	Development of bone and teeth, necessary for hemoglobin production Important in energy metabolism and many enzymes
Phosphorus	Development of bone and teeth, structure of cell membranes (phospholipids) Involved in nucleic acid production, energy utilization and many enzymes
Potassium	Necessary for the function of nerve and muscle, important in protein synthesis Essential for acid/base, osmotic pressure, and fluid control in the blood
Sodium	Necessary for the function of nerve and muscle Essential for acid/base, osmotic pressure, and fluid control in the blood
Sulfur	Is in the amino acids methionine and cystine Involved in many enzymes and hormone production

TABLE 12–6 Microminerals

Mineral	Functions
Cobalt	Involved in many enzyme reactions
Copper	Involved in bone and hemoglobin formation, necessary for proper iron metabolism
	Included in many enzymes—especially those of protein production
Iodine	Portion of thyroid hormone, necessary for proper reproductive cycle
Iron	Portion of hemoglobin and oxygen metabolism, many enzymes
Manganese	Activates many enzymes, involved in bone and connective tissue development
	Carbohydrate and fat metabolism
Selenium	Antioxidant, necessary for healthy immune system
Zinc	Many enzyme systems, especially protein synthesis
	Necessary for healthy skin, hooves, and immune system
	Important role in blood cell formation and wound healing

The concentration of a nutrient in a feed or diet can be expressed either on an as-fed or dry matter basis. An as-fed basis measures the concentration just as it is delivered to the animal. This is affected by the amount of water in the feed. It is difficult to compare two feeds with different moisture contents when the concentrations are reported on an as-fed basis. To make the comparison easier, concentrations are often reported on a dry matter basis. A dry matter basis calculates the concentration of the nutrient with all the water removed from the feed (Figure 12–3).

Table 12–7 shows a forage analysis report of an alfalfa haylage. Alfalfa is commonly fed to dairy cows. In haylage, alfalfa is put in a storage structure with a relatively high amount of moisture. The two columns of the report show the concentrations on an as-fed and dry matter basis. The numbers are describing the same feed, with and without water. A formula can be used to convert the numbers.

$$\frac{\text{Concentration as fed}}{\text{Percent dry matter}} = \text{Concentration on dry matter basis}$$

or

Concentration on dry matter basis × Percent dry matter = Concentration as fed

Using the numbers in Table 12–6:
Crude protein as fed = 5.3
Percent dry matter = 28.6

$$\frac{5.3\% \text{ Crude protein}}{0.286 \text{ Dry matter}} = 18.5\% \text{ Crude protein on dry matter basis}$$

FIGURE 12–3 *A.* Koster tester positioned for heating a sample. *B.* Koster tester with sample being weighed. Measuring the amount of water removed allows the dry matter to be evaluated.

TABLE 12–7 Forage Analysis Report: Alfalfa Haylage

	As Fed	Dry Matter
% Moisture	71.4	
% Dry matter	28.6	100
% Crude protein	5.3	18.5
% Acid detergent fiber	10.3	35.9
% Neutral detergent fiber	12.8	44.6
% Crude fat	1	3.4
Net energy of lactation (Mcal/lb)	0.17	0.58
% Calcium	0.43	1.5
% Phosphorus	0.08	0.29
% Magnesium	0.08	0.28
% Potassium	0.65	2.27
% Sodium	0.003	0.01
Iron ppm	27	95
Zinc ppm	7	24
Copper ppm	2	8
Manganese ppm	13	45
Molybdenum ppm	0.2	0.7

The dry matter percentage of the feed is critical in evaluating how much food the animal is actually consuming. Consider a cow consuming 100 lb of a complete ration with a dry matter of 50%. The cow is actually consuming 50 lb each of dry matter and water. If the ration becomes wetter (lowering the dry matter to 45%), the cow will consume less feed and more water (45 lb of dry matter and 55 lb of water). Because a farmer wants to feed a cow the same amount of dry matter every day, in this case the cow must be fed more of the total ration.

To calculate the amount of total ration that must be fed to a cow, producers must know how much dry matter is required (50 lb in this example) and the percent dry matter of the total ration (45% in this example).

$$\text{Pounds of total ration} = \frac{\text{Pounds of dry matter}}{\text{Percent dry matter}}$$

$$\text{Pounds of total ration} = \frac{50}{0.45}$$

$$\text{Pounds of total ration} = 111 \text{ lb}$$

CLINICAL PRACTICE

OBJECTIVE

■ *Discuss the Clinical Significance of the Academic Material Learned in the Chapter*

Copper is an essential mineral in the diet of animals. However, any species can develop toxicity if excess is fed. Some species of animals, such as sheep, are much more sensitive to excess copper. Sheep require much less copper (per pound of body weight) than do cattle. Because of this variation, feeding sheep with grain formulated for cattle can result in toxicity. Improper formulation of a sheep diet can lead to similar problems.

Copper is stored in the liver. With chronic exposure to excess copper, the levels continue to increase in the liver. When the animal is put under stress (such as trucking the sheep to a show), large amounts of copper are released into the bloodstream. As a result, the red blood cells begin to break down in the vessels (**hemolysis**). The animal becomes anemic, and large amounts of hemoglobin spill into the urine, which can damage the kidneys.

Animals showing clinical signs begin with depression, weakness, and a loss of appetite. As it progresses, the animal often develops diarrhea, has difficulty breathing, and becomes ataxic. Death usually occurs quickly after the animal begins hemolysis. The sheep discussed at the beginning of this chapter both died. Postmortem examination confirmed the diagnosis. The level of copper in the sheep's liver was 458 ppm. The normal range for this test is 25 to 100, with the toxic range starting at 250 ppm. The postmortem was not helpful in saving any of the clinically affected animals. With a positive diagnosis, the diet of the remaining flock was adjusted and no further animals were affected. This is a common procedure in herd health medicine. The loss of an individual animal can be very helpful in treating or preventing disease in the remainder of the herd or flock.

Cattle are generally allowed free access to salt. They can regulate intake based on their need. The kidneys excrete any excess that is consumed. This works well as long as cattle have free access to water. If the animal consumes a normal amount of salt (sodium chloride) but water is restricted, the concentration of sodium and chloride become elevated to dangerous levels. The water restriction can occur during times when electricity fails, a water pump quits, or troughs freeze. The animals from the introduction and Table 12–1 show evidence of severe hypernatremia (elevated sodium).

Salt toxicity is much more common in swine but does occasionally occur in ruminants. I failed to make this diagnosis until after the blood results had returned. Every source that I found listed salt toxicity in combination with some water restriction. These cattle had not been restricted in their water intake. In this farm situation the cattle had not had access to salt for a long period. The farmer had gotten salt that day and supplied them with a large amount of loose granular salt. Many animals consumed a large amount very quickly, with the two affected animals consuming even more.

I visited the farm the following day to check on these animals, and the first cow had already died. It was obvious that the herd in general had been consuming large amounts of water, and there was an excessive amount of urine in the gutter behind the cow. The elevated sodium in the blood had stimulated the thirst center, and the animals were consuming large amounts of water. The cattle were not eating well the following day.

The two most severe cases had obviously consumed quite excessive amounts. A high level of sodium increases to extremely high levels in the brain. The high sodium level stimulates animals to drink. Unfortunately, the osmotic force of the sodium in the brain draws large amounts of water into the brain tissue, causing swelling (edema). The swelling then results in the clinical signs that I had observed: ataxia, inability to rise, muscle twitching, and death. The normal mechanism of providing water, which cured the problem for the majority of the herd, actually worsened the signs for the two worst animals. This incident proved a valuable lesson for the farmer and myself. The farmer now keeps salt available at all times.

Polioencephalomalacia (PEM or polio) occurs with a lack of thiamine in ruminants. (It is important that polio in ruminants not be confused with an infectious disease in humans that is also called polio.) Ruminants do not require a large source of thiamine in the diet, because the organisms in a normally functioning rumen produce adequate amounts. With certain feed changes, such as a high-grain diet, certain organisms that produce an enzyme called thiaminase can flourish in the rumen. Thiaminase destroys thiamine, leaving the animal deficient.

Thiamine is involved in the chemical reactions that produce energy for cells. When deficiency occurs, the cells become starved for energy. The lack of energy to the cells of the central nervous system leads to the classic signs that these animals were exhibiting. Ataxia, inability to rise, and blindness are

FIGURE 12–4 A sheep showing evidence of polioencephalomalacia. *(Courtesy Ron Fabrizius, DVM, Diplomat ACT.)*

all typical signs with PEM (Figure 12–4). Treatment is straightforward and involves injection of high levels of thiamine. If caught early enough in the disease, it can be cured.

Although many of our domestic species have tremendous similarities, small differences can be quite significant. The diets of cats and dogs may appear similar, but there are differences. Taurine is an amino acid that can be produced from other amino acids. Cats are unable to produce adequate amounts, and therefore it is considered an essential amino acid. Cats eating only dog food would become taurine deficient.

Several serious health problems can occur when a cat becomes taurine deficient. Blindness is one possible outcome, as the retina of the eye becomes damaged. Female cats will be unable to maintain a pregnancy and young animals will have growth problems. Most serious is a disease condition of the heart muscle. The muscle becomes weakened and damaged, leaving the heart functioning poorly. This condition can be fatal, although it is reversible if taurine is supplied early enough in the course of the disease.

SUMMARY

Carbohydrates, protein, fats, vitamins, minerals, and water comprise the six nutrients of animal diets. All working in conjunction allow the body to function efficiently and effectively. Deficiency in any of the six, especially water, will result in low performance or even death. To maintain a healthy animal, producers must ensure that all six nutrients are presented in correct amounts in the daily ration.

REVIEW QUESTIONS

1. Define the following terms:
   autopsy
   postmortem
   constipation
   flatulence
   calorie
   hydrolysis
   rodenticide
   free radicals
   dry matter
   hemolysis

2. Carbohydrates provide _____ to maintain blood sugar levels.

3. Adding _____ to the diet is helpful in both diarrhea and constipation.

4. Proteins are composed of chains of _____.

5. One kilocalorie equals _____ calories.

6. Minerals are divided into two categories, _____ and _____.

7. Name the simplest carbohydrates.

8. Which remain solid at room temperature, fats or oils?

9. Addition of which nutrient generally adds palatability to food?

10. What proportion of the body is composed of water?

11. Does sweating cool or heat the body?

12. Which common barnyard ruminant is quite sensitive to cooper toxicity?

13. List the three classes of nutrients used for energy supply and cellular structure.

14. List the three attributes of animal fat deposits.

15. List four fat-soluble vitamins.

ACTIVITIES

Materials needed for activities:
   forage samples
   Koster tester or gram scale and microwave oven
   paper plates

1. A very important technique in animal nutrition is the ability to measure the dry matter content of the feed. There are several techniques available to measure dry matter. A commercially available method is the Koster tester. This is an electrical unit that heats the sample and has a scale to measure the changes in weight as water is removed. These units come with complete instructions on proper usage.

2. These units are relatively expensive, and the same technique can be accomplished with a gram scale and a microwave oven. A note of caution: Because the samples become extremely dry, it is possible for them to catch on fire! This is dangerous, plus the burning causes a loss of dry matter and an inaccurate reading. Always maintain observation on the sample—DO NOT leave drying samples unattended.

3. Obtain a forage sample from a local farm; ideally this will be silage, such as corn silage or haylage, in which there is a significant amount of moisture. Dry hay can also be tested in the same manner, but the amount of moisture is quite low. Weigh a paper plate. Add a sample of the forage (approximately 50 g). Record the actual weight of forage. Spread the sample out thinly on the plate.

4. Heat the sample in the microwave for short periods. Table 12–8 lists approximate times for various forages and moisture. An important warning is that determining dry matter with a microwave

TABLE 12–8 Approximate Microwave Times for Dry Matter Analysis

	Corn Silage (<40% dry matter)	Haylage (<40% dry matter)	Haylage (>40% dry matter)
First heating	90 seconds	60 seconds	50 seconds
Second	45 seconds	35 seconds	40 seconds
Third	35 seconds	25 seconds	25 seconds
Fourth	30 seconds	15 seconds	15 seconds
Subsequent heatings	10 to 15 seconds	5 to 10 seconds	5 to 10 seconds

Heating times are approximate. Do not leave the sample unattended.
In general, the drier the sample the shorter each heating time should be.
Do not allow the sample to burn, which will result in a loss of dry matter from the sample.

does produce a distinctive cooked odor. Parents generally don't appreciate this procedure being done in the kitchen with the good microwave! Following each heating, weigh and stir the sample. Use care not to lose any of the forage. As the forage becomes drier and more brittle, heat for shorter periods. Repeat the procedure until the weight does not change between samples. The goal is that all the water be released as steam, leaving only the dry matter of the sample without any loss from combustion.

5. Record the final weight of the sample. Divide the final weight by the initial weight and multiply by 100 to calculate the dry matter percentage:

$$\text{Percent dry matter} = \frac{\text{Final weight}}{\text{Initial weight}} \times 100$$

6. A farmer recently purchased a Koster tester and has found that the moisture of available haylage has changed. You examine the ration, which says to feed 45 lb of haylage as fed, which is 45% dry matter. Using the Koster tester, the producer has determined that the haylage is now only 32% dry matter. Show the farmer how to calculate how much haylage should be fed.

7. A producer calls you to ask advice. An opportunity has presented itself to buy haylage from two neighbors. Both neighbors are asking $40 per ton. Haylage 1 has a crude protein of 6.3% on an as-fed basis and a dry matter of 32.3%. Haylage 2 has a crude protein of 7.8% on an as-fed basis and a dry matter of 44.3%. Which haylage would you recommend and why?

13

Species Comparison

INTRODUCTION

Veterinary science presents the challenge of having to understand the anatomy and physiology of many species. Nutrition emphasizes these differences. A tremendous variation exists among the anatomy and type of diet of domestic species. It is well beyond the scope of this text to describe the nutritional requirements of all the domestic species. This chapter describes many of the considerations used in the nutrition of dogs, cats, horses, and cows. In this chapter the cow represents the classic ruminant digestive system and the horse provides illustration of a different means of fiber digestion.

A DAY IN THE LIFE

Some Days the Odor Is Overwhelming...

The veterinary profession is well respected, but it is not always a glamorous job. Many days when I return to the office, the office staff will tell me that I stink! Unfortunately, smelling badly is just part of my job. When working around cows, getting dirty and picking up distinct odors are routine.

I recently visited a farm where we had to laugh about one of these episodes. It occurred a number of years earlier, but we still talk about it. I was working on a cow with an obstruction in her teat. The farmer could not get milk from this teat, and I was attempting to relieve this obstruction. The procedure does cause some degree of pain, and the cow was not happy about my work. The farmer was attempting to control the cow and keep her from kicking me.

I bent over and placed my head firmly into the cow's flank. This kept me somewhat safe, because if the cow tried to move my way, it just shoved me backward. In this position my backside was dangerously close to the cow in the stall behind me. I was concerned about her kicking me as well, but I had to concentrate on the task at hand.

I was making progress, but it was a bit of a struggle. The cow was quite nervous, and all the activity was making the neighboring cows excited as well. Anyone that has worked around cows will tell you that when cows get nervous they have a tendency to defecate. I had definitely made the cow behind me nervous. I was in the middle of the procedure, when I noticed a wet, warm feeling on my backside. I finished as quickly as possible and stood up to find the problem. The cow had filled my back pocket with feces! The farmer and I just looked at each other and began to laugh. I have these kinds of days.

I was visiting Dr. Baker, my co-author, when a very large black Lab walked into the house. This dog had obviously been eating quite well and had become down-

FIGURE 13–1 An obese cat.

right fat. Dr. Baker's parents had adopted this former stray. How did this dog become so heavy? Moreover, the dog's housemate, a former barn cat, is morbidly obese, weighing a hefty 30 pounds (Figure 13–1). Only a few weeks ago, I saw another cat that had developed a skin infection around its hindquarters. This cat had become so heavy that it was unable to groom itself in that region. Obesity is a very common problem in small animal medicine.

So often it is the emergency calls that I remember. This may be because being called out of bed and losing sleep is so traumatic. It may also be because the cases are so dramatic as well. I remember the night in veterinary school when I had to see a horse showing signs of colic. When I arrived at the farm, I discovered that the horse had gotten into the feed room and had consumed a huge amount of grain. What is a normal part of the horse's diet in excess had caused a severe illness. The digestive system of this horse, designed to digest forages, had become quite upset with the overload of such a large amount of grain.

ANIMAL NUTRITION

OBJECTIVE

■ *Explain the General Principles in Animal Nutrition*

Many of the factors considered in nutrition are consistent among the species. The basic goal of nutrition is to meet all the needs of the animal and maintain good body weight and condition.

Many factors influence the nutritional demands on an animal. In developing a diet a natural starting place is to consider a resting animal in a comfortable environment. Dietary needs in this condition are referred to as maintenance requirements. Any change from these factors alters the needs of the animal. Increasing the level of activity immediately increases the animal's needs. The activity level can be dramatic for working animals. Racehorses, draft horses, and hunting dogs are just a few examples of animals that have nutritional needs much higher than the maintenance levels. The environment also influences requirements. The same animal in cold, wet conditions must expend more energy to maintain its body temperature. This increase in energy must be supplied in the diet or the animal will mobilize fat to provide this energy.

Animals in other stages of life also require different levels of nutrients. Growth, pregnancy, and lactation also increase the demands that must be supplied through the diet. A dog in peak lactation may have an energy requirement two to four times that needed for maintenance. It is also important to recognize that there is tremendous variation between individuals. Even littermates sharing the same environment can have quite different needs. It is common to see, in two pets in the same household, one that is dramatically overweight and the other in good body condition. Finally, the health status of an animal can dramatically change the requirements of the animal. Certain disease conditions can increase the metabolic needs of an animal.

There are several methods available to deliver the feed. One of the simplest methods is to provide a **free choice diet**. The goal is to have good quality feed available for the animal at all times. Using the dog as an example, free choice feeding requires the least work and helps to eliminate any competition between animals. The pet can eat at any time. This can be helpful in quieting a confined dog. There are drawbacks to this type of feeding. In multiple pet households, it can be difficult to determine when a pet stops eating. The most common problem, however, is that obesity is a high risk. Many animals eat much more feed than is required. The excessive energy is converted to fat and deposited in the fat tissue.

The diet can also be delivered by controlling the time allowed for eating or the amount of feed supplied. This allows much greater control over the amount consumed by the animal. A pet that begins to gain too much weight can be fed less or allowed less time to eat. This does require more time commitment as meals are supplied multiple times throughout the day. The pet may also show more food hunting or begging behavior between meals.

Most animals should also have free access to water throughout the day. The diet should then provide all the necessary nutrients (carbohydrates, fats, protein, vitamins, and minerals) in the proper amount and proportions. The caloric need of the animal can be used to determine how much food is required. The **resting energy rate** (RER) is the amount of energy required by an animal at rest in a comfortable environment (temperatures not requiring the animal to heat or cool itself). Numerous formulas exist to calculate RER, based on the animal's body weight. An example of such a formula is:

$$RER \text{ (kcal)} = 70 \times (\text{weight in kg}^{0.75})$$

The **maintenance energy requirement** (MER) accounts for the RER, plus any additional energy required for the normal activity of the animal. The MER for the typical or average dog is usually twice the RER. Cats, on the other hand, typically have an MER that is 1.4 times the RER. Although inactivity decreases the requirements, many of the other factors discussed create an extra need for calories. For example, a dog in peak lactation with a large litter may have energy needs two to four times that of a typical MER.

Every animal has a limit to the volume or weight of food that can be consumed in a day. To meet the high energy needs of the lactating or working dog, each amount of food may need to contain a higher amount of calories. Energy density describes the calories supplied by each weight of food (for example, kcal per pound). Whereas the lactating dog may benefit from a higher energy density food, the overweight, inactive dog will benefit from a lower energy density food. With the lower energy density, the fat dog will feel satisfied with the volume consumed, while obtaining fewer calories.

PET FOOD LABELS

OBJECTIVE

■ *Describe the Important Features Found on Pet Food Labels and Compare and Contrast the Nutritional Requirements for Dogs and Cats*

Anyone that has walked down the pet food aisle in the grocery store will recognize that there is a tremendous variety in the types of foods available for pets (Figure 13–2). In addition to purchasing commercial feeds, it is possible to make diets for pets at home. For the sake of convenience, the vast majority of pet owners elect to purchase commercially available foods. The three basic feed types available are moist or canned, semi-moist, and dry foods (Figure 13–3).

FIGURE 13–2 A tremendous variety of foods are available. Making a selection can be difficult. *(Courtesy Giant food stores.)*

FIGURE 13–3 Three major varieties of food: canned, semi-moist, and dry.

Canned foods in general are about 75% water. These moist foods do tend to be more palatable and more digestible than dry food. Palatability describes how well the animal likes the food. Palatability is affected by several factors: odor, texture, nutrient content, and habit. Canned foods tend to have a higher level of water, protein, and fat, which also influences the odor. Fat plays a large role in the palatability of a food. The role of smell in the sense of taste is very important. An animal with an upper respiratory tract infection and a congested nose may become anorexic (have a poor appetite). Cleaning the nose and warming the food to increase the smell released can often help to improve the appetite of such an animal.

Dogs in general prefer canned food to dry. Cats are not so consistent. Cats often develop a preference for one type or shape of food. Manufacturers attempt to take advantage of this habit of cats, by producing foods with a distinctive shape or texture. Cats can become such creatures of habit that it can be difficult to get them to switch foods if the need arises.

Canned foods have the advantage that any type of feed ingredient can be used (wet or dry). Once opened, canned foods should be refrigerated, because they can spoil. In general, canned foods are the most expensive on a dry matter basis. Because of the high palatability,

canned foods are not often used on a free choice basis. Pets often eat food well beyond their needs, and obesity is a common problem if intakes are not regulated.

Semi-moist or soft-moist foods are generally packaged in a sealed foil or plastic package. This type of food generally contains about 30 percent water and is usually highly palatable. The small pieces of food are coated with a carbohydrate and treated with an acid, which helps to retain moisture and prevent spoilage. Because of the external treatment, this type of food does not require refrigeration.

In general, similar ingredients are used in semi-moist and dry foods. Some wetter protein (meat) sources can be used in semi-moist food because of the higher water content. The cost of semi-moist food can approach the cost of canned food.

Dry food is most commonly used and is generally the least expensive of the three types of food. For many pets, dry foods are used free choice. The hard food particles also provide an abrasive action on the teeth. This helps to slow the accumulation of tartar on the teeth. Tartar is a hard mineral plaque that builds up on the teeth. Excessive accumulation allows for bacteria to invade the gums and can eventually lead to tooth loss.

Most commonly the dry foods are packaged in paper bags or cardboard boxes. These foods are very stable, have a long shelf life, and do not require refrigeration. The fat in the food can become oxidized, lowering its nutritional value. It is recommended that the food be used within 6 months of when it is made. Because this date may not be known, it is ideal to only purchase amounts that can be used within a month or two.

The ingredients are limited in dry food, because the moisture content must be controlled. Dry foods typically have 10% to 14% moisture. This limits the amount of fat and fresh meat that can be used in manufacturing dry food. With the type of ingredients needed, dry food tends to be the least palatable and least digestible of the three food types.

The government establishes rules for the labeling of pet foods (Figure 13–4). Obviously the label needs to list the name of the product, the words *dog* or *cat food*, and the net weight. In addition, the name and address of the manufacturer or distributor must be identified. The label must also include a description of the designed usage or purpose of the food. This may describe various life stages, such as puppy, adult, inactive and overweight, or senior.

Also on the label is the guaranteed analysis of the pet food. The guaranteed analysis does not define the absolute quantity of nutrients, but rather certain minimums and maximums. The label lists the minimum for crude protein and fat. The actual amount may be higher than that listed on the label. Maximum values are listed for moisture and crude fiber. For these

FIGURE 13–4 An example of a pet food label.

nutrients, the actual content may be less than that listed on the label. The minimum and maximum values assure the consumer of a certain quality of the food. The manufacturer can, for example, formulate the food with a higher level of protein. If the quality of the ingredients used in the food declines, the manufacturer is still safe in producing an acceptable product.

The label must also include the ingredients, listed in descending order. This can be somewhat confusing, however. Consumers may find it appealing to find a pet food that lists meat sources as the first ingredient, but the ingredients are listed based on weight. Meat products may be listed as the highest ingredient, but may contain much higher water content than the second ingredient, which might be a grain source. Multiple forms of a similar ingredient may also be listed (such as corn grain, flaked corn, corn byproduct, ground corn, and so on). Again this gives the appearance that the grain source is much lower in the list. Meat byproducts may have quite a variation in their nutritional content, based on what sources are used.

Veterinarians are commonly asked, How much should I be feeding my pet? This is a very difficult question to answer completely and directly. The labels on pet foods give an estimate of how much to feed. However, this recommendation needs to be adapted to the individual pet. Variations in the energy requirement of an individual have already been discussed.

The goal of the diet is to provide enough food to maintain an ideal body weight and body condition. *Body condition* is a term that describes the outer appearance of an animal. Body weight is very helpful, but it is important to realize that muscle tissue is more dense than fat. So two animals with the same body weight could have quite different appearances and body composition. Body condition is used in many species to judge the amount of fat on the animal.

In pets a simple evaluation of body condition is to feel the flesh over the ribs. It should take only gentle pressure to feel the individual ribs. If firm pressure is required to feel the ribs, the animal is becoming too fat. There should be enough flesh over the rib cage that the ribs are not visible. The animal is too thin if the ribs are visible. (It is important to realize that long hair can mask this sign.) Pets should also have a waistline (Figure 13–5). That is, the abdomen should be smaller than the ribs. When the abdomen protrudes beyond the ribs, the animal is becoming too heavy.

Although cats and dogs share a very similar digestive system, they do have distinct differences in their nutritional demands and habits. Several generalizations can be made; however, many exceptions exist. Dogs do not require a change in foods to provide variety. This may actually create a finicky eater. Often the dog can train the owner! When the dog learns to expect feed changes often, it eats poorly for a few days and the owner responds with a new food. It is better to maintain the dog on a consistent diet. If changes do need to be made, it is best to gradually make the transition to the new diet over several days to prevent upsetting the digestive system.

Cats can become very fixed on a particular food type. It is best to provide a diet with multiple protein sources, which makes transitioning to a new diet easier if the need arises. Cats can also be made into finicky eaters by frequent feed changes. Cats often eat small amounts frequently, unlike many dogs that eat a large meal. Typically cats even prefer to eat alone and tend to be erratic in their appetite. One day they may eat a large amount and eat very little the following day.

More important than the behavior involved, cats and dogs have distinct nutritional differences. Cats are true carnivores (meat eaters), whereas dogs fit more into the omnivore (eating both meat and plant) category. Dog foods are not designed to be nutritionally sound for cats. Cats have a higher demand for protein and require much higher levels of the amino acids arginine and taurine than the dog. Cats cannot convert linoleic acid to arachidonic acid like the dog. Arachidonic acid and taurine are only found in animal tissues and therefore must be included in the cat's diet. Cats cannot survive on a complete vegetarian diet without appropriate supplementation.

Dogs can produce the B vitamin niacin from the amino acid tryptophan and can produce vitamin A from betacarotene. The cat is unable to make either of these conversions. As a result, the cat's requirements for niacin and vitamin A are much higher than that of the dog.

To ensure proper nutrition for a dog or cat, it is important to feed a good quality food designed for that animal in that stage of life and activity level. Cat and dog foods are not interchangeable. The amount of food to provide is the amount necessary to keep an animal in good body condition and at ideal body weight and maintain its health. Supplements are generally not needed for healthy animals on a balanced diet.

EQUINE NUTRITION AND FIBER DIGESTION

OBJECTIVE

■ *Discuss the Horse's Ability to Digest Fiber and its Role in Equine Nutrition*

The discussion in Chapter 7 has already illustrated the horse's ability to digest fiber. This section reviews the basic structures of the horse's digestive tract and discusses in greater depth the considerations involved in equine nutrition.

Many of the principles found in the nutrition of dogs and cats also fit in the feeding of horses or any species. The main goal of a nutritional program is to maintain the ideal body weight and condition while maximizing performance. There is a tremendous variation in the demands placed on horses. It is easy to see that a backyard pleasure horse ridden infrequently requires much less energy than a thoroughbred training for a major race. Likewise, a massive Percheron draft horse plowing all day has a much higher nutritional requirement than a pony giving rides at a party. The nutritional requirements are also influenced by the health, age, condition, and temperament of the horse.

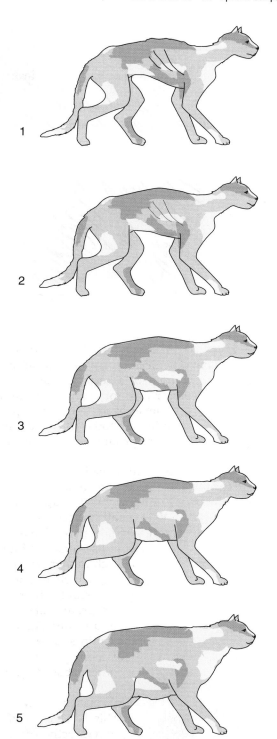

FIGURE 13–5 Evaluating body condition in cats on a scale of 1 to 5. The illustrations show cats ranging from emaciated (1) to obese (5). Number 3 is ideal. This cat will still have a visible waistline, and the ribs can easily felt but not seen. Long hair can make visual evaluation difficult.

The environment also influences the demands on the animal. A horse kept outside on a wet, cold day burns much more energy maintaining its body temperature than a similar sized animal kept in a warm barn.

It is useful to consider a horse in the wild to understand the design of the equine digestive system. These horses spend the majority of their day (up to 18 hours) **grazing** (Figure 13–6). The grazing provides a regular input of small meals, delivering new nutrients into the intestinal tract throughout the day. These horses live almost entirely off the leafy portion of grasses or roughages. Horses in the wild have minimal grain in their diet. Over the course of a year, the diet does change significantly in quality. Typically horses gain weight, depositing reserves of fat in the spring and early summer. As the seasons progress, the quality of the plants declines and the horse loses much of its fat over the winter. Wild horses' diets must support them in maintenance and provide adequate nutrients to support reproduction.

Our domestic horses do not share many of these challenges. In general a domestic horse has a consistent diet throughout the year and is often protected from the harsh weather. Many horses have increased nutritional demand due to the work that they are required to do. To compensate for these needs, grain, or **concentrates,** have been added to their diet. Grain feeds have been called concentrates because the nutrients have a higher concentration in these feeds compared with roughages. Grazing plays a much smaller role in domestic horses than in those in the wild. The diet is controlled by the owner and often comes in larger meals fewer times a day.

An inactive horse or one at only light work (for example, the occasional pleasure ride) needs little to no concentrate. As the workload increases, the need for concentrates in the diet also increases. Concentrates are divided into two categories: energy or protein sources. There can be overlap as well; a high-protein feed may also be high in energy. Grains such as oats, wheat, corn,

and barley are all considered energy concentrates, high in carbohydrates. Other seeds, such as linseed and soybeans, are a high-protein source. Often several ingredients are combined in one concentrate to increase the energy and protein.

Roughages or forages, such as hay, should comprise at least half the diet for horses. As discussed, horses are designed to digest the fiber found in plant material. Adequate levels of roughage in the diet helps to maintain a healthy digestive system. The fiber also causes the full feeling that limits the appetite of the horse. Eating forages is more time consuming that eating grain. This can help to occupy the horse's time, which may be important for confined animals.

On average, feed entering a horse takes around 70 hours to completely pass through the digestive tract. The stomach of an average adult horse is relatively small, able to hold only 2 to 4 gallons. This reflects the natural diet of horses, where small amounts are taken frequently throughout the day. Grain feeding needs to be limited to less than 3 to 4 pounds at a feeding because of this small stomach size. Feeding larger amounts can force grain into the intestine too early, disrupting the normal fermentation. Colic is just one side effect of excessive grain intake.

Much of the concentrate portion of the diet and a small portion of the roughage are digested and absorbed in the stomach and small intestine. The small intestine averages 60 to 80 feet in length and can hold approximately 10 to 15 gallons. Much of the protein and carbohydrate not bound in fiber is digested and absorbed in this region.

The fiber remaining then passes into the large intestine. This is the site of fiber fermentation. *Fermentation* describes the digestion of the fiber by the large number of bacteria housed in the large intestine. More than 40 species of bacteria have been identified in the gut of the horse. This normal flora (the collection of microorganisms naturally found in an area) varies with the type of diet being fed. A sudden feed change can kill millions of bacteria. It is important to change a diet slowly over the course of a week or 10 days to allow the necessary bacteria to increase.

The first portion of the large intestine is the large blind sac, the cecum. The cecum averages 3 to 4 feet in length and can hold about 6 to 8 gallons of material. The material then flows into the colon, which further allows for fermentation and water absorption. The colon is about 10 to 12 feet in length and can hold up to 25 gallons.

The bacteria are able to break down the beta bonds in the plant fiber. The bacteria utilize the simple sugars and then release volatile fatty acids (VFA), which are absorbed by the horse. The bacteria gain nutrients from the fermentation process, allowing them to flourish. The horse provides a warm, liquid environment for the

FIGURE 13–6 Horses grazing pasture in early spring.

bacteria to live and supplies them with nutrients. In exchange, the bacteria provide nutrients for the horse. This beneficial relationship is called symbiosis.

The volatile fatty acids are short-chain fatty acids. The three most common VFAs are acetic, butyric, and propionic acids. These VFAs are absorbed in the cecum and colon. The horse then utilizes them as an energy source. The intestinal bacteria also produce many B vitamins and vitamin C, which are subsequently absorbed by the horse. As the bacteria continue to reproduce, many are passed to the external environment in the feces.

Acetic Acid	$C_2H_4O_2$
Propionic Acid	$C_3H_6O_2$
Butyric Acid	$C_4H_8O_2$

Many common principles exist in feeding horses, just as in dogs and cats. The first is that the ration needs to be adapted to the needs of the horse. This includes the influence of the size of the horse and the work done. The diet should provide nutrients to optimize the output of the horse but should not result in overconditioning. An overweight horse usually has decreased performance in whatever type of work it performs. The added fat has an insulating effect that creates an increase in body temperature during exercise. The added body size also increases the oxygen and circulatory demands on the heart and lungs. In addition, the excess weight puts more stress on the musculoskeletal system. This can be important as a horse runs and jumps.

Horses are very sensitive to the quality of feed. Moldy feed can contribute to respiratory problems and colic. Concentrates should only be fed based on the horse's need and, equally important, should be fed in small meals. With the small size of the stomach, no more than 3 to 4 pounds should be fed at any one time. At most, concentrates should contribute 50% or less to the total diet. Horses are creatures of habit, and maintaining a consistent feeding schedule can be beneficial.

A horse that becomes too heavy may have problems, but a loss of weight is also of concern. A horse that is extremely excitable may always stay thin. A common problem as a horse ages is that the margins of the molars become very sharp. With the motion of chewing, the teeth wear in such a way that a sharp edge forms. This can cut the cheek or tongue and cause the horse discomfort. When this condition progresses, the horse may drop whole grain out of its mouth as it chews. This problem can be corrected by having the horse's teeth **floated**. In this procedure a special rasp is used to file down the sharp edges of the teeth (Figure 13–7).

Intestinal parasites are also a common problem in horses. Any grazing animal has a high risk of intesti-

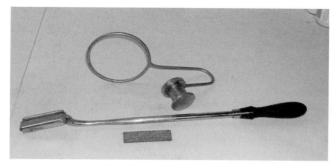

FIGURE 13–7 A tooth float used to file the teeth of horses. A gag is used to hold the mouth open for the procedure.

nal parasites. These parasites pass eggs in the feces, which then contaminate the ground and plants. When animals graze these eggs are ingested and new parasites develop in the intestinal tract. Regular treatment with medications to kill these parasites is very important.

Any animal losing weight should have its diet evaluated. This is just as true in horses. It is possible that there is inadequate concentrate being supplied for the level of work. Another possibility is that the **forage** is of poor quality. As plants mature, the quality as a feed declines. Much of the fiber can become indigestible if the plant is too old.

Horses, especially those confined for long portions of the day, can develop bad habits. **Cribbing**, or wood chewing, is one of those habits (Figure 13–8). The horse stands and chews on the wooden stall for long periods. Another problem occurs when horses eat their own feces. Both habits can be difficult to break. However, some basic principles can be useful

FIGURE 13–8 A horse cribbing. Note the destruction of the board fence.

in controlling the behaviors. First it is essential to ensure that the horse is fed a well-balanced diet, with good quality feed and adequate fiber. Having long-stemmed hay to eat can provide the horse with activity to occupy its time. Maintaining a regular feeding schedule is also helpful. Providing regular exercise can prevent boredom in the horse.

Some horses **bolt** their concentrate; that is, they eat extremely rapidly, with very little chewing. This increases the risk of choking and colic in these horses. Feeding on regular schedules and having adequate fiber in the diet can help keep the horse from becoming extremely hungry at grain-feeding times. In addition, preventing competition between horses is also helpful in avoiding the problem. If the bolting continues, a large, smooth stone can be placed in the feed bucket, so that the horse has to eat around it, slowing his ability to consume the grain.

RUMINANT NUTRITION AND FIBER DIGESTION

OBJECTIVE

■ *Detail the Ruminant's Ability to Digest Fiber and the Role in Ruminant Nutrition*

Cattle, sheep, and goats are all ruminants. In this section the cow is used as an example to show the considerations involved in feeding a ruminant. As already discussed, a ruminant has a large four-compartment stomach containing the rumen, reticulum, omasum, and abomasum. The abomasum is called the true stomach because it functions like that of a monogastric. The large rumen serves as a fermentation vat. It occupies the majority of the left side of the abdomen in adult cattle. The size of the rumen varies with the size of the animal, but on average it can hold 30 to 40 gallons.

Young ruminants actually function as a monogastric, relying on the mother's milk for their initial nutrition. Over the next few months of life, the rumen begins to develop until the animals become complete ruminants. Cattle initially eat quickly, swallowing forage in relatively large pieces. Just like the symbiotic relationship in the lower intestinal tract of horses, the rumen houses a large number of bacteria and protozoa that are responsible for digesting fiber.

The cow provides the bacteria with warmth, water, and nutrients. In exchange, the bacteria digest fiber, supporting their own replication, but also release VFAs for the cow to utilize. Cattle have a higher efficiency in digesting fiber than horses. In horses the bacteria use many of the nutrients and eventually are passed in the feces. By having the rumen before the glandular abomasum, bacteria that pass through the rumen are then

digested. The large numbers of bacteria that are digested are a significant source of protein for the cow. In designing rations for cattle, we must consider that we are actually feeding the rumen organisms, which in turn feed the cow.

The rumen and reticulum are active stomachs, frequently contracting. On average, healthy cattle have a rumen contraction between 1 to 3 times a minute. These contractions can easily be heard with a stethoscope as a rumbling noise on the left side of the abdomen. Evaluating the motility of the rumen is very important in the physical examination of the cow.

The contractions of the rumen keep the contents stirred. The contents in a healthy rumen divide into three layers. On the very bottom is a very liquid fraction with the finest particles. Above this is a firmer layer with longer fiber particles. Floating to the very top is a gas layer. As much as 8 to 10 gallons of gas can be produced every hour by the rumen organisms. The rumen contractions bring this to the appropriate location to allow it to be **eructated,** or belched. If eructation cannot occur, the cow will bloat (a dramatic distention of the rumen with gas). If not corrected, this condition can become life threatening.

The rumen contractions are also responsible for rumination, where boluses of food are regurgitated into the mouth. The cow at rest spends time chewing this food again, making much smaller pieces. This is called cud chewing. In addition to making the food particles much smaller for bacteria to utilize, cud chewing stimulates the production of saliva. It is estimated that cattle produce between 40 to 50 gallons of saliva every day. This actually contributes a large portion of the liquid in the rumen. In addition, it helps to buffer the pH of the rumen fluid. Maintaining a consistent pH is very important in protecting the large number of rumen microorganisms.

Ruminants rarely vomit. If the ingesta is completely vomited to the outside, it indicates a problem. On many occasions, goat owners have called our clinic concerned because their goats are all vomiting. The classic presentation is that the goats were just introduced to a new section of woodland that the owner wanted to have cleared. When examined more closely, these sections have rhododendron or mountain laurel. Unfortunately, these plants are toxic to ruminants and usually cause vomiting. This can be fatal if the goat ingests enough of the plant.

For the rumen organisms to grow and divide they must be supplied with carbohydrate and protein. Crude protein is commonly measured in a forage analysis. The total nitrogen level multiplied by 6.25 defines the crude protein. (Feed protein averages 16% nitrogen; dividing the nitrogen content by 0.16 determines the protein. Dividing by 0.16 is the same as multiplying by 6.25.)

Not all the nitrogen in the plant is found in protein. Urea contains nitrogen and is included in the nitrogen amounts.

When the feed enters the rumen, a portion of the crude protein, especially the urea enters the liquid portion very quickly. This soluble protein is rapidly available to support the growth of the rumen microbes. Another fraction of protein also dissolves into the rumen fluid but at a slower rate. On average, 60% to 70% of the crude protein is degraded in the rumen and utilized by the rumen microbes (this includes the soluble protein). The remaining 30% to 40% moves into the abomasum to be digested. This is called rumen bypass protein. The huge number of bacteria and protozoa that pass into the abomasum are also digested as a source of protein.

The carbohydrate portion of plants also divides into fragments. A small portion of simple sugars and starch exists, which also rapidly dissolves in the rumen. This provides the energy supply, which organisms need to utilize the soluble protein. A much larger supply of the carbohydrate is found in fiber. Not all the fiber is digestible by the cow or the microorganisms. Cellulose, hemicellulose, and lignin all comprise the plant's fiber. Lignin is indigestible (not able to be used). The older the plant, the higher the level of lignin.

The rumen microbes are able to digest much of the cellulose and hemicellulose. The organisms utilize the energy from the plant fiber and the protein to support their own growth. These organisms release large quantities of VFAs into the rumen fluid, which are then absorbed by the cow. Very little glucose is absorbed in the intestinal tract of the cow. The majority of the simple sugars are utilized in the rumen. The VFAs are the source of energy for the cow.

Acetic acid, propionic acid, and butyric acid are the three major VFAs utilized in the cow. The liver removes most of the propionic acid and converts it to glucose. Acetic acid is used to create energy and is also used in the synthesis of lipids. Butyric acid is used in many tissues throughout the body for energy.

Cows during their dry period are fed a maintenance diet. In general, dry cows are not expected to gain much weight and their workload is relatively low. They need only support themselves and the developing fetus. Following calving, the cows are placed on a diet that is much richer in energy and protein. This diet is designed to supply nutrients to support large amounts of milk production. With time the rumen develops longer papillae (tiny fingerlike projections) on the lining of the rumen. These papillae are very important in absorption of VFAs.

The formation of these papillae takes several weeks. During that time, the quantity that the cow can eat (dry matter intake [DMI]) gradually increases. Typically the energy demands for milk production exceed what the cow can consume for the first 6 to 8 weeks. During this time, the cow is in negative energy balance (that is, it is using more energy than it is consuming) and loses weight. Finally its intake is adequate to meet its needs, and with time, as production declines, the cow is able to gain weight. Figure 13–9 shows graphs describing the relationship among milk production, dry matter intake, and body weight.

It would seem that a straightforward question would be, Why not increase the concentrates in the ration to meet the needs? Feeding excess grain may cause health problems in cattle. The rumen organisms need adequate fiber to survive. By increasing grain and decreasing fiber, the pH in the rumen begins to decline. This can kill millions of microbes. Remember that the organisms are essential to supply energy to the cow and they are a source of protein. Feeding excessive grain can often make a cow go off feed (have a poor appetite) and actually result in a more negative energy balance.

Even in diets in which normal amounts of grain are fed, feeding too much at one time can cause the same pH changes in the rumen. This is called slug feeding. To prevent this, many farmers feed a **total mixed ration** (TMR). In a TMR, all the feed ingredients are combined in a mixer and blended together (Figure 13–10). The goal of this feeding method is to supply a uniform feed to the cattle throughout the day. With each mouthful, the cow is consuming the proper balance of fiber, protein, and energy.

CLINICAL PRACTICE

OBJECTIVE

■ *Link the Clinical Significance of the Academic Material Learned in This Chapter to Veterinary Practice*

Obesity is the most common nutritional disorder that veterinarians see in dogs and cats. Certain disease conditions, such as hypothyroidism and hyperadrenocorticism, may contribute to an animal's obesity. However, most cases are associated with an excessive intake of calories relative to the animal's needs. Several factors can contribute to this problem. The animal may be fed too much or fed a diet not designed for its activity level. Many pets confined to a kennel or to the house just do not get enough exercise to utilize all the calories consumed. Additionally, many pets are given too many snacks or treats (both animal treats and table food) that add additional calories.

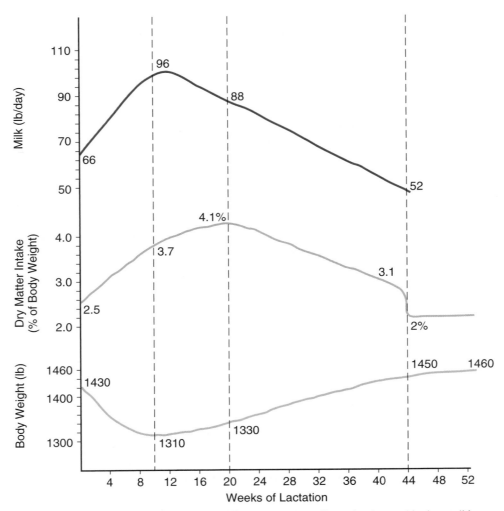

FIGURE 13–9 Graphs illustrating the relationship among dry matter intake, milk production, and body condition over lactation.

Weight regulation in healthy pets requires strong willpower by the owner. Often when treats are taken away or less food is given, the pet begs for food. The owners need to recognize that they are trying to help the dog. Increasing the pet's activity level also requires a significant time commitment by the owner.

FIGURE 13–10 A TMR mixer being loaded with feed.

Obesity may contribute to or worsen other diseases. A common presentation is an overweight dog that is having difficulty rising or getting around. Many elderly pets develop arthritis, and adding extra weight increases the stress on the joints. As muscles weaken and the arthritis worsens, the extra weight can become quite crippling.

The added weight also puts more stress on the circulatory and respiratory systems. The extra fat puts a higher demand on cardiac output and oxygen usage. This can worsen an existing problem, such as heart failure. These problems also make obese pets more of an anesthesia risk if surgery is required. Obesity may also have a relationship to controlling diabetes. Diabetes is much more readily regulated when their weight is brought under control.

As a veterinarian, it is common to hear about the huge variety of food items that pets are fed. Many owners are quite proud when they can say that their pet will eat anything. Unfortunately, many items that humans consume regularly can be toxic to small pets. Chocolate is one such food. Pets often find chocolate very tasty and will consume large quantities if they have the opportunity. Chocolate contains a toxin called theobromine. This toxic component has effects on the kidney and central nervous system. Signs often begin with an increased thirst and urine production, vomiting, diarrhea, and urinary incontinence. If the dosage is high enough, the signs progress to excitability, muscle twitching, seizures, and coma. Chocolate toxicity can be fatal.

Cats happen to be very sensitive to a toxin in onions. If consumed in high enough levels, the onions cause a defect in the red blood cells. The red blood cells break down in the bloodstream, producing hemolytic anemia. These animals develop darkened urine from the free hemoglobin and develop a fever. The combination of anemia and damage to the kidneys can result in death if sufficient amounts are eaten.

The role of fiber in the nutrition of both horses and ruminants has been emphasized. It is possible to overfeed concentrates and cause digestive disorders. But animals can accidentally gain access to grain and eat extremes amount in a short period. The result is grain overload. In the horse, grain overload results in a dramatic decline in the pH within the colon. Many of the natural organisms die and are replaced by other bacteria that can survive in those conditions. The death of such a large number of bacteria releases toxins that are absorbed into the bloodstream. In moderate levels the result is typically diarrhea and anorexia.

In more severe cases the toxins can have a damaging effect on the blood supply to the hooves. The third phalange (P3) is suspended within the hoof by a tissue rich in blood supply, called laminae. The laminae become inflamed, resulting in a disease called laminitis. With laminitis the horse has a great deal of pain with each step and there is swelling at the top of the hoof. The gait of these horses is often described as "walking

on eggshells." An interesting side note is that horses bedded with black walnut shavings may also develop laminitis. The walnut shavings have a toxin that can cause laminitis.

Horses with mild laminitis improve with rest and medications to reduce the inflammation. More severe cases can result in permanent lameness. The disease can be so severe that the point of P3 rotates downward. The laminae are so inflamed that they are unable to suspend the bone. It is possible that the entire hoof can become detached. Many of these horses do not recover.

Nutrition is often used in combination with treatment or as a follow-up to treatment in many diseases. Bladder stones and urinary blockage in cats was discussed in Chapter 6. Diet can play a significant role in preventing recurrences of these diseases. The diet change must be adapted to the specific cause. Many different types of bladder stones may develop. Some are more likely to occur in low pH, whereas others occur in high pH. Each has specific mineral composition. Diets can be adapted to control both the pH and the mineral content of the urine. This must be carefully balanced to prevent deficiency and yet restricted enough to minimize crystal formation. Initially these types of diets were only sold by prescription through veterinarians. Now many foods are available over the counter that help to control such problems. Many diets are now labeled with the claim of promoting urinary health in cats.

Other diseases that can benefit from nutrition include kidney and heart failure. In kidney failure, nitrogen-containing waste from protein metabolism builds up in the blood. Diets used to control kidney failure contain a very high quality protein source in limited amounts. Other minerals, such as phosphorus, are also restricted to help limit the accumulation in the blood. In heart failure the diets often restrict the amount of sodium, which helps prevent excessive blood volume from accumulating.

SUMMARY

A review of Chapter 12 will provide information about the six nutrients and their affects on the body. This chapter discussed how to interpret pet labels and differentiate between the nutritional needs of cats and dogs. In this chapter a horse's ability to digest

large amounts of fiber and how the horse and ruminant differ in the digestion of fiber were also covered. Understanding differences in species digestive tracts gives veterinarians the ability to successfully develop species-specific rations.

REVIEW QUESTION

1. Define the following terms:

   free choice diet   float
   resting energy rate   forage
   maintenance energy   cribbing
       requirement   bolt
   grazing   eructation
   concentrates   total mixed ration

2. True or False: To develop a diet, a veterinarian first considers the needs of a resting animal in a comfortable environment.

3. True or False: The adult rumen in cattle can hold 40 gallons.

4. The maintenance energy requirement is typically _____ times that of resting energy requirement for an animal.

5. Filing the sharp edges of a horse's teeth is called _____.

6. Does a lactating animal have more or less nutritional needs than when not lactating?

7. Which feeding system can help quiet confined dogs?

8. What type of food do dogs typically prefer?

9. In general, what is the water content of semi-moist pet food?

10. Do horses prefer to graze throughout the day or eat one large meal daily?

11. Where are simple sugars utilized in the cow?

12. Are cats carnivores or omnivores?

13. Where does fiber fermentation occur in the horse's digestive tract?

14. How are ingredients listed on a pet food label?

15. List the three types of feed available for dogs and cats.

ACTIVITIES

Materials needed to complete activities:
    samples of canned, moist, and dry pet food
      accompanied by their respective labels
    copies of the pet food labels

1. Compute your resting energy rate by multiplying your body weight times 10 (keep in mind that variations will exist, but this simple equation will give a ball park figure). Example: $155 \times 10 = 1550$ calories. The resulting number gives your RER.

List activities that would increase your energy needs. Expected answers could be sports or physical labor. If you have an animal, explain how much feed you provide to the animal and why you choose that amount.

2. Based on the display of pet food samples, guess the percent of water in each. Suggest an ingredient list for each. Suggest a favorite from a pet's perspective.

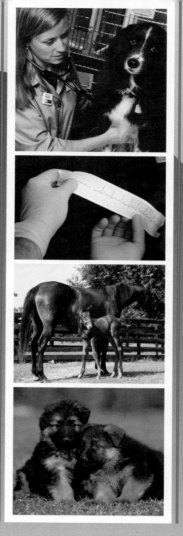

UNIT III

Diseases

14

Principles of Infectious Disease

OBJECTIVES

Upon completion of this chapter, you should be able to:

- *Describe Koch's postulates.*
- *List the important distinguishing features and give examples of major disease agents and discuss the resulting diseases.*
- *Relate text material to common presentations.*

KEY TERMS

coliform	fomite	prodromal phase	antimicrobial
botulism	vector	anemia	bacteriostatic
anthrax	eukaryotic	systemic	antiseptics
Koch's postulates	prokaryotic	Lyme disease	disinfectants

INTRODUCTION

A disease is a change that occurs in the body and prevents normal function. Some diseases occur as a result of other organisms' invasions. Numerous organisms are necessary to keep our bodies functioning normally (for example, intestinal bacteria). Only a small percentage of microorganisms are capable of causing disease.

A DAY IN THE LIFE

Sometimes I Worry...

While writing, I often come across cases that relate to the text subject matter. Often these cases interfere with my writing progress. Last night one of those situations occurred. I was on call for large animal emergencies, so it wasn't surprising when the phone rang at 8:30 P.M. A young farmer reported that he had a cow down and she was unable to rise. This was not a classic milk fever case, because the cow had calved 2 months earlier.

I went to the farm and found the down cow to be obviously quite ill. Her eyes appeared sunken, and she had severe diarrhea. I began my physical examination and found that she was running a fever. While checking her milk, I discovered the source of the problem: mastitis. Mastitis is an infection of the udder that causes abnormal milk to be produced (Figure 14–1). The milk from infected quarters contains a very high number of white blood cells that have migrated into the milk from the blood system.

This cow had a yellowish, watery secretion from the infected quarters (see Figure 14–1). This type of secretion is classic for what is called **coliform** mastitis. The word *coliform* describes a related group of bacteria that cause this type of mastitis. Fortunately, this cow responded well to treatment and soon was able to rise. Many cows that reach this degree of severity with coliform mastitis do not respond as positively.

I remember one demonstration quite clearly, although it happened many years ago in veterinary school. A horse was presented for weakness. The veterinarian on the case set a bucket of water in front of the horse. The horse immediately placed its muzzle in the water and apparently began to drink. After several minutes the horse raised its head, and the water level had not changed. The students observing the case were all surprised. It appeared that the horse was trying to drink but was not able. Next

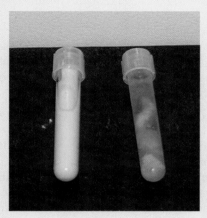

FIGURE 14–1 Normal white milk is shown in the left tube. Abnormal secretion from a cow with coliform mastitis is shown on the right.

the clinician held the horse's head and pulled gently on the tongue. The horse did not resist. Normally a horse has such strong muscle tone in its tongue that it is difficult to hold. These two tests confirmed the diagnosis; this horse was showing the earliest signs of **botulism**.

Our clinic often receives calls from cat owners reporting that their pets have developed swellings. When examined, these cats are usually also feverish. The swellings often occur around the base of the tail or on the head and neck. Swelling such as this typically happens after a cat fight. After clipping the cat's hair, I usually find a small puncture or punctures in the skin caused by the bite or scratch from another animal. These punctures have begun to heal, sealing the opening. Unfortunately, the teeth or claws introduce bacteria under the skin. The bacteria then multiply inside the warmth of the cat. As the body's immune system fights the infection, pus accumulates under the skin. These lumps can break and drain before the cat is presented.

KOCH'S POSTULATES

OBJECTIVE

■ *Describe Koch's Postulates*

In the fall of 2001, current events brought the disease **anthrax** to the forefront. Frequent news reports taught us of the skin, intestinal, and inhaled forms of

this disease. Nearly 125 years earlier, Dr. Robert Koch, a German physician, was investigating the same disease. At that point, the medical profession was attempting to discover why certain diseases were occurring.

Dr. Koch's investigation led to the development of certain foundation principles about infectious diseases. Dr. Koch studied anthrax in cattle in an attempt to prove that an agent was responsible for the disease. This research occurred before bacteria and associated disease conditions were clearly understood. Dr. Koch knew that blood taken from an infected cow would cause the same disease in another cow. He was able to isolate the bacteria (*Bacillus anthracis*) from the blood.

His work led to the formation of a set of principles that help to define an infectious disease. These principles have come to be known as **Koch's postulates**. To prove the cause of a disease, he postulated the following requirements:

1. The infectious agent should be detectable in sick animals but not healthy animals.
2. It should be possible to isolate and culture the organism.
3. Organisms taken from the culture and introduced into a healthy animal should cause the same disease.
4. The same organism should be isolated from this second animal as well.

Dr. Koch's theory helped to establish germ theory. Physicians finally recognized that infectious agents caused many diseases. To this day, Koch's postulates are helpful in understanding infectious diseases. The basic principles still hold true, although not every disease is so clearly defined. For example, an infectious agent may be isolated from a healthy animal. In other situations, more than one infectious agent can be involved in causing an illness.

Infectious diseases are caused by microorganisms that gain entry into the animal's body. Some infectious diseases are also contagious. This means that a healthy animal exposed to a clinically infected animal will also develop the disease. The organism is transmitted from one animal to the next. It is important to realize that not all infectious diseases are contagious. In Chapter 11, tetanus was discussed. This is an infectious disease; the animal picks up the organism from the environment. However, other animals in contact with the infected animal are not at risk of becoming infected from the exposure.

Tetanus is an example of a disease in which the organism is introduced into the body through a wound. For many other diseases, organisms come into contact with a mucous membrane. This can be in the nose, mouth, eyes, respiratory, gastrointestinal, or urinary tracts. The organism must first adhere to the surface of the mucous membrane and then begin to replicate.

The animals may become exposed to the organism from several sources. In contagious diseases the animal may become sick following exposure to another clinically ill animal. Some animals, following recovery from an infectious disease, continue to shed the organ-

FIGURE 14–2 A cow suffering from pinkeye. Pinkeye is an infectious disease that is commonly spread between animals by flies.

ism. These carriers may be a source of infection for others.

Although contact with infected animals may be direct, some infectious diseases can be spread through the air. Sneezing and coughing can make the infectious organisms airborne, and other animals subsequently inhale them. The organisms can often be found in high numbers in the secretions of the animal (for example, nasal discharge, saliva, tears, urine, and feces). These discharges can contaminate inanimate objects, or **fomites**. These fomites can then be a source of infection for other animals. Sharing water bowls or feeding utensils are common examples of how this type of contamination may occur. Different organisms can survive for variable periods outside the animal. This is influenced by the outside temperature and the availability of moisture. Extremes of temperature and dryness generally shorten the ability of an organism to survive.

Arthropods (such as insects, mites, ticks, and mosquitoes) may also play a role in transmitting disease (Figure 14–2). These arthropods are then called **vectors**. Some diseases actually require the arthropod to be involved. In these diseases the arthropod serves as a host necessary for the development of the infectious agent. In other cases the arthropod is merely contaminated and serves as a vector to carry the organism to other animals.

Some organisms are commonly found in the soil and require some means of being introduced into the body (again tetanus serves as an example). Food and water also may become contaminated with infectious organisms. Listeriosis is a disease caused by bacteria infecting the brain. Contaminated silage is a common source of this infection in cattle.

DISEASE AGENTS

OBJECTIVE

■ *List the Important Distinguishing Features and Give Examples of Major Disease Agents and Describe the Resulting Diseases*

Infectious diseases are divided into four major classes of agents: bacteria, viruses, fungi, and parasites. Each type of organism has features that distinguish it from the others. Within each class exists tremendous variation. This discussion begins with a description of bacteria.

Until this point, all the text's discussion of cells has described **eukaryotic** cells. These cells have membrane-bound organelles, such as the nucleus, mitochondria, and endoplasmic reticulum. Bacteria are one-celled **prokaryotic** organisms. Prokaryotic cells lack membrane-bound organelles. Each bacterium has a single chromosome that is not surrounded by membrane.

Bacteria reproduce through cell division, one cell dividing into two. The process begins with a replication of the chromosome. Cell membrane and cell wall grow across the cell, thus dividing it into two. This process can occur in less than 20 minutes. At this speed, millions of bacteria can be formed in the matter of hours. Replication of bacteria is limited by the availability of nutrients and the accumulation of waste products released by the bacteria.

Bacteria are also affected by the surrounding air, including gases such as oxygen, nitrogen, and carbon dioxide. Bacteria vary considerably in their need for oxygen. Many bacteria, called aerobes, require oxygen to be present in the environment to allow for growth. Other bacteria, such as *Clostridium tetani*, only grow in environments lacking oxygen. These bacteria are termed anaerobes. A third class of bacteria, known as facultative anaerobes, grow in either type of environment.

Bacteria also have a cell wall that provides a rigid framework and maintains the cell's shape. The cell wall is composed of short polypeptides and sugars combined into one large macromolecule that surrounds the cell. The plasma membrane lies immediately beneath the cell wall. Many bacteria must be able to survive in a hypotonic environment (such as water). A normal animal cell would burst in such an environment, because the osmotic pressure of the cell would draw water into the cell. The cell wall prevents this flow of water into bacteria.

The cell wall is significant in the classification of bacteria. Differences in the cell wall influence various bacteria's ability to pick up stain. In the late 1800s, Christian Gram, a Danish bacteriologist, developed a staining technique for bacteria in which the sample is first treated with crystal violet, a purple dye. The sam-ple is then treated with alcohol or acetone, to remove any free crystal violet. The last step is to use a different colored stain, such as pink safranin.

In this technique, some bacteria absorb the crystal violet and are stained blue. Another class of bacteria does not retain this stain when decolorized and subsequently are stained red with the safranin. The Gram stain then distinguishes two classes of bacteria. Gram-positive bacteria are stained blue. They pick up the crystal violet. Gram-negative bacteria do not retain the crystal violet and are visible as red or pink under the microscope.

Some bacteria have a slime layer or capsule surrounding the cell wall. This additional layer makes it more difficult for phagocytes to capture and engulf them. Some bacteria are mobile with the help of flagella. The flagella do have a different structure than those found in eukaryotic cells. Also found on the surface of the cells are small protein filaments projecting from the cell membrane through the cell wall. These are called pili, and they play a role in the bacteria's abilities to attach to other cells. The pili may also function to transmit DNA material between bacteria.

In addition to the main chromosome, many bacteria have small fragments of DNA called plasmids. The plasmids can replicate independently of the bacteria. Plasmids can provide the genetic code for certain enzymes, for the exchange of genetic material, and for resistance to antibiotics. Resistance to antibiotics is a very serious concern in human and veterinary medicine.

Three methods are available for bacteria to transfer genetic material. The first method, conjugation, requires direct contact between two bacterial cells. The pili come into contact, and a bridge of cytoplasm forms between the two cells. The DNA then transfers across this cytoplasmic bridge. In transformation, the second method, a fragment of DNA is released by one bacterium and then engulfed by another. The final technique, transduction, requires the transmission of the DNA by a virus. Viruses are discussed later in this chapter. The virus infects one bacterium, picks up genetic material, replicates, and then infects other bacteria. By infecting other bacteria, the new genetic material is introduced into them.

When nutrient supplies are scarce or environmental conditions harsh, many bacteria form endospores. Only one endospore is formed from each bacterium, this is not a form of replication. The bacterium loses water and shrinks into a very durable cell. Endospores are capable of surviving very hot and dry conditions or even freezing. When conditions improve, the endospore is capable of becoming active again and functioning as a normal bacterium. *Bacillus anthracis*, the organism responsible for causing anthrax, is capable of forming an endospore.

TABLE 14–1 Common Bacterial Infections

Disease	Strangles	Pneumonia	Greasy pig disease
Species	Equine	Equine	Swine
Causative organism	*Streptococcus equi*	*Rhodococcus equi*	*Staphylococcus hyicus*
Signs	Usually affects young horses, anorexia, fever, nasal discharge, swollen lymph nodes	Usually affects young horses, fever, nasal discharge, respiratory difficulty, lung damage and inflammation, may spread to joints	Reddened skin, anorexia, fever early in disease, thickened skin with purulent discharge
Disease	Erysipelas	Leptospirosis	Swine dysentery
Species	Swine	All	Swine
Causative organism	*Erysipelothrix rhusiopathiae*	*Leptospirosis* sp.	*Treponema hyodysenteriae*
Signs	Acute form can cause sudden death, high fevers, reddened skin progressing to a diamond appearance, arthritis	Abortions, fever, anemia, jaundice	Diarrhea, often with blood and mucus, dehydration, weakness, anorexia, may cause death
Disease	*E. coli* diarrhea	Contagious equine metritis	Kennel cough
Species	All	Equine	Canine
Causative organism	*Escherichia coli*	*Hemophilus equigenitalis*	*Bordatella bronchiseptica*
Signs	Usually young animals less than 2 weeks old, severe diarrhea, dehydration, anorexia, fever, may cause death. (The same organism is one of the causes of coliform mastitis.)	Disease seen in mares, purulent discharge from the vulva, spread by stallions	Can be secondary to viral infection, causes a dry, hacking cough, anorexia
Disease	Salmonellosis	Thromboembolic meningoencephalomyelitis (TEME)	Pinkeye
Species	All	Cows	Cows
Causative organism	*Salmonella* sp.	*Hemophilus somnus*	*Moraxella bovis*
Signs	Severe diarrhea, often with blood, high fever, anorexia, dehydration, weakness, may cause death	Fever, severe depression, hind limb paralysis when it invades the spinal cord, may cause death	Increased tear production, inflamed conjunctiva, cloudy cornea, pain to eye in bright light

Note: This is only a partial list of bacterial diseases commonly found in domestic species. The complete list is well beyond the scope of this text. In general, when a bacterial infection is diagnosed, an appropriate antibiotic and supportive treatment are used.

There are a tremendous number of bacteria species in the world, many of which are normal inhabitants of animals. The importance of bacteria in fiber digestion has already been discussed. The normal bacteria actually compete with invading pathogens, helping to prevent disease. However, a small percentage of bacteria are pathogens that cause disease (Table 14–1).

Only when pathogens are able to compete with the normal flora of bacteria are they able to cause disease. Infectious diseases can be divided into several stages. The first stage, infection, occurs when the organism invades the host animal. The number of bacteria may actually be quite small at this point, and the bacteria must multiply. It is during this incubation phase when the pathogen number increases dramatically.

The signs of disease only occur after sufficient numbers of pathogens are present. In the case of bacteria, the signs are a result of toxins released by the bacteria. The effect is quite variable, depending on the toxin. Some organisms, such as in tetanus, release a toxin from the cell. These exotoxins are produced and released by secretion. Endotoxins are the other form of toxin found in bacteria. Endotoxins are actually a portion of the cell wall of Gram-negative bacteria. These toxins are only released on the death of the bacteria. In these situations the destruction of large numbers of bacteria by the immune system causes a release of large quantities of endotoxin. The sudden release of endotoxin is responsible for the clinical signs of the disease.

The first signs of illness occur in what is called the **prodromal phase**. This is a short-lived period

TABLE 14–1 Common Bacterial Infections—cont'd

Disease	Wooden tongue	Contagious mastitis	Lockjaw
Species	Cows, sheep	Cows	Horses, cows, sheep, goats
Causative organism	*Actinobacillus lignieresi*	*Staphylococcus aureus, Streptococcus agalactiae*	*Clostridium tetani*
Signs	Severe inflammation of the tongue and local lymph nodes. In sheep usually does not affect tongue	Infection of the mammary gland that can be transmitted between cattle during milking	Spasms of muscles, jaws held shut tightly, sawhorse stance, stiffness, progressing to recumbency and eventually death
Disease	Foot rot	Anthrax	Botulism
Species	Cows, sheep, goats	All	All
Causative organism	*Fusobacterium necrophorum*	*Bacillus anthracis*	*Clotridium botulinum*
Signs	Deep infection between the claws of the hoof, foul odor, discharge, lameness	Cutaneous form: painful skin swelling. Inhalant form: difficulty breathing, ataxia, weakness and death	Muscle paralysis, difficulty in swallowing and chewing, weakness, recumbency, often death
Disease	Circling disease	Rain scald	Lyme disease
Species	Cows, sheep, goats	Horses, cows, sheep, goats	All
Causative organism	*Listeria monocytogenes*	*Dermatophilus congolensis*	*Borrelia burgdorferi*
Signs	Fever, anorexia, begin pressing head against objects, progressing to ataxia and circling in one direction, can be fatal	Crusting of skin at base of hairs, often on the top of the back, most common in animals housed out of doors in damp conditions	Spread by ticks, primarily arthritis, can be chronic and intermittent, may also show lethargy, anorexia
Disease	Johne's disease	Enzootic pneumonia of pigs	Pneumonitis
Species	Cows, sheep, goats	Pigs	Cats
Causative organism	*Mycobacterium paratuberculosis*	*Mycoplasma hyopneumoniae*	*Chlamydia psittaci*
Signs	Chronic diarrhea, weight loss, anorexia, death	Often effects young pigs, dry cough, can be chronic, poor weight gain	Ocular discharge and inflammation, sneezing tearing, fever, cough

and may have such signs as fever and muscle aches. The disease quickly progresses into the acute period, when the signs of the disease are maximized. Hopefully the animal's immune system (possibly with the aid of medication) is able to control the infection. The disease enters the decline period, when the symptoms begin to improve. Finally the animal enters the convalescent phase, when it begins to return to normal. During this phase the animal regains strength, begins eating well, and becomes more active.

If the immune system and any treatment are unsuccessful in clearing the infection, the animal may enter a phase of chronic illness. The length of time for the chronic phase can be extremely variable. Some chronic illnesses can last a lifetime.

In addition to being classified as Gram-positive or Gram-negative, bacteria are classified based on shape (Figure 14–3). Spherical bacteria are called cocci. If the cocci bacteria cluster in groups of two, they are called diplococci. Streptococci are individual cocci bacteria grouped in long chains. Bacteria that are cylindrical are called rods or bacilli. Some bacteria take on a spiral shape. *Spirochete* describes bacteria that have a flexible spiral.

The classification of bacteria is always being evaluated, and at times the names of bacteria are changed. Many more factors are involved in classification, but the Gram stain characteristic and the shape of the bacteria are two important features. It is well beyond the scope of this text to list all the disease-causing bacteria. Only a few examples of each are given.

Example

Staphylococci (stahf-ih-lō-kohck-sī) are grapelike clusters of round bacteria; **coccus** (kohck-uhs) means round.

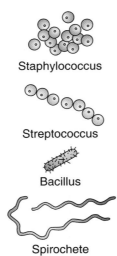

Staphylococcus

Streptococci (strehp-tō-kohck-sī) are round bacteria that form twisted chains.

Streptococcus

Bacilli (bah-sihl-ī) are rod-shaped bacteria.

Bacillus

Spirochetes (spī-rō-kētz) are spiral-shaped bacteria.

Spirochete

FIGURE 14–3 The shape of bacteria.

Gram-positive cocci are one such class. This group includes the *Streptococcus* (they grow in chains) and *Staphylococcus* (clusters of cocci bacteria). Many species in this group infect the mammary gland, causing mastitis. Other species can cause skin infection (*Staphylococcus aureus*) and infection in the respiratory tract of horses (*Streptococcus equi*).

Gram-positive rods include the *Clostridium* organisms and *Bacillus anthracis.* In addition to tetanus, which has already been discussed, *Clostridium* species also cause botulism. *Clostridium botulinum* is a toxin-producing organism that is often found in spoiled food. The same disease can occur in humans who consume improperly canned food. The toxin blocks acetylcholine, which is released by motor neurons. The result is severe muscle weakness. The disease can be fatal if the muscles that control respiration are paralyzed.

The Gram-negative rods include *Escherichia coli*, commonly abbreviated *E. coli*. In mastitis the Gram-negative rods are grouped as coliforms because of their similarities to *E. coli*. There are a large number of disease-causing Gram-negative rods (such as *Klebsiella, Pseudomonas, Salmonella, Yersinia, Pasteurella,* and *Hemophilus*). These species commonly cause intestinal, respiratory, and mammary gland infections. The disease-causing spirochetes are also Gram negative. The most common spirochetes in veterinary medicine are species of *Leptospira. Leptospira* can cause damage to the kidneys or abortions as a result of a disease called leptospirosis.

Not all bacteria fit neatly into the Gram-positive and -negative classification. *Mycoplasma* species are some of the smallest bacteria and lack a cell wall. Technically these organisms are Gram negative be-

cause of the lack of cell wall. However, other staining techniques are necessary to identify these organisms. These organisms commonly cause respiratory, joint, and mammary gland infections in farm animals.

Another class of disease-causing organisms is the viruses (Table 14–2). Viruses are neither eukaryotic nor prokaryotic cells. Viruses have no cell wall, no organelles, and no enzymes capable of producing energy. Viruses are not even able to replicate on their own. Viruses can only replicate by forcing infected cells to produce more virus. The viruses are so simple in form that it has been debated whether or not they can be considered living things.

Viruses have a central core of nucleic acid, either RNA or DNA, not both. Surrounding this strand of nucleic acid is a protein coat called the capsid. The capsid is responsible for protecting the nucleic acid (Figure 14–4). In addition, the protein of the capsid determines to what kind of cell the virus can attach. Once attached, the capsid plays a role in the insertion of the nucleic acid into the infected cell.

Viruses can be classified as either RNA or DNA. DNA viruses contain a double strand of DNA. Infection begins when the virus locates a host cell and attaches to the surface. Viruses can be very specific in the type of host and the type of cell to which they can attach. Some viruses may only attack one species, whereas others can infect all mammals. Following the attachment phase, the protein of the capsid aids in the penetration and insertion of the DNA into the host cell.

Once the penetration phase is complete, the genes coded in the virus's DNA use the host cell's ribosomes to produce enzymes (Figure 14–5). These enzymes can actually shut down the host cell's DNA while using the cell to produce multiple copies of the virus's DNA. This replication phase also has the cell producing proteins for the capsid. The cycle then enters the assembly phase, in which the DNA and capsid protein are combined into a new virus. In the release phase, lysosomes are then released to destroy the host cell. The assembled viruses are then released, ready to attack new cells. This entire process from the initial attachment to the final destruction of the host cell and release of virus is called the lytic cycle.

RNA viruses have a similar cycle upon infecting host cells. The host cell is once again used to create new strands of RNA and capsid. Upon assembly of the virus, the cell can be lysed and new virus released.

Not all viruses result in the death of the cell. The retroviruses actually incorporate genetic material into the host cell's DNA. The feline leukemia virus (FeLV) is a common example of this type. FeLV is an RNA virus that infects the cells of cats. The RNA codes for a strand of DNA that is formed in the host cell. The newly produced DNA inserts into the chromosomes of the host cell. With every division of the cell, the viral

TABLE 14–2 Common Viral Infections

Disease	Shipping fever complex	Calicivirus infection	Panleukopenia
Species	Cows	Cats	Cats
Causative organism	IBR, PI-3, BVD, BRSV	Calicivirus	Feline parvovirus
Signs	High fevers, nasal discharge, ocular discharge, coughing, severe pneumonia	Inflamed eyes, ocular discharge, sneezing, pneumonia, ulcers in mouth	Young cats, fever, vomiting, diarrhea, anorexia, often fatal, can damage bone marrow
Disease	Feline viral rhinotracheitis	Feline leukemia	Feline calicivirus
Species	Cats	Cats	Cats
Causative organism	Herpesvirus	Retrovirus	Calicivirus
Signs	Coughing and sneezing, discharge from eyes and nose, fever, anorexia	Chronic weight loss, anemia, anorexia, tumors	Inflammation of nose and eyes, ulcers in mouth, fever, usually self limiting
Disease	Viral rhinopneumonitis	Equine viral arteritis	Equine infectious anemia
Species	Horses	Horses	Horses
Causative organism	Equine Herpesvirus	Herpesvirus	Retrovirus
Signs	Fever, cough, inflammation of nose and throat, nasal discharge	May attack upper respiratory tract with fever, nasal discharge, or coughing or may cause abortions	Fever, hemolytic anemia, icterus, weight loss
Disease	Equine influenza	Equine encephalomyelitis (3 forms: Eastern, Western, and Venezuelan)	Canine distemper
Species	Horses	Horses	Dogs
Causative organism	Equine influenza virus	Togavirus	Canine distemper virus
Signs	Coughing, fever, anorexia, tearing, cloudy cornea	Fever, mental depression, anorexia, central nervous system signs	Respiratory: fever, nasal and ocular discharge, pneumonia. Nervous: seizures, can be fatal
Disease	Infectious canine hepatitis	Parvo	Rabies
Species	Dogs	Dogs	All
Causative organism	Canine adenovirus	Canine parvovirus	Rhabdovirus
Signs	Fever, lethargy, enlarged liver, anorexia, bleeding disorders, enlarged lymph nodes	Usually young animals, severe vomiting, diarrhea, anorexia, fever, often fatal	Fatal infection of the central nervous system: quite variable signs, paralysis, inability to swallow (foaming at the mouth), aggression, stupor
Disease	Kennel cough	Coronavirus infection	Pseudorabies
Species	Dogs	Dogs	Primarily pigs (can infect many species)
Causative organism	Adenovirus and parainfluenza	Canine coronavirus	Herpesvirus
Signs	Dry, hacking cough, fever, anorexia. Can lead to secondary infection with bacteria (*B. bronchiseptica*)	Generally a mild case of vomiting and diarrhea, usually self limiting	Affects the central nervous system, shaking, ataxia, convulsions, seizures, fever, anorexia
Disease	Swine influenza		
Species	Pigs		
Causative organism	Influenza type A		
Signs	Sudden onset, high fever, anorexia, coughing, respiratory distress		

IBR = Infectious bovine rhinotracheitis, PI-3 = parainfluenza 3, BVD = bovine viral diarrhea, BRSV = bovine respiratory syncytial virus.
Note: This is only a partial list of the viral diseases found in domestic species. The complete list is beyond the scope of this text. In general, no specific treatment is available to cure viral infections. Supportive treatment is necessary. Vaccination is critical in preventing viral diseases.

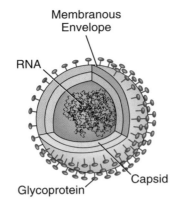

FIGURE 14–4 Classic virus structure.

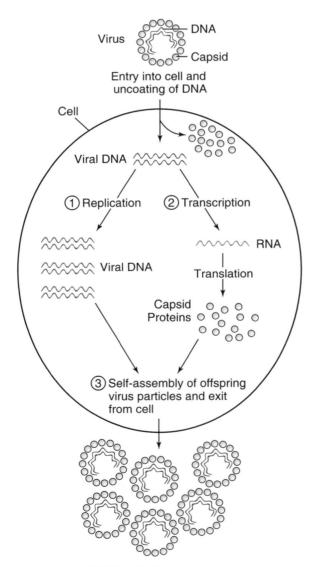

FIGURE 14–5 Life cycle of virus.

genetic material is passed on to the daughter cells. Once the virus has inserted into the host DNA, the infection will last for the life of the animal.

The DNA codes for the production of proteins specific for the virus. Some of these proteins can be detected within the serum of the cat. Enzyme-linked immunosorbent assay (ELISA) tests have been developed to detect these proteins. These tests are available in self-contained kits and can be performed in the practitioner's office.

Feline leukemia virus can eventually result in fatal diseases. Many cats can live for long periods while infected with FeLV. However, the virus can result in a suppressed immune system. As a result, the cats are more susceptible to other bacterial and viral infections. FeLV may also suppress the bone marrow, resulting in **anemia**. Infected cats can also develop tumors of the immune system or bone marrow.

Viruses are classified by several distinguishing features. As already discussed, the type of nucleic acid is the first major classification. Other features include how large a nucleic acid is included, the shape of the capsid, and the type of host infected (Figure 14–6). Some viruses require a vector to transmit them between hosts. This characteristic is also included in the classification.

The third class of infectious microorganisms is called fungi (singular is fungus). Fungi have eukaryotic cells. They have cell walls, as do plants. Fungi, however, do not have chlorophyll, the substance that plants use to capture the energy from the sun (Figure 14–7). Fungi gain their nutrition by absorbing nutrients from the surroundings. Fungi can use inorganic nitrogen to create protein.

Most fungi are spore producers. The spores provide the ability to spread the organism to other areas and also to survive harsh conditions. Fungi have an appearance of filaments. Each filament is called a hypha (plural is hyphae). Cells from the end of the hyphae produce an asexual spore. The spore in this case is a cell budding off the end of the filament. These spores are very resistant to drying. Once adequate moisture is present, these spores germinate into an active growing fungus. Sexual spores can also be produced when two specialized hyphae fuse and combine genetic material. The sexual spores tend to be much more resistant to heat than are the asexual spores.

Fungi can affect the health of animals in several ways (Table 14–3). A common fungal infection is called ringworm. It is important to realize that ringworm is caused by a fungus and has nothing to do with worms! The infection often appears in a circle and expands outward. In humans, ringworm appears as a red circle. It is this presentation that gave the disease its name.

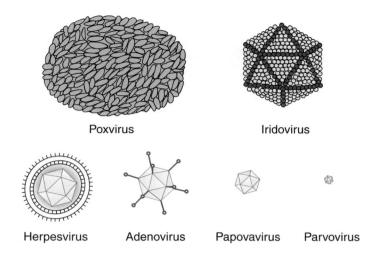

Poxvirus   Iridovirus

Herpesvirus   Adenovirus   Papovavirus   Parvovirus

DNA Viruses

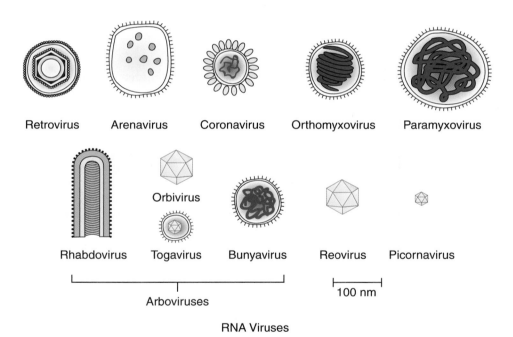

Retrovirus   Arenavirus   Coronavirus   Orthomyxovirus   Paramyxovirus

Orbivirus

Rhabdovirus   Togavirus   Bunyavirus   Reovirus   Picornavirus

Arboviruses   100 nm

RNA Viruses

FIGURE 14–6 Examples of the shape of different viruses.

In animals, ringworm often has a very crusty, flaky appearance (Figure 14–8).

Several different fungi are capable of causing ringworm (such as species of *Microsporum* and *Trichophyton*). These fungi are often introduced into the skin through abrasions. They invade the outer layers of the skin and work into the hair follicles. Damage to the hair follicles results in loss of hair in the infected region. A special culture medium is available to detect ringworm. Hair from infected regions is placed on the medium. The fungus absorbs nutrients from the medium and grows. A color reaction occurs to help provide the diagnosis of ringworm (Figure 14–9).

Fungi can also infect the internal organs. The fungi can enter through the respiratory or the intestinal tracts. The infection can begin in those areas, but it is then possible for it to spread to other regions. Any organ system can become involved. The skin, eyes, bones, and lymph tissue are all common sites of infection. These **systemic** (affecting the entire body) infections are much more

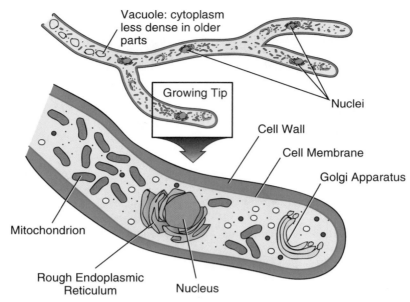

FIGURE 14–7 Illustration of a fungus.

common in animals with suppressed immune systems. The same occurs in human medicine, with patients who are immune suppressed following organ transplantation or as a result of diseases such as acquired immunodeficiency syndrome (AIDS).

Other fungi that do not directly infect animals may also cause diseases. Fungi can grow on poorly stored feeds. Some of these fungi can produce toxins (mycotoxins) that are harmful to the animals consuming these feeds. In very high levels, these mycotoxins can result in sudden illness and even death. Often the problems result from smaller levels over long periods. One toxin (zearalenone) can mimic the hormone estrogen. Swine are quite sensitive to this toxin. Affected pigs have swollen vulvas, enlarged mammary glands, and decreased fertility.

The final class of infective organisms is parasites. There are many different types of parasites. Parasites may be single-celled organisms or much larger arthropods that are visible with the naked eye. Parasites may

---

TABLE 14–3 Common Fungal Infections

Disease	Blastomycosis	Valley fever	Sporotrichosis
Species	Dogs	Dogs, cats	Dogs, cats
Causative organism	*Blastomyces dermatitidis*	*Coccidioides immitis*	*Sporothrix schenckii*
Signs	General signs of anorexia, fever, weight loss, usually begins with a severe pneumonia and may spread to lymphatics, eyes, bone	Most prevalent in Southwest U.S., may infect the skin with lumps and abscesses, also infects lungs, eyes and bone	Usually introduced into skin by trauma, which causes nodules, and these lumps in skin often drain
Disease	Histoplasmosis	Cryptococcosis	Ringworm
Species	Cats, dogs	Cats, dogs	All
Causative organism	*Histoplasma capsulatum*	*Cryptococcus neoformans*	*Microsporum* sp. and *Trichophyton* sp.
Signs	Often begins as respiratory infection with dyspnea, weight loss, anorexia; may also invade intestinal tract with diarrhea, weight loss	More common in cats, may affect upper airways with sneezing and discharge, open mass in skin, also affects eyes	Superficial infection of skin, areas of hair loss, with crusting and flaking

Note: This is only a partial listing of fungal diseases found in domestic species. A complete list is beyond the scope of this text.

FIGURE 14–8 A Holstein heifer with ringworm.

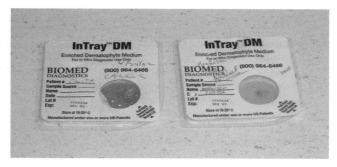

FIGURE 14–9 A culture for ringworm. Note the fuzzy growth of fungus on the culture media.

be within the body (internal) or on the surface of the body (external). In general, parasites use the host animal as their source of nutrients and protection. This description may sound very much like the symbiotic relationship that describes the rumen bacteria and the cow. The difference is that the host is actually harmed by the parasite (Table 14–4).

Parasites can damage the host in several ways. Some parasites actually compete for nutrients. For example, parasites living in the intestinal tract can utilize the nutrients before the host has the opportunity to absorb them. The parasites can also damage the intestinal mucosa, hindering the ability of the host to absorb nutrients. Certain intestinal worms and external parasites actually consume the blood of the host. If the number of parasites is excessive, they can consume such a large amount of blood that the host becomes anemic. Parasites may also have stages that migrate through the body, resulting in further tissue damage.

Many parasites have complex life cycles that allow them to survive and infect other animals. In many instances this requires more than one host to complete the life cycle. The animal that carries the adult of the parasite is called the definitive host. If another animal is required to transmit the immature stages of the parasite, it is called the intermediate host.

The tapeworm (*Dipylidium caninum*) in dogs and cats offers an example of a life cycle requiring an intermediate host. The tapeworm is a flatworm joined together in small segments (Figure 14–10A and B). The head of the worm attaches to the lining of the intestine. Over time, individual segments containing eggs are passed in the feces. Owners often see these segments on the hair coat around the anus of the animal. The individual segments are flat, often moving, and the size of a rice grain.

These segments rupture, releasing the eggs. The larvae (immature stages) of a flea consume these eggs.

As the flea matures, it jumps onto the skin of another animal. As this animal grooms itself or bites at the skin, the flea is consumed. The tapeworm is released from the flea and establishes itself in the new host. In this life cycle the dog or cat is the definitive host and the flea is the intermediate host. The definitive host contains the stage in which the parasite is sexually mature and able to produce eggs to infect others. The intermediate host is infected with larval stages.

The tapeworm is just one example of the flat-segmented parasitic worms (Cestodes) that infest animals. Another type of flatworm is not segmented (Trematodes). These worms are also called flukes. Many flukes leave the intestinal tract to invade the liver or lungs. A third type of worm is the roundworm (Nematodes). Roundworms commonly invade the intestinal tract. In dogs the heartworm is a roundworm that invades the bloodstream. This parasite requires a mosquito as the intermediate host. The mosquito ingests blood from one dog and then infects another dog with a subsequent bite.

Another common parasite is a single-celled organism called protozoa. Protozoa lack a cell wall and obtain their food through phagocytosis. Many types of protozoa are parasites in the intestine or blood of animals. Intestinal protozoa often form a cyst, a structure that is resistant to dry conditions. The cyst is usually passed in the feces, which can then be consumed by other animals.

In general, intestinal parasites pass an infective form from the host in the feces. These may be eggs, cysts, or an immature stage of the worm (larva). To carry the infection to other animals, the parasite relies on an intermediate host (such as the tapeworm) or some means of fecal contamination that is ingested by another host.

In grazing animals (such as horses, cows, sheep, and goats) it is obvious how easily an infection can be spread. Eggs passed in the feces of an infected animal contaminate the ground and plants on the pasture. Rain can contribute to spreading the eggs to new

---

TABLE 14–4 Common Internal Parasitic Infections

Roundworms of dogs and cats
CAUSED BY: *Toxocara canis, Toxocara cati,* and *Toxascaris leonina*
CLINICAL SIGNS: diarrhea, vomiting, pot-bellied appearance, dull coat, poor weight gain, coughing
LIFE CYCLE: Ingested eggs develop into larvae in the intestinal tract. The larvae migrate through the liver and lungs. Eventually migrate up the trachea and are swallowed to develop into adults. Eggs are then shed in the feces.
Four means of transmission exist: 1. Larvae can cross from the mother to the fetus through the placenta (*Toxocara canis*). 2. Larvae can pass through the milk to infect newborns. 3. Animal can ingest the eggs with feces contamination of food. 4. Ingestion of certain animals carrying the larvae (e.g., rodents, rabbits).

Hookworms of dogs and cats
CAUSED BY: *Ancylostoma sp.* and *Uncinaria sp.*
CLINICAL SIGNS: diarrhea, anemia, anorexia, weight loss, may cause death
LIFE CYCLE: Four means of transmission exist: 1. Ingestion of larvae. 2. Across the placenta. 3. Through the milk. 4. Larvae may penetrate through the skin of the foot pads to enter the host.
The larvae enter the intestinal tract and develop into adults. The adults attach to the lining of the intestines and suck blood from the host. Adult hookworms produce eggs which are passed in the feces. These eggs develop into larvae, which are then infective to new hosts.

Whipworms of dogs
CAUSED BY: *Trichuris vulpis*
CLINICAL SIGNS: chronic diarrhea often with blood or mucus, weight loss, dehydration
LIFE CYCLE: Eggs are passed in the feces. Larvae develop within the eggs, which are then ingested. The larvae hatch out of the egg and develop infection in the intestines. Adults then shed eggs.

Tapeworms of dogs and cats
CAUSED BY: *Dipylidium caninum, Taenia pisiformis, Taenia taeniaeformis*
CLINICAL SIGNS: Usually minimal, severe infections may cause weight loss or low blood sugar in small young animals. Often a concern to owners because the segments are visible.
LIFE CYCLE: Requires intermediate host (Dipylidium—flea, Taenia—rabbits, rodents). The intermediate host eats the eggs; the definitive host (dog or cat) eats the intermediate to ingest larvae. The larvae establish infection in the intestines. The adult tapeworm forms; segments or eggs are passed in the feces. These are consumed by the flea or rodent.

Heartworm of dogs
CAUSED BY: *Dirofilaria immitis*
CLINICAL SIGNS: coughing, exercise intolerance, fluid accumulation in abdomen (ascites), cardiac or respiratory failure
LIFE CYCLE: Adult worms live in the major vessels close to and in the heart chambers. Adult worms produce larvae (called microfilariae) that travel in the blood stream. A mosquito picks up the larvae when consuming a blood meal from the dog. The larvae mature in the mosquito (the intermediate host). At this stage if the mosquito bites another dog, the larvae are introduced. The larvae finish their development in the dog, a process requiring up to five months. Adult worms then mate and produce more microfilariae. The entire cycle requires six to seven months for microfilariae to be detectable in the blood.
Diagnosis of heartworm disease requires a blood test. Tests are available to detect the microfilariae, or antigens released from the adult worms.

Note: This is only a partial listing of the parasitic infections found in domestic species. The complete list is beyond the scope of this text.

---

areas. The next host then consumes the egg by eating contaminated plants. Dogs and cats are also infected in similar ways. The pet walking on contaminated soil gets eggs on the skin or footpads. As the pet grooms itself, it can become infected as well. It is important to note that some intestinal parasites of dogs and cats can cause infections in humans. Young children who place things in their mouth are at highest risk. Children playing in contaminated soil can also develop infections of these parasites. This risk emphasizes the importance of controlling parasites in pets.

Animals are susceptible to external parasites as well. These external parasites are all arthropods. Arthropods are the class of organisms that includes insects and spiders. There are a tremendous number of arthropods in the world. They constitute the largest group of organisms. Only a relatively small number are parasites in domestic species. These external parasites include ticks, fleas, mites, lice, mosquitoes, and biting flies (Figure 14–11). As already mentioned, these parasites can be responsible for transmitting other parasites and diseases.

TABLE 14–4 Common Internal Parasitic Infections—cont'd

Strongylosis of horses
CAUSED BY: *Strongylus sp.* (numerous species exist)
CLINICAL SIGNS: colic, weight loss, diarrhea
LIFE CYCLE: Eggs passed in the feces develop into larvae. The larvae crawl up blades of grass, which are then ingested by grazing horses. The larvae migrate through the blood stream of the horse. Damage done to the blood vessels can result in colic. At a stage in their development the larvae enter the large intestines and mature into adults. The adults then pass eggs in the feces.

Roundworm infection of horses
CAUSED BY: *Parascaris equorum*
CLINICAL SIGNS: coughing, anorexia, poor weight gain; large numbers can cause impaction and obstruction of the intestines
LIFE CYCLE: The horse ingests eggs, which then release a developing larva. The larvae migrate through the liver and lungs. Eventually the larvae are swallowed, where they develop into adults in the small intestine.

Bot infection in horses
CAUSED BY: *Gasterophilus intestinalis* and *Gasterophilus nasalis*
CLINICAL SIGNS: mild stomach irritation, often no clinical signs
LIFE CYCLE: see text portion of this chapter

Trichostrongyles of ruminants
CAUSED BY: species of *Hemonchus, Ostertagia, Trichostrongylus, Cooperia, Bunostomum*
CLINICAL SIGNS: diarrhea, weight loss, low blood protein resulting in fluid accumulation under the skin (e.g. bottle jaw—the accumulation of fluid, edema, under the jaw)
LIFE CYCLE: These parasites have a direct life cycle. Eggs passed in the feces contaminate pastures and are then consumed. The larvae mature in the intestines and eventually produce adults that release eggs into the environment. The larvae of different species do migrate into tissues to varying degrees.
   All of the developing larvae have the ability to enter a stage of arrested development when conditions are bad (e.g. during winter). The larvae halt development until conditions improve (e.g. spring) and then mature into adults.

Coccidiosis in ruminants
CAUSED BY: numerous species of protozoa, many *Eimeria sp.*
CLINICAL SIGNS: chronic diarrhea, rough hair coat, poor weight gain
LIFE CYCLE: An egg, called oocyst, is passed in the feces. The oocyst is consumed by another animal. Immature stages are released from the oocyst. These immature stages penetrate the lining of the intestine and go through several stages of development. Each step damages lining cells of the intestines and is responsible for the clinical signs. Once mature, the coccidia releases more oocysts to continue the life cycle.

One insect, the botfly, is also the source of an internal parasite. The botfly lays eggs on the legs of horses. As the horse licks these eggs, larvae hatch and penetrate into the gums and tongue of the horse. As development continues, the next stage of larvae move down the esophagus and enter the stomach. Large numbers of these larvae may develop in the stomach, causing irritation to the lining (Figure 14–12). Each larva can be up to 20 mm in length. The larvae develop in the horse for 10 to 12 months and then pass in the feces, usually in the spring. The larvae then develop into the botfly, which lays egg to continue the life cycle.

Fleas may be the most common external parasite seen in veterinary medicine. The flea is a blood-sucking insect that can cause significant skin irritation, transmit tapeworms, and cause anemia. Small kittens and puppies may develop a life-threatening anemia when infested with a large number of fleas. The eggs of fleas develop in the environment, off the infected animal. This makes control difficult, because eliminating all the fleas from the animal does not prevent reinfection.

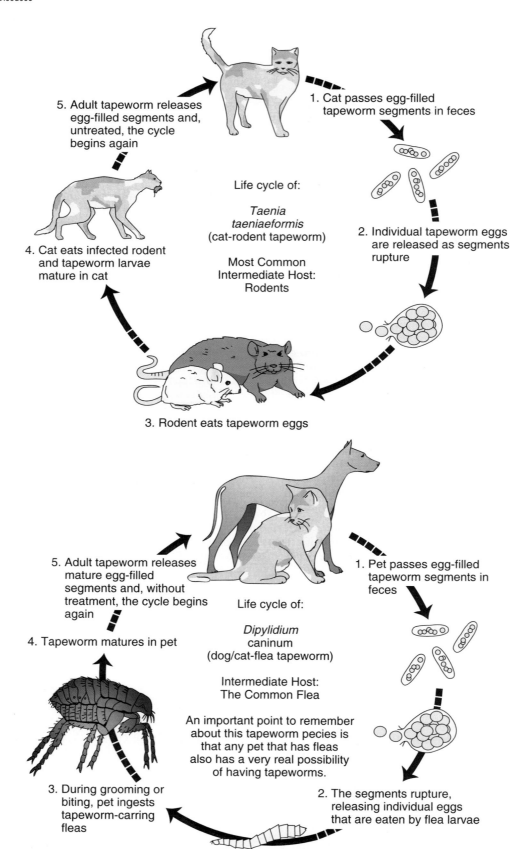

5. Adult tapeworm releases egg-filled segments and, untreated, the cycle begins again

1. Cat passes egg-filled tapeworm segments in feces

Life cycle of:

Taenia taeniaeformis (cat-rodent tapeworm)

Most Common Intermediate Host: Rodents

2. Individual tapeworm eggs are released as segments rupture

4. Cat eats infected rodent and tapeworm larvae mature in cat

3. Rodent eats tapeworm eggs

A

5. Adult tapeworm releases mature egg-filled segments and, without treatment, the cycle begins again

1. Pet passes egg-filled tapeworm segments in feces

Life cycle of:

Dipylidium caninum (dog/cat-flea tapeworm)

Intermediate Host: The Common Flea

An important point to remember about this tapeworm pecies is that any pet that has fleas also has a very real possibility of having tapeworms.

4. Tapeworm matures in pet

3. During grooming or biting, pet ingests tapeworm-carring fleas

2. The segments rupture, releasing individual eggs that are eaten by flea larvae

FIGURE 14–10 *A.* The life cycle of two tapeworms that commonly infect dogs and cats. *(Reprinted with permission of Miles, Inc., Animal Health Products.)*

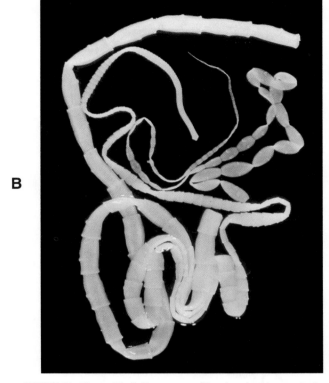

B

FIGURE 14–10 cont'd *B.* Tapeworm. *(Reprinted with permission of Miles, Inc., Animal Health Products.)*

Flea shampoos can be very useful in ridding the animal of existing fleas. The effect of the shampoo is eliminated once the animal is rinsed. If fleas exist in the environment, they will quickly reinfest the animal. Many forms of insecticides exist to kill fleas. Flea sprays, dips, collars, and powders have all been helpful in controlling flea infestations. Unfortunately, the treatment had to be consistent to prevent reinfection. Recent advances have provided treatments that can be effective for a month or longer. These products have been a tremendous aid in controlling flea problems.

Many pets are allergic to fleabites. As the flea sucks blood from the animal, some of its saliva enters the skin. If the pet is allergic to the saliva, scratching, biting, and skin irritation result. The animals often lose hair in these areas, and the skin can develop a bacterial infection. Large numbers of bacteria live on the surface of healthy skin. Once the skin becomes irritated, the bacteria can enter and establish an infection. The infection in this situation is not what started the problem, but it does make it worse. The flea is the primary problem and must be controlled. The bacterial infection is called a secondary infection.

Ticks are blood-sucking external parasites. Ticks have gained notoriety for transmitting **Lyme disease** in humans. Lyme disease is a bacterial infection that can

result in many disease symptoms in humans, including fatigue and joint pain. The same disease may infect pets as well. In addition, ticks can transmit a number of bacterial, viral, and protozoal diseases. It is the concern of these other diseases that makes control of ticks so important.

Numerous insecticides are available for controlling ticks. Just as in fleas, long-lasting products have been developed. The tick attaches to the skin for long periods, engorging itself with blood. Once full, the tick detaches and falls from the animal. It is during attachment that diseases can be transmitted. Even with some of the long-lasting control products, it is still possible to find ticks on the animal (the product kills the tick but does not repel it). The important benefit of these products is that the tick is killed before it has time to transmit the disease.

Many chemicals exist that hinder the growth or kill microorganisms. **Antimicrobial** is the general term for these types of agents. A natural antimicrobial is called an antibiotic. The suffix *–cidal* describes antimicrobials that kill microorganisms, whereas *–static* describes those that slow the rate of growth of microorganisms. For example, a germicide is capable of killing a variety of microorganisms. An antibiotic that only slows the rate of growth of bacteria is called **bacteriostatic**.

Antiseptics and **disinfectants** are two types of germicides. Antiseptics are germicides that may be used on the skin of animals. These products are not mild enough to be taken orally. Disinfectants are too harsh to be used safely on the skin and are used as germicides on inanimate objects. An antiseptic soap is used to kill organisms on the skin, whereas a disinfectant would be used to clean the examination table in the veterinarian's office.

Antibiotics are products produced by one microorganism that kill or slow the growth of another microorganism. (Modern technology has allowed for the production or synthesis of new antibiotics without the need for a microorganism.) Penicillin was one of the first antibiotics discovered. Dr. Alexander Fleming, a physician doing research in London, first wrote about penicillin in 1929. Dr. Fleming noticed that bacteria failed to grow around a mold that had accidentally contaminated his cultures. Dr. Fleming was able to isolate a product produced by the *Penicillium* fungus, which he called penicillin. Dr. Fleming showed how penicillin was able to kill many bacteria. It was almost 10 years until penicillin was produced in quantities capable of being used to treat humans and animals. Penicillin is credited with saving many lives of soldiers wounded in World War II.

Antibiotics are capable of killing bacteria but not viruses. Many antibiotics are only effective on a limited range of bacteria. For example, penicillin is most effective for the Gram-positive bacteria. Penicillin

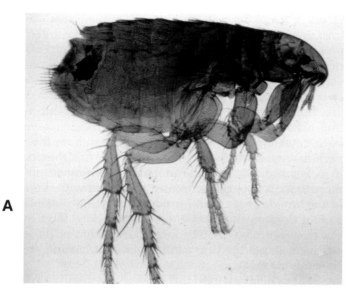

A

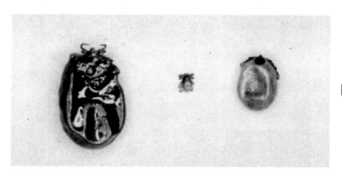

B

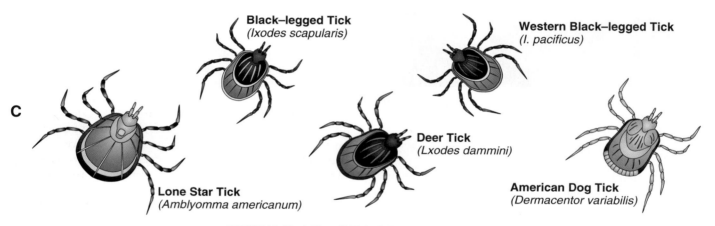

C

Black–legged Tick
(Ixodes scapularis)

Western Black–legged Tick
(I. pacificus)

Lone Star Tick
(Amblyomma americanum)

Deer Tick
(Lxodes dammini)

American Dog Tick
(Dermacentor variabilis)

FIGURE 14–11 *A.* Flea. *B.* Tick. *C.* Several species of ticks.

FIGURE 14–12 Larvae of the horse bot fly (*Gastrophilus*) attached to the stomach lining of a horse.

blocks the formation of units of the cell wall that are specific for Gram-positive organisms. *Broad spectrum* describes an antibiotic that is effective against a large variety of bacteria. Broad-spectrum antibiotics are effective against many Gram-positive and -negative organisms. No antibiotic has been found that can kill all the different bacteria. Selection of the appropriate antibiotic is important in treating infectious diseases.

CLINICAL PRACTICE

OBJECTIVE

■ *Relate Text Material to Common Presentations*

Culturing is the technique used to isolate and identify microorganisms causing infectious diseases. Because microorganisms are prevalent in the environment, it is important that proper sampling and handling techniques are used to isolate the actual organism causing the disease. For example, milk is often cultured to identify the bacteria causing mastitis. The skin of the teat must be thoroughly cleaned before obtaining the milk sample. Otherwise bacteria present on the skin will contaminate the sample. From contaminated samples, many different bacteria will grow in the culture and the actual disease-causing organism may be missed.

It is not practical to swab the nose of a horse with pneumonia to identify the source of the infection. Many bacteria that normally inhabit that region would be detected. The sample must be obtained from deep in the trachea or bronchi. A transtracheal wash is often used to obtain these samples. In this procedure a special catheter (the catheter has a needle and a long plastic tubing that can be threaded into the trachea) is slipped between the rings of the trachea (Figure 14–13). The tubing is passed into the trachea, and a small amount of saline (0.9 % sodium chloride in water) is pushed through the catheter. The liquid is then drawn back into a syringe. This saline is then used as the source for the culture.

The sample is added to a medium (solid or liquid) that provides all the necessary nutrients to allow for the growth of bacteria. The nutrients include sugars, amino acids, vitamins, and minerals. Culturing anaerobes requires that oxygen be eliminated from the surrounding air by adding other gases such as carbon dioxide. Many different media exist. Differences in the nutrients added allow for specific organisms to grow.

Selective media favor the growth of one type of organism. By using a selective medium, an organism that is typically found in small numbers can be isolated by slowing the growth of the predominant bacteria. For example, *Salmonella* bacteria can cause a serious diarrhea in animals and humans. Feces have a tremendous number and variety of normal bacteria. By culturing feces on a selective media, disease-causing organisms such as *Salmonella* can be isolated.

Once the sample is applied to the culture medium, it is incubated at 37°C. When using a solid medium, the sample is streaked across the surface. The goal is to distribute individual bacteria on the medium (Figure 14–14). Each bacterium can then replicate, using the nutrients supplied. A single bacterium can divide to such an extent that within 12 to 24 hours, it has produced a colony that is visible to the naked eye (usually about the size of a pin head). Some bacteria are much slower growing and require much longer culture times. The appearance of the colony is distinct for each type of bacteria. When applying a specific amount of sample (for example, 0.01 ml), the number of colonies can provide an estimate of the concentration of bacteria within the sample.

Once a specific bacterium is isolated, antibiotic sensitivity testing can be performed. The goal of this testing is to identify the appropriate antibiotic for this strain of bacteria. Again *Salmonella* offers an excellent example of why sensitivity testing is performed. Although the same organism is causing the disease on two different farms, it does not mean that the same antibiotic will be effective in each case.

Small disks saturated with antibiotics are placed on a culture plate coated with bacteria. The antibiotic within the disk diffuses out into the culture medium. If the antibiotic is effective, a ring will develop where there is no bacterial growth. The size of this ring is influenced by several factors: the sensitivity of the

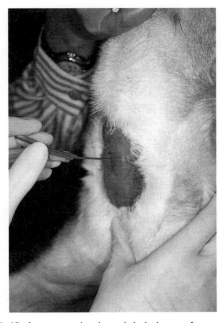

FIGURE 14–13 A transtracheal wash is being performed on a dog. The catheter inserted into the trachea is used to deliver and recover a small amount of saline. This fluid can then be cultured.

FIGURE 14–14 A bacterial culture on a plate with three selective media. The growth pattern helps to identify the type of bacteria present. The other plate shows the organism being tested for antibiotic sensitivity. A clear zone appears around the small disks when the organism is sensitive to the antibiotic on the disk.

organism, the concentration of the antibiotic, and how it diffuses on the culture medium. Because other factors exist, it is important to interpret the size of the ring based on the specific antibiotic in question (that is, a disk with a small ring may be the most effective because it diffuses poorly and is being used at a very low concentration).

Choosing the appropriate antibiotic is important in the success of treatment. Plasmids within bacteria can carry the genetic code that provides the resistance to a given antibiotic. Treating with an inappropriate antibiotic will destroy many bacteria, leaving the resistant bacteria to thrive. This can make subsequent treatment even more difficult. The plasmids can also be exchanged between other bacteria, creating even more resistant strains.

For treatment to be successful, several important factors must be considered. Choosing the correct antibiotic is only the first step. In addition, the antibiotic must be used at a dosage that delivers adequate amounts of the drug to the affected tissue. For example, antibiotics often do not penetrate into bone tissue very easily. In cases of bone infection, the dosage of antibiotic is often much higher than if the same infection was present in the skin. A third factor is the length of time that the antibiotic is used. Some bacteria may be only slightly resistant to the antibiotic and can be destroyed by treating for long enough time periods. If treatment is stopped too soon, the mildly resistant organisms can survive and continue the infection. The next time treatment is required, this infection may be more difficult to cure, because it stemmed from these resistant organisms.

Proper usage of antibiotics and client cooperation is a very important issue in veterinary and human medicine. It is not unusual for clients to see significant improvement in their pet's condition and stop giving the antibiotic. The infection can then return. It is stressed to clients that they must use the entire course of antibiotics that is prescribed.

Antibiotics are not effective against viral infections. Administering antibiotics to these animals may destroy many of the normal bacteria and allow resistant bacteria to survive. This inappropriate use of antibiotics can create strains of bacteria that are becoming more difficult to cure. An excellent example is actually the common cold in human medicine. A virus causes the common cold, and antibiotics will not alter the course of the disease. Doctors should not prescribe antibiotics for this type of condition. The challenge arises because a bacterial infection can develop secondary to the virus. Veterinarians and physicians must attempt to determine if an antibiotic is necessary.

Many intestinal worms shed eggs or larvae in the feces. The presence of these eggs is necessary to iden-

tify infected animals. The simplest technique is to smear a small amount of feces on a microscope slide and examine it under the microscope. The drawback of this technique is that only a small amount of feces is used and eggs can easily be missed. Other techniques exist that concentrate the eggs, making a diagnosis more likely. Fecal flotation is a concentration technique that uses the differences in specific gravity between the egg and other fecal material.

Specific gravity is basically a measure of how particles will float. Parasite eggs will not float in tap water. Various salts or sugars (such as table sugar, sodium nitrate, and zinc sulfate) are added to the water to raise the specific gravity of the solution to a point that the eggs will float. Particles in the feces remain too heavy to float and settle to the bottom. Kits are available to make this procedure quick and convenient.

Feces are placed in the container, and the salt solution is added (Figure 14–15). The feces are stirred, and the container is completely filled with solution. Enough solution is added that it actually domes above the surface of the container. A microscope cover slip is placed on top of the solution. Over the next 5 to 20 minutes, the eggs float to the surface and adhere to the cover slip. The eggs are being concentrated from the larger sample into one small region. The cover slip is lifted and placed on a microscope slide. The slide is then examined for the presence of parasite eggs or larvae (Figure 14–16).

It is the egg passed from the adult worm that allows for a diagnosis. There is a period when the ani-

FIGURE 14–15 The steps in doing a fecal flotation on a stool sample. 1. Feces is added to the sample container. 2. A sleeve is slid over the container to hold solution. 3. Flotation solution is added, and a sieve is placed to trap large particles. 4. The tube is filled completely, and a cover slip is placed on top of the solution. With time the eggs float to the surface and adhere to the coverslip. The coverslip is examined under a microscope.

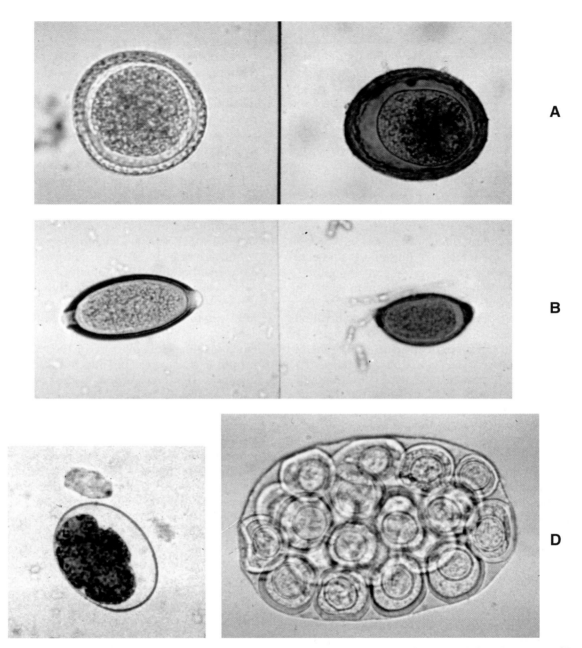

FIGURE 14–16 Parasite eggs commonly found on fecal flotation. These eggs are used to diagnose an infection. *A.* Roundworm eggs (*Toxocara canis,* left; *Toxascaris leonina,* right). *B.* Whipworm, left (*Trichuris vulpis*), lungworm, right (*Capillaria aerophila*). *C.* Hookworm (*Ancylostoma*). *D.* Tapeworm (*Dipylidium caninum*).

mal has developed an infection, but adults have not yet developed. This is called the prepatent period. This is significant because there are times when clinical judgment allows us to treat an animal for the potential of having a parasitic infection, even though a positive diagnosis cannot easily be made.

The cow from the introduction, suffering from coliform mastitis, was showing severe signs secondary to a bacterial infection in the udder. The bacteria gain en-

trance into the udder through the teat canal and begin to rapidly divide, using the milk as a source of nutrients. The cow's immune system attacks the bacteria, and many begin to die. Coliform mastitis is caused by Gram-negative bacteria, which contain endotoxins in their cell wall. The sudden death of large numbers of bacteria releases a large amount of endotoxin.

The endotoxin causes many signs in animals. Typically the cow shows signs of fever, rapid heart rate, diarrhea, and an inactive rumen. Treatment usually includes fluid therapy and medications that block the effects of the endotoxins. A very useful technique in treating cows with coliform mastitis is frequent milking of the infected quarter. By removing the milk every 1 to 2 hours, much of the endotoxin is eliminated before it can be absorbed into the bloodstream. In addition, many bacteria are physically removed. The combination of frequent milking and treating the effects of the endotoxins keeps the animal supported until its immune system can eliminate the infection.

Botulism, on the other hand, is the result of an exotoxin. With botulism, the animal is not infected by the bacteria but ingests the exotoxin that is produced. Spores of *Clostridium botulinum* are commonly present in the soil. Often the source of the toxin is a decomposing animal carcass that is caught in the animal's hay. The carcass provides an environment in which the spore can germinate and begin dividing. Exotoxin is produced and released into the carcass and surrounding hay.

When ingested, the botulism toxin blocks the release of acetylcholine at the nerve muscle synapse. The result is the profound muscle weakness. In horses the earliest signs are muscle weakness and the difficulty chewing and swallowing. When examined, these animals have very weak tongue tone. If not caught early, the weakness becomes generalized and so severe that the animal is unable to rise.

When botulism is this severe, the horse may die from dehydration or from paralysis of the respiratory muscles. Beginning treatment early in the disease is very important in improving the outcome. The specific treatment is botulism antitoxin (antibodies specific against the toxin). The antitoxin is extremely effective against any circulating toxin but does not reverse the effect of any toxin that has already bound at the synapse. Therefore the more quickly the treatment is initiated, the better the results.

Supportive treatment is also essential. Supplementing water and feed is necessary in those horses that cannot eat. Fluids can be given intravenously through a catheter. A slurry of protein-rich food can be pumped into the stomach through a tube passed through the nose and esophagus. Down horses must also be rolled side to side to prevent pressure damage in the muscles and nerves from lying too long.

Cats that spend a portion of their time outdoors have the potential to encounter other cats. Often this can result in a fight between or among animals. This is especially true of male cats that are fighting to protect their territory. A bite can introduce bacteria that are common in the mouth of cats or introduce bacteria that are present on the skin and hair of the bitten animal.

An abscess, the accumulation of a pocket of pus, is a common result. *Pasteurella multocida* is a very common organism that causes these catfight abscesses. This type of interaction can also introduce viruses (such as feline leukemia virus). The viruses do not contribute to the abscess but can result in a more serious infection. Treatment of the abscess does generally include antibiotic therapy. A number of antibiotics are effective for the common bacteria found in abscesses. Recovery is also speeded by draining the pus from the abscess. This may be done through the initial bite wound or by making a new incision.

In large abscesses, a drain may be placed. A Penrose drain is soft rubber tubing that is placed in the abscess cavity and exits through the skin. The purpose of the drain is to keep the opening from healing too quickly. As more pus accumulates, it is able to drain out of the hole kept open by the Penrose drain. As the amount of discharge declines, the drain can be removed.

SUMMARY

Koch's postulates lay the foundation for investigation of infectious disease. The following four statements paraphrase these postulates: (1) The infectious agent should be detectable in sick animals but not healthy animals. (2) It should be possible to isolate and culture the organism. (3) Organisms taken from the culture and introduced into a healthy animal should cause the same disease. (4) The same organism should be isolated from this second animal as well. Knowing these postulates and possessing understanding of the four major infectious disease agents (bacteria, virus, fungi, and parasites) help veterinarians identify, prevent, and treat infectious diseases.

REVIEW QUESTIONS

1. Define any 10 of the following terms:

coliform	prodromal phase
botulism	anemia
anthrax	systemic
Koch's postulates	Lyme disease
fomite	antimicrobial
vector	bacteriostatic
eukaryotic	antiseptics
prokaryotic	disinfectant

2. True or False: Horses typically have poor muscle tone in their tongues.

3. True or False: Bacteria are multicelled organisms

4. True or False: Fungi contain chlorophyll.

5. Transduction requires the transmission of bacterial DNA by a _____.

6. The suffix *–cidal* means the antimicrobials that _____ microorganisms.

7. A Gram-_____ bacterium causes coliform mastitis.

8. How many principles are found in Koch's postulates?

9. Give an example of an arthropod.

10. What causes ringworm?

11. What external parasite transmits Lyme disease?

12. Is botulism the result of an endotoxin or an exotoxin?

13. Who developed the commonly used bacteria staining technique?

14. List the three types of anthrax.

15. List the four major classes of infectious disease-causing agents.

ACTIVITIES

Materials needed for completion of activities:
   tryptic soy agar media plates
   sterile cotton tipped applicators
   incubator at 37°C (local veterinarians may be able to incubate the plates)
   assorted disinfectants
   hole punch
   tweezers
   filter paper
   marker or wax pencil
   incubated cultures from step 1 or 2 or prepared bacterial cultures (allow for a known Gram-positive or -negative organism to be used)
   microscope (with oil immersion lens)
   immersion oil
   sterile swabs
   microscope slides
   Gram stain kit
   Bunsen burner
   tongs for holding slides
   latex or vinyl gloves
   protective eyewear

1. This experiment is designed to culture bacteria from humans and the natural environment. Take a sterile, cotton-tipped applicator and swab the area being investigated. The swab should be wet when swabbing dry surfaces. Potential surfaces include skin, mouth, under the fingernails, desktops, soil, and soles of shoes, or other surfaces if approved by the instructor. Use the swab to inoculate the surface of the culture media. The swab should be held at approximately a 45-degree angle and placed gently against the media. Too much pressure will cut the surface of the media. The swab should be taken in a zigzag fashion across one half of the plate. This swab should then be discarded (Figure 14–17).

A new sterile swab should then be taken across the already inoculated section one time. Inoculate the remaining half of the plate with this swab in a zigzag manner. By using the second swab, the number of bacteria in the second half will be significantly lower. The goal is to spread the bacteria, so that an individual cell can produce a colony. If the tested surface is highly contaminated, the colonies may grow too close to identify individual types.

When a single bacterium produces a colony, it will have a distinct appearance. The colony will be consistent for each type of bacteria. Colonies can be judged on size (tiny or pinpoint, small, medium, large, or very large), shape (round, irregular, spreading), color (white, yellow, creamy, red, metallic) and cross-section view (flat-topped, domed, concave).

The plates should be labeled with your name and the surface tested. The plates are turned upside

Swab 1

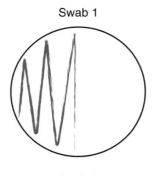

Swab 2

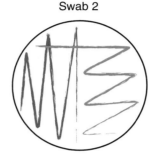

Sensitivity Plate

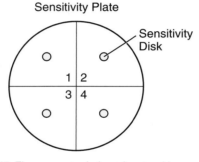

FIGURE 14–17 The proper technique for streaking a culture plate.

down (culture media on the top) and placed in an incubator at 37°C. The media plates can be left at room temperature; however, results may be inconsistent and growth may be poor. The culture plates should be examined at 24 and 48 hours for growth of microbial colonies.

For each surface tested, answer the following questions:
■ How many different colony types can be identified?
■ Describe the shape, color, and size of each type of colony. Sketch each type of colony.
■ Compare the different media plates. Which surfaces seemed to have the largest number of bacteria? Which surfaces seemed to have the largest variety of bacterial types (that is, different colonies)?

2. Individual colonies from question 1 can be used in this experiment to evaluate the effectiveness of various antiseptics. Gather various antiseptics and disinfectants from home. (This may include items such as mouthwash, liquid soap, topical wound sprays, and household cleansers listed as disinfectants).

Identify a single isolated colony from one of the cultures performed in question 1. Take a new sterile swab through this colony. Be careful to only touch one colony. Take this swab and cover the entire surface of a new culture plate. Turn the plate approximately 60 degrees and repeat the swabbing using the same swab. Turn another 60 degrees and repeat. The goal is to evenly distribute the bacteria over the entire surface of the culture plate.

Using a standard hole punch, make numerous circles or disks from a piece of filter paper.

Using tweezers, dip a circle of filter paper in a disinfectant or antiseptic. Hold until all excess has dripped from the paper. Place the filter paper disk on the surface of the newly streaked culture plate. Distribute a number of circles around the perimeter of the culture plate. It is important to divide the bottom of the plate into sections with a marker or wax pencil. Place a number in each section. Make a key to identify each number with the product used (for example, 1 = mouthwash, 2 = hand soap, and so on).

Gently press each disk against the culture media with a probe. Invert the plate and incubate for 24 hours. If the disks do not stick to the surface of the media, the plate can be cultured with the media on the bottom. If incubated in this fashion, it is important to check frequently because the media will dry out quickly in this position.

Examine the plate for bacterial growth. Observe for any clear zone surrounding the individual disks, where no bacteria have grown. Measure each zone and record.
■ Were any of the disinfectants or antiseptics effective in preventing growth of bacteria?
■ Which product produced the largest zone of inhibition?
■ Were the products consistent with all different bacteria?

3. Perform the Gram stain. Take a sterile swab and lightly touch a colony of bacteria. Place a drop of distilled water on a slide. Smear the surface of the slide with the swab. Allow to dry completely. The bacteria must be fixed to the surface of the slide.

This is accomplished by holding the slide with the tongs and passing the slide (bacteria side up) through the flame of the Bunsen burner two to three times. This prevents the bacteria from being washed off the surface of the slide.

Stain the slide following the Gram stain procedure. Specific times and techniques will vary with the stain purchased. The basic procedures are to saturate the surface of the slide with the stain and rinse with water between steps. One step is to decolorize the initial stain. This is what distinguishes Gram-negative from Gram-positive bacteria. Following the staining, examine the slides under a microscope.

■ Is the bacteria Gram negative (pink) or Gram positive (blue)?

■ Are the bacteria cocci or rods?

When dealing with bacterial cultures, wear latex or vinyl gloves and protective eyewear. Wash your hands thoroughly following each procedure, and thoroughly clean laboratory equipment and tables as well.

4. Research a disease and answer the following questions in your report:

■ What are the features of the causative organism (such as Gram stain characteristics, type of virus [RNA or DNA])?

■ How is the disease transmitted? Is it contagious? How long is the incubation period or life cycle?

■ How is the disease diagnosed?

■ What is the treatment for the disease? Is it often fatal?

■ How can it be prevented?

■ Is it a threat to infect humans?

15

Disease Prevention

OBJECTIVES

Upon completion of this chapter, you should be able to:

- *Name the basic components of disease prevention.*
- *Describe the types of vaccines available and their roles in disease prevention.*
- *Link the clinical significance of the academic material learned in this chapter to veterinary practice.*

KEY TERMS

antioxidant	wet dewlap	equine infectious	quarantine
ventilation	biosecurity	anemia	fibrosarcoma
tunnel ventilation			

INTRODUCTION

Treating infectious diseases is a critical part of a veterinarian's life. However, the goal of the profession is to prevent disease. With prevention, the animal and owners are not faced with the losses associated with disease. Many factors must be considered in preventing disease.

A DAY IN THE LIFE

Here We Go Again...

Over the years I have seen this case too many times. I receive a call from a farmer reporting several cows off feed. During the visit, it is obvious a respiratory infection outbreak is spreading through the herd. Often a quarter to half of the cattle show clinical signs. These animals cough, breath heavily, and have nasal discharge. Typically they present with high fevers and very poor appetites (Figure 15–1).

The classic history reveals that a few animals were purchased recently and introduced to the herd. These new cattle do not show any clinical signs of disease. Additionally, the herd has not been recently vaccinated. The purchased animals are typically healthy, but with the stress of transport and a new environment, they shed bacteria or viruses. However, they have adequate immunity because of prior exposure and therefore remain healthy. Unfortunately, the existing herd is not immune and the disease spreads rapidly.

In any herd with such problems, the economic costs can be quite staggering. Expenses include veterinarian examinations, treatment (medication costs can be very high for large animals), death loss, and a decrease in productivity as inflicted animals produce little milk or gain little weight.

Many kennels in our area require dogs to be current on all their vaccines, including kennel cough. The kennel cough vaccine is not one that I recommend for every animal. However, when many dogs are combined in a confined area such as a kennel, the risk increases for kennel cough, therefore causing owners to require the use of the vaccine.

I remember graduating from veterinary school. For the first time, I heard the word *doctor* in front of my name. All the years of hard work had finally paid off. Actually, there was one more step. In addition to obtaining a doctorate, veterinarians are required to meet additional licensing requirements before practicing medicine.

I had already passed the National Board Examination. The majority of states require a minimum standard on this test before permitting the veterinarian to enter practice. Many states have additional requirements beyond the National Board Examination. When I graduated, Pennsylvania required the successful completion of an oral comprehensive examination for all new veterinarians. The candidate entered a room

FIGURE 15–1 Heifer suffering from pneumonia. This heifer has a thick nasal discharge and labored breathing.

and sat before four veterinarians on the licensing board. The candidate could be asked anything about veterinary medicine by the board.

While waiting for my turn to enter the examination, I felt like I was being sent into an interrogation. My mind raced with the thought of bright lights shining into my eyes, sweat beading on my forehead, and the interrogators prying answers from me. Actually, the entire process was quite intimidating. Over the 4 years of veterinary school, a tremendous amount of information was covered. I wondered what were they going to ask.

When I finally entered the room, the board was quite nice. They welcomed me and told me to make myself comfortable. The honest truth was that I was not going to be comfortable until I knew if I had passed. My entire career had come down to the next few questions.

The first question concerned a condition called gastric torsion in dogs. In this situation the dog's stomach twists, trapping gas inside. This is a life-threatening problem, requiring surgical and medical treatment. I had just been through the emergency service rotation, and this disease and treatment were fresh in my mind. At this point I was feeling quite smug. Maybe this oral exam was not so tough after all.

The next question burned in my mind. I was asked to discuss the causes and treatment of wool eating, wet dewlap, and cannibalism in rabbits. My jaw dropped and heart stopped. I'm sure the panel immediately knew I was

continued

in trouble from the expression on my face. Exotic animals were not a significant part of the core veterinary school curriculum. Plenty of opportunity existed for studying this kind of information, but it was elective, and up until that moment, I did not have a lot of interest in the subject.

I honestly could not remember hearing of these diseases before that moment, and I sure did not know any specifics. But one thing that veterinary school does teach is how to use the knowledge that you do have. After a few seconds of stuttering, I began to discuss some very important principles in disease prevention. I generally discussed proper nutrition, proper sanitation,

and minimizing stress. They asked me to explain a few other points about the topics before moving to the next question. (Fortunately, the remaining questions related to familiar material and I was granted a license!)

As I left the room, I did not know if I had passed. I just had to investigate the unfamiliar diseases. Even though I had not learned about these specific diseases, I had learned the important principles of disease prevention. My answers, although more general than the board may have wanted, were quite correct. In this chapter the discussion centers on important principles in preventing infectious diseases.

DISEASE PREVENTION

OBJECTIVE

■ *Name the Basic Components in Disease Prevention*

Many factors contribute to an animal's resistance to disease, such as immunity and nutrition, which have already been discussed in depth in previous chapters. A healthy animal remains resistant to most diseases. The skin and mucous membranes serve to prevent invasion of pathogens. The normal bacterial flora play a role in maintaining the health of an animal. These bacteria secrete substances that inhibit the growth of other organisms. In addition, pathogens must compete for adhesion sites. The pathogen must adhere to the animal's cells before an infection can begin.

It is an obvious conclusion that maintaining the normal bacterial flora helps to maintain the health of an animal. Stress, nutrition, medications, and other diseases influence the normal bacteria. Humans can think of factors that produce stress (for example, a major examination pending). In animals, stress does not require conscious thought. Stress is any factor considered a threat by the animal. An animal under stress releases higher levels of epinephrine and cortisol (see Chapter 10). In addition, stress increases the activity of the sympathetic nervous system.

A wide range of factors can contribute to stress. Examples include overcrowding, competition for feed, extremes of weather, rough handling, noise, and transport (Figure 15–2). Some factors that are designed to help an animal can also contribute to stress. Hospitalization and surgery both stress the animal. Following major surgery, an animal can be more susceptible to contagious diseases.

The increase of cortisol in response to stress has specific benefits. Cortisol helps to increase carbohydrate metabolism and shifts glucose to the brain. The cortisol also minimizes inflammation in damaged tis-

FIGURE 15–2 These individually tied cows have feed delivered directly to them. This feeding method greatly decreases competition.

sue. As a side effect, the elevated cortisol suppresses the immune system, increases the risk of diabetes, and weakens muscle tissue. These signs are quite evident in hyperadrenocorticism (Cushing's disease), where cortisol levels remain chronically elevated.

Nutritional problems can also be considered a stress or may have a direct effect on diminishing the immune response. Many vitamins and minerals (such as vitamin E and selenium) are crucial in maintaining a healthy immune system. All the **antioxidants,** oxidation inhibiting vitamins and minerals, play a role in the animal's immune system. Other deficiencies may weaken the defense of the animal in more general ways. For example, a protein and energy deficiency may weaken the skin and mucous membranes, increasing the likelihood that pathogens will invade.

Sudden changes in diet are also an important stress in animals. A major effect of diet changes is the decline in the normal bacterial flora in the intestinal tract. This greatly increases the risk of pathogen invasion and consequential disease. Medications, primarily antibiotics, pose the same threat to the animal's defense system. When treating a bacterial infection, antibiotics also cause a decline in the normal flora—a particularly high risk if the pathogen has resistance to that antibiotic. This is another key reason why antibiotics should be used only in situations where they are actually required.

Some factors that increase the stress of the animal also contribute to increasing exposure to pathogens. Overcrowding was mentioned as a stress. Overcrowding increases the competition for feed and comfort, in addition to increasing the contact between animals, which aids in the spread of contagious diseases. The high density of animals typically increases the fecal and urine contamination in the environment as well. Humidity is often higher in confined areas, which aids in the survival of pathogens outside the body. Pathogens that are spread through the air have less physical distance to travel when contacting another susceptible individual.

Ventilation describes the exchange of air from within a building to the outside. As animals exhale, they add moisture, heat, and potential pathogens to the air. Ammonia is also released from the urine and feces in the building. Excessive ammonia can cause irritation to the mucous membranes, increasing the risk of a pathogen's invasion of the respiratory tract. Most organisms require some degree of moisture to survive, so exhaled moisture contributes to the success of pathogens.

To help prevent disease, fresh air from the outside needs to be brought into the building and the stale air exhausted. This helps to lower the moisture, heat, ammonia, and pathogen load in the air. Some facilities are built to allow for natural airflow to make this exchange. Other buildings rely on fans to move the air efficiently. **Tunnel ventilation** has become quite popular in recent years (Figure 15–3). In tunnel ventilation the fans are placed at one end of the barn, and all the air inlets are on the opposite end. With the fans running, the air is brought into the barn over a broad area and an even flow of air is swept over the animals. In warm weather, enough air is exchanged to produce a breeze of 3 to 5 miles per hour.

In cold weather, less air can be exchanged. In these conditions, enough body heat must be maintained to prevent freezing within the barn. Even in the coldest weather, at least four air exchanges per hour should be accomplished. This means that the exhausted air should be equivalent to four times the amount of air contained within the building. Unfortunately, in many barns there are areas that do not completely exchange air. These areas can be regions where confined animals are at a higher risk for disease.

In general, older animals have a higher level of immunity than the very young. They have been exposed to more organisms and vaccines over the years. They also have a higher likelihood of being carriers for pathogens. Therefore when designing animal facilities, it is important for the airflow to move from the youngest to the oldest. In this way the young animals are not exposed to the pathogens that could be exhaled from the mature animals. It is ideal that young animals do not commingle with adults for the same reasons. Having separate areas of confinement is very helpful.

Urine and feces not only release ammonia into the air but also contribute pathogens. Proper sanitation and providing clean dry bedding are extremely

FIGURE 15–3 Barn equipped with tunnel ventilation.

important in reducing pathogen load. Physically this helps to lower the number of organisms in the environment. In addition, it helps to keep the animal's hair clean and dry. Hair has a very important role in insulating the animal and conserving body heat in cold weather. When the hair coat becomes wet and matted with feces, the animal loses higher levels of heat. In addition, prolonged exposure to moisture, urine, and feces increases the risk of developing skin disorders (such as **wet dewlap** in rabbits, an infection in the skin of the lower neck).

VACCINES

OBJECTIVE

■ *Describe the Types of Vaccines Available and Their Roles in Disease Prevention*

The previous details discussed are designed to minimize the exposure of the animal to pathogens and to limit any factors that decrease the animal's resistance to disease. Another component of disease prevention is to increase the animal's immunity with vaccination. Much has been discussed about vaccines in Chapter 11. Many factors are involved in determining a vaccination program.

The goals of a vaccination program vary with the type of animal and the disease. Most vaccines do not offer complete disease protection. Properly vaccinated animals can still become ill when overwhelmed with exposure to a pathogen. Therefore vaccination programs must be established with realistic goals in mind. Successful vaccination requires that an effective vaccine be given to an animal that is capable of responding well. In addition, the vaccine must be given long enough before exposure to allow immunity to develop.

Consideration must be given as to what vaccines should be included in the program. The risk of acquiring an infection is one important factor. Certain organisms have a specific geographic location. Therefore vaccination is not considered in areas where the organism is not found. Other factors such as age, sex, and functional purpose of the animal must also be evaluated. Some diseases may only be common in young animals, and some infections are associated with reproductive problems that require females to be vaccinated. Certain diseases are spread from bulls to cows during natural breeding. Artificial insemination and the use of semen from bulls free of these diseases prevent the need for vaccination against these organisms.

The severity of the disease must also be evaluated in terms of medical and economic severity. A disease that is very mild and self-limiting usually does not require vaccination. Diseases such as distemper in dogs

and panleukopenia in cats are so deadly that they are included in all vaccination programs for these species. In food-producing animals a disease may be self-limiting, but if it attacks a large percentage of animals, it may still be a very costly disease, due to loss of production. This type of disease may be included in the vaccination program.

Vaccines vary considerably in how effective they are at preventing disease. Some vaccines only provide a short-lived immunity. Others may only decrease the incidence slightly or may lessen the severity of the disease. The success must be evaluated against the cost of the vaccine. A costly vaccine that has only limited success may not be used. A vaccine may also have adverse effects (such as suppressed milk production in dairy cattle), which is another factor considered in establishing a program.

In companion animals, vaccination programs are established with the goal of preventing the disease in every animal vaccinated. The emotional attachment to pets makes the individual animal of extreme significance. The vaccines are given at frequent intervals in young pets to maximize their protection at an early age. Even the most thorough program may not prevent every case.

In the farm setting, vaccination programs help establish herd immunity. To establish herd immunity, the vaccination program is designed to decrease the number of susceptible animals in an attempt to prevent a herd outbreak. Ideally every animal would develop complete immunity. However, it is quite possible for an individual to develop clinical disease. If herd immunity is established, the disease will not spread throughout the remainder of the animals.

Herd immunity is maximized when all the at-risk animals are vaccinated. The vaccination programs are established to minimize the potential losses experienced with an outbreak of clinical disease. Producers also attempt to keep the cost of the program as low as possible. Therefore only animals with a reasonable threat of exposure are vaccinated, and they are only vaccinated as often as is necessary to maintain immunity.

In review, immunity may be active or passive. Passive immunity is acquired with the transfer of antibodies. This may occur with the ingestion of colostrum by a newborn. A well-vaccinated mother generally has better-quality colostrum. The higher quality of colostrum improves the amount of antibody that is absorbed by the newborn. The amount and how quickly after birth the colostrum is delivered directly influence the passive transfer.

Certain vaccines also provide passive immunity by supplying antibodies. The immunity is only temporary, and therefore the timing of the vaccination is critical. An example is tetanus antitoxin, which may be administered at a time when an animal is at high risk.

This may include a horse with a deep puncture wound or an animal that is being castrated. If the animal was not vaccinated previously for tetanus, these antitoxins have the distinct advantage of delivering immediate immunity.

For an animal to develop active immunity, a vaccine must be administered that allows the animal to produce its own antibodies. Natural infections also create an active immunity. Natural infections often provide the strongest and longest lasting immunity. Vaccines must provide enough antigens to stimulate the immune system. The two major types are killed (commonly virus or bacteria) and modified live vaccines (typically virus). In a killed vaccine, enough antigens are provided for the immune stimulation. The organism within the vaccine is completely inactivated.

With killed products, a primary vaccination initiates the immune response. A booster shot must be given in three to four weeks to stimulate the memory response. It is only after this second dose that effective levels of antibodies develop. Relative to modified live vaccines, the killed products provide a short immunity. Periodic boosters are then required to maintain immunity (which typically ranges from several months to a year).

Modified live vaccines (MLV) stimulate both antibody production and cell-mediated immunity. Initially, MLV deliver a small amount of antigen. The virus then undergoes replication once injected into the animal. The amount of antigen then increases to a level adequate to stimulate the immune response. The virus has been modified so it can replicate without causing

clinical disease. In general, MLV provides a higher level and longer lasting immunity than do killed products. Many MLV do not require a booster vaccination in the 3- to 4-week period, as the killed products do. The immunity may not be indefinite, and a booster is required every 1 to 3 years.

Numerous factors can lead to a failure of a vaccination program. Modified live vaccines require very careful handling to ensure that the organism will replicate once injected. These products come as a dehydrated powder and a liquid to mix with it (Figure 15–4). Once rehydrated, the vaccine should be used immediately. The organism can be killed with extremes of temperature, direct sunlight, and exposure to many disinfectants. Proper storage at controlled refrigerated temperatures is essential. (These factors may also damage killed products.) If the organism does not replicate, there will not be enough antigen present to stimulate an immune response.

With killed vaccines, a failure to give a booster shot also prevents acceptable levels of immunity. The timing of the booster is also important. Administering the booster in less than 2 or more than 8 weeks greatly decreases the level of immunity developed. It is essential to remember that immunity only develops following the vaccination with either type of product. Failure may occur if the animal is exposed before or shortly after the vaccination.

The animal must be able to respond to the vaccine. Very young animals may have levels of maternal antibody that prevent the immune system from responding. The maternal antibodies quickly bind to the antigen and therefore prevent the need for the immune system to respond. Animals that are under stress or are ill may not be able to mount an effective response. Poor nutrition or certain medications also decrease the level of immunity.

Even a well-vaccinated animal with a good level of immunity can be overwhelmed if exposed to high levels of pathogens. In effect, poor management can defeat a good vaccination program. Any failure in management (such as poor sanitation, poor ventilation, or overcrowding) can result in the animal being exposed to such high numbers of pathogenic organisms that the immune system cannot defeat them.

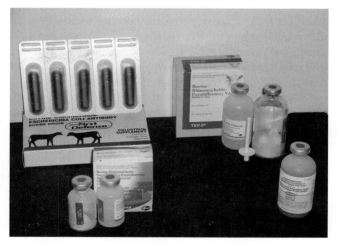

FIGURE 15–4 Examples of vaccine selection (clockwise from lower left): (1) Modified live vaccine. The freeze-dried powder must be reconstituted with liquid before usage. (2) An oral vaccine (bolus) used to provide passive immunity to newborn calves. (3) A modified live vaccine used to administer intranasally. The white nasal canula shown in front of the bottles is applied to the end of a syringe to administer the vaccine into the nasal passages. (4) A killed vaccine that is ready to be administered.

CLINICAL PRACTICE

OBJECTIVE

■ *Link the Clinical Significance of the Academic Material Learned in This Chapter to Veterinary Practice*

Kennel cough in dogs represents an excellent example of how vaccination programs are adapted. Several dif-

FIGURE 15–5 This farm has implemented biosecurity measures to limit the risk of disease entering the farm.

ferent viruses or the bacteria *Bordetella bronchiseptica* may cause kennel cough. Dogs with kennel cough typically present a severe dry hacking cough. Dogs that develop kennel cough generally respond to medication (cough suppressants and possibly antibiotics) and improve within 1 to 2 weeks.

Because the disease is generally self-limiting and the incidence of kennel cough is quite low in the average house pet, routine vaccination is not always required. Kennel cough is highly contagious and is able to be spread through the air or by direct contact. Therefore dogs that are housed in close quarters with others are at a higher risk. This is especially true in a boarding kennel, where dogs from multiple households are brought into close confinement. Because the risk is higher in these situations and the disease can spread so rapidly, vaccination for kennel cough is often required before boarding.

Biosecurity is a term used to describe practices that protect the health of the animals on a farm by preventing the spread of pathogens (Figure 15–5). Ideally these practices prevent the introduction of disease into the herd. For example, horses are often screened for **equine infectious anemia** (EIA), prior to a show or purchase. EIA is a viral disease causing fever, an anemia from the breakdown of red blood cells, depression, and weight loss. Some horses die from EIA; others develop a chronic infection and become carriers of the virus. These horses may appear normal but, when under stress, may shed the virus. These horses can then be a source of infection for other animals on the farm. The testing of animals for infectious dis-

eases before introducing them onto a new farm is an important biosecurity practice.

The other aspect of biosecurity is to minimize the spread of disease if it is introduced onto the farm. The previously mentioned dairy herd with the outbreak of respiratory infection emphasizes the importance of biosecurity. The farm in this case had poor biosecurity. The purchased cattle represented a real threat to the health of the existing herd. The purchased animals had not been screened for any infectious diseases, and the existing herd was not well vaccinated. By recognizing that the purchased cattle were a potential threat to introduce infectious respiratory diseases, the farmer should have vaccinated the herd. Even with proper vaccination, a small number of cases may have developed. The goal of the vaccination program is to establish herd immunity and minimize the spread of the disease.

Diseases can be introduced onto a farm by many vectors. Visitors, trucks (feed, livestock), rodents, birds, and water supply are all potential sources for new diseases. Veterinarians represent visitors to a farm. Veterinarians take precautions by sanitizing boots and hands between farm visits.

The greatest risk of new disease is introducing new livestock. The newcomers have the greatest potential to introduce large enough numbers of pathogens in close contact to the existing animals. An excellent procedure for minimizing this risk is **quarantine**. Quarantine confines the animal in a location that prevents contact with the existing herd. A separate set of utensils (such as pitchforks and feed buckets) should be used only in the quarantine area.

The goal of quarantining is to ensure that the purchased animals are not incubating a disease. The new addition may appear perfectly healthy at the time of purchase but be in the earliest stage of infection. Separating the animals for 2 to 4 weeks helps to minimize this risk. Unfortunately, many farms do not have facilities to keep animals completely separated.

Vaccination programs are developed to maximize protection of the animal. However, consideration must be placed on the cost of the vaccine and the potential for side effects. In farm animals the cost of the vaccine is balanced against the potential for loss if the disease occurs. The effectiveness of the vaccine must also be considered.

Any vaccine has the potential for causing side effects. Soreness and swelling at the site of injection are very common. Many animals also develop mild fevers, lethargy, and poor appetites. In general these side effects are short lived and disappear without treatment. Some vaccines may increase the risk of abortion in pregnant animals. The most serious side effect is anaphylaxis. This allergic reaction may be so severe that it is life threatening. Prompt treatment with epinephrine is essential to reverse the allergic response.

In companion animals, vaccination programs generally begin with a series of vaccines in the young. Recommendations vary significantly, but vaccination often begins in dogs and cats at 6 to 8 weeks of age. Booster vaccinations are given at 3- to 4-week intervals until there is confidence that maternal antibodies have declined to a level that allows the pet to respond (recommendations vary from 12 to 18 weeks of age). The long-held standard is that the pets are then given yearly booster vaccinations to maintain a high level of immunity.

In recent years it has been discovered that certain vaccines may increase the risk of developing a cancerous tumor of connective tissue (**fibrosarcoma**) in cats. This discovery has raised significant controversy in vaccination recommendations. New protocols are being developed and tested in which vaccines are not repeated as frequently.

SUMMARY

Disease prevention begins with sound biosecurity practices such as maintaining a sanitary environment. Employing a comprehensive vaccination program tailored for each client's needs also helps prevent the spread of disease. Ultimately, disease prevention proves cost effective for both the large-scale producer and the pet owner alike.

REVIEW QUESTIONS

1. Define the following terms:
   antioxidant
   ventilation
   tunnel ventilation
   wet dewlap
   biosecurity
   equine infectious anemia
   quarantine
   fibrosarcoma

2. True or False: Vaccinations all offer complete protection from the intended disease.

3. A _____ shot must be given a month or so after administration of a killed vaccine to elicit a memory response.

4. Which have higher immunity levels, young or older animals?

5. How long should a new animal be quarantined when brought home?

6. Name a common side effect of vaccination

7. Does stress in animals require conscious thought?

8. Does a vaccine exist for distemper in dogs?

9. Can a vaccine affect milk production?

10. Should modified line vaccine be refrigerated?

11. How does allowing an animal's hair coat to become wet and matted contribute to disease conditions?

12. Describe the cough in kennel cough.

13. Name two types of immunity.

14. List at least two factors used in developing a vaccination program.

15. List the symptoms of equine infectious anemia.

ACTIVITIES

1. Low stress levels in cattle increase milk production. Many companies are designing low-stress facilities for dairy cattle. Do a Web search to investigate such items. Share information with the class.

2. Interview a livestock producer and a pet owner regarding their vaccination protocols. Compare and contrast the responses. Report findings to the class.

3. Pretend to be a livestock producer of one specific species. Develop a biosecurity plan for your intended farm.

16

Classification of Diseases

OBJECTIVES

Upon completion of this chapter, you should be able to:

- *Classify diseases, match them with the domestic species in which they occur, and discuss their clinical significance.*

KEY TERMS

schistosomus reflexus	arthritis	peritonitis	neoplasm
congenital	pneumothorax	idiopathic	metastasis
hemophilia			

INTRODUCTION

In the most general terms, a disease is any disorder that creates a problem in the normal function of the body. During the discussion of the individual organ systems, a variety of diseases have been introduced. This chapter discusses the major classifications of diseases and the underlying causes.

A DAY IN THE LIFE

All I Want Is a Full Night's Sleep...

A common part of the job is to be called in the middle of the night. With the first ring, I begin to wonder what the call is going to concern. The calls have such variety; it could be anything from someone canceling an appointment to a distraught owner whose five dogs have porcupine quills in their mouths. When awakening from a deep sleep, it takes me a few minutes to pull my thoughts together. I find it difficult to think clearly when people begin to immediately ask questions.

One particular call a few years ago really caught me off guard. It was 2:47 A.M. when the phone rang. On the other end was a woman who was quite upset. She immediately began to talk; the conversation went something like this (my thoughts are in parenthesis):

"Hello, Dr. Lawhead, my husband and I are getting ready to go on vacation. (Well, thanks for calling to tell me.) My dog is on heartworm preventative (Oh no, she is out of medication and needs some more. But at this hour?), and I laid out his pill along with my husband's heart medication. Unfortunately, my husband took the dog's pill. Will it hurt him?"

At this point I crawled from bed and went to look through my pharmacology text. I tried to think about what potential problems could develop. I just could not remember what effects it had in humans. Finally it struck me—had the dog taken the husband's medication, I could help. I was not qualified to give human medical advice. I referred her to a poison control hotline. This call was quite unusual. However, we do receive many calls from owners whose pets have ingested items they should not have.

Most cows are able to calve on their own. When there is difficulty, a majority of farmers are capable of assisting in the delivery. When that fails, they call the veterinarian for help. One notable calving call came 3 years ago. The farmer told me that she had reached into the cow in an attempt to determine the problem. She told me that she was confused because all she felt was this *thing*.

Now I was confused. Her description did nothing to tell me what she was feeling. I drove to the farm, wondering what I was going to find. Even someone with limited experience is able to tell me if they can feel a hoof or a nose. She did not recognize what she was feeling. I cleaned the cow and reached through the

FIGURE 16–1 A deformed calf (schistosomus reflexus). This calf developed inside-out along the length of its spine. *(Photograph courtesy Jeff and Joanne Walker.)*

vulva and into the uterus. I have attended hundreds of calvings, so I was confident that I was going to explain to her what was occurring. As my hand finally reached the calf, my first impression was exactly the same—all I felt was this *thing*.

She was right; it did not feel like a hoof, a nose, a rump, or a back. I had to investigate more thoroughly. Reaching in as far as my arm could go, I was able to feel more parts. Unfortunately, I now knew what was occurring, but I also knew that I was in for a lot of work. I was finally able to feel a leg, but I also felt loops of intestine and a heart! When I felt the heart, I knew that the calf was inside-out. This calf had a condition called **schistosomus reflexus** (a good gravy splasher to try!) (Figure 16–1).

The deformity of the calf prevented a natural delivery. With the assistance of the farmers, I performed a cesarean section, surgical removal of the calf from the cow. We removed the calf through a large surgical incision on the left side of the cow (Figure 16–2). The cow stood through the entire procedure. It is difficult to visualize what the calf looked like. The calf was basically turned inside-out, lengthwise. Instead of the legs pointing downward from the spine, they were pointing upward. The skin was folded upward as well, and the internal organs were all exposed. The calf was dead at birth, but the cow did well and actually had a normal calf the next lactation.

continued

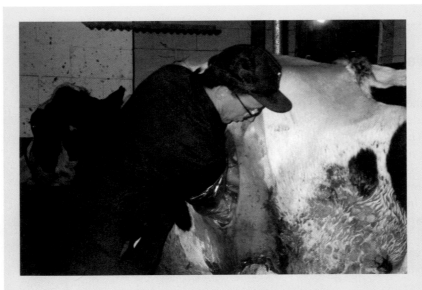

FIGURE 16–2 Dr. Lawhead in the process of a C-section. The surgery is performed with the cow standing and with the use of a local anesthetic. *(Photograph courtesy Jeff and Joanne Walker.)*

Traumatic diseases are often the easiest to diagnose. Tonight I received a call from an owner who was obviously quite shaken. The neighbor's dog had just attacked her dog. She was concerned with how badly the dog was still bleeding, so I asked her to bring the dog to my office. Lady, a 40-lb mixed-breed dog, had obviously tangled with a much larger dog. The dog had multiple bite wounds over her neck, face, and ears.

There was no problem making the diagnosis in this case. The challenge then came as my partner and I tried to repair the damage. The punctures on the neck were quite deep, with extensive muscle damage. The skin had been torn loose from the underlying muscle tissue. We sutured many of the wounds and placed a drain to allow the tissue fluid to escape. These wounds were severe, but we hoped that Lady would make a full recovery.

DISEASE IN CLINICAL PRACTICE

OBJECTIVE

■ *Classify Diseases, Match Them with Domestic Species in Which They Occur, and Discuss Their Clinical Significance*

Diseases can be divided into several major classes. The letters of the words *MAD TIN* will help the student to remember the major classes (Table 16–1), but not every

TABLE 16–1 Major Classifications of Diseases
M– Metabolic
A– Anomalies (birth defects)
D– Degenerative
T– Trauma
Toxins
I– Infectious
Immune
Iatrogenic
Idiopathic
N– Neoplasm
Nutrition

disease fits neatly into one class. For example, an infectious agent or toxin may cause a **congenital** (birth) defect. Nutritional problems can result in many metabolic diseases. Diseases are grouped in classes to help aid the veterinarian in considering a diagnosis when examining a sick animal. Although classification may appear somewhat arbitrary, it is designed to help in the diagnostic process.

Typically the clinical signs that an animal presents could be the result of several different diseases. Following the physical examination, the veterinarian considers a list of different diagnoses. This list includes all the diseases that may result in those clinical signs. By considering each class of disease, the veterinarian can efficiently consider a wide range of diseases. The veterinarian attempts to list the most likely diseases first. This list helps to guide the diagnostic tests that will be performed.

An example may help to illustrate how this thought process occurs. Benji, a 25-lb, 10-year-old mixed-breed dog, is presented because he will no longer sleep through the night. Benji is a house dog and only goes outside while on a leash. Benji eats only dry dog food and never gets table scraps. While giving the history, the owner explains that Benji has to get up several times during the night to urinate. She also explains that he has been drinking more water than he ever had before. She mentions that his appetite has declined over the past few days. The physical examination showed that Benji had lost 3 lb but did not provide any further clues to help in the diagnosis.

These signs are quite common, and a large number of diseases may fit this presentation. Each class of disease can be evaluated. Although many diseases can be eliminated, many must be considered:

■ Anomalies: Such problems are very unlikely given the age of the dog.
■ Trauma: The history helps to rule this out.
■ Infection: A kidney or bladder infection is a possibility.
■ Nutrition: Benji's diet has not changed, so this is not a likely problem.
■ Toxin: Toxins such as antifreeze can cause signs like this, but the history helps to rule out the possibility.
■ Neoplasm: A tumor must be considered. In general, the risk of tumors increases with age. Benji's age increases this possibility.
■ Metabolic: Diabetes and kidney failure are very likely with this history and clinical presentation.

Following consideration of these possibilities (the complete list is actually much longer), the list of differential diagnoses would include:

■ Kidney failure
■ Diabetes
■ Tumor
■ Kidney infection

By preparing this list, the veterinarian can then plan to confirm a diagnosis.

Metabolism describes all the processes occurring within the animal. This includes the chemical reactions occurring within the cell and all the functions of the body, such as urine production and hormone control. *Metabolic* diseases occur when there is a disruption in this natural process.

Many metabolic diseases have already been discussed throughout this text. All the endocrine diseases are considered metabolic disorders. Diabetes, hypothyroidism, and hyperadrenocorticism are just a few examples. Milk fever, the low blood calcium that occurs in cows at calving, is also a metabolic disease. Failure of organs, such as the liver or kidney, disrupts the normal metabolism of the animal. These diseases are also included in this class. The complete list of metabolic diseases is very long.

Pregnancy toxemia is a metabolic disorder common in sheep and goats in the last month of pregnancy. Typically this occurs in animals carrying multiple fetuses. These animals become weak, depressed, and anorexic. As the disease progresses, the animal is unable to rise and eventually dies if left untreated.

The multiple fetuses take up a large space in the abdomen. This limits the size of the stomach and how much feed the animal can eat. In addition, the multiple fetuses increase the demand on the mother. If inadequate nutrients are consumed, the animal mobilizes fat to produce energy. As lipids are broken down, small molecules called ketones are produced. A series of chemical reactions occur in the liver, using the ketones in combination with simple sugars to produce energy. In pregnancy toxemia the carbohydrate metabolism is disrupted and the ketones accumulate in the bloodstream.

Acetone is one of three major ketones. Acetone (commonly found in many fingernail polish removers) has a distinct odor. The ketones accumulate in the blood but also pass into the milk and urine and are exhaled from the lungs. The breath of animals with pregnancy toxemia develops a sweet smell, which is diagnostic for these ketones. Not all people are sensitive to the smell of ketones. Ketones can be detected in urine with a simple color reaction test strip. (Note: Many sheep can be stimulated to urinate by holding their nostrils closed!)

Treatment for pregnancy toxemia includes replenishing the energy supply and decreasing the energy demand on the animal. Energy can be delivered by giving intravenous glucose or drenching with an oral product (propylene glycol) that increases the blood sugar. The fetuses continue to be a demand on the mother. A complete cure often requires that the fetuses be removed from the mother. This may be accomplished by inducing labor or by performing a cesarean section (surgical removal of the fetuses). Prevention is not always possible, but proper nutrition is essential for minimizing the number of cases.

Seizures in dogs can be a result of liver disease. Ammonia that is absorbed from the intestines is removed from the bloodstream by the liver. In liver disease the ammonia and certain other toxins accumulate in the blood. These products are toxic to the central nervous system and can produce seizures.

This type of liver disease, a metabolic disease, may occur in animals as they age. However, the same symptoms can be a result of an anomaly or birth defect. This anomaly may be a portocaval shunt in which blood from the intestinal tract is diverted around the liver. (Review Chapter 8 and the material on fetal circulation.) This shunt, a normal structure in the fetus, should close shortly after birth. If it remains open, neurologic signs may develop early in life.

A wide range of *anomalies* can occur. In general, anomalies are detectable in young animals. Some defects may be so severe that abortions and stillbirths are a result (Figure 16–3). Other defects may never cause clinical problems in the animal (for example, six toes

FIGURE 16–3 A calf born with a portion of two heads. The calf was born alive but did not survive. *(Photograph courtesy John Lynn.)*

on a foot of a cat or a short tail). Clinical signs may not be present at birth but may develop as an animal grows. The demands on the body increase rapidly with the growth of the young animal. It is the increasing demand that brings on the clinical signs. An example is a heart murmur in a dog. The animal may be able to grow and crawl as a young pup. As it matures and becomes active, the heart may not be able to compensate for the increased demands that come with greater size and activity. Therefore a condition that was present at birth may not produce signs until the animal is 6 months old.

Anomalies can be inherited; that is, a chromosome has a defect, resulting in the abnormality. A classic example of an inherited disease is **hemophilia**. In this disease a genetic defect results in a deficiency of one of the clotting factors. As a result, the animal is at high risk for serious bleeding problems. This problem may not be evident at birth and may only become obvious when there is significant trauma or during surgery.

Other anomalies are a result of a defect in the development of the fetus. One of the most common defects detected is an umbilical hernia. The umbilicus (belly button) is where the vessels from the placenta enter the fetus. Following birth, the umbilical vessels begin to shrink and the ring of muscle surrounding them close. With an umbilical hernia, the ring does not close and connective tissue or intestines pushes out through the opening (typically the contents remain under the skin).

Often animals live with an umbilical hernia without any complications. The risk is that a loop of intestine will slip through the hole and become twisted.

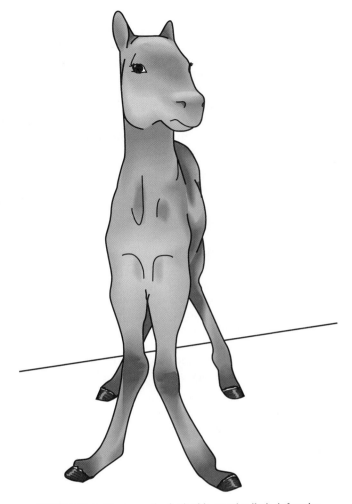

FIGURE 16–4 The legs of a foal with angular limb deformity.

The twisting will cut off the blood supply and trap gas within the intestinal loop. This causes a painful swelling and is called a strangulated hernia. If left untreated, the loop of intestine can die due to a lack of blood supply. A strangulated hernia requires immediate surgical correction. Many hernias are repaired to prevent this occurrence.

Angular limb deformity is a very obvious defect that can occur in foals (Figure 16–4). When observed from the front, these foals' legs have a deviation (often at the carpus) either to the outside or the inside. This can occur because of positioning within the uterus, weak ligaments, or a failure for normal bone to form.

In mild cases, keeping the foal confined with controlled exercise can be very helpful. This protects the animal as the muscles and ligaments around the joint strengthen. In more severe cases, splints and even surgery may be required. Some cases may be too severe to correct.

Degenerative diseases are often associated with aging. These can occur with the normal wear and tear that occurs to the body over the years. Musculoskeletal problems are often degenerative. **Arthritis** and disk disease are two examples of degenerative diseases. In arthritis the normally smooth cartilage that lines the joints becomes rough and irregular. Bony spicules are deposited around the joint in response to chronic irritation. The end result is pain and restricted movement in the affected joint.

In disk disease the disk deteriorates with age and eventually protrudes from the normal site between the vertebrae. Pain and paralysis may result from the disk applying pressure to the nerves or spinal cord. All the clinical signs are secondary to the degeneration of the disk.

Trauma and *toxic diseases* can be some of the easiest to diagnose and treat. Trauma often damages specific regions, and the effect can be quite localized. The veterinarian's attention is directed to one location. Also, many animals present with clinical signs that could have been a result of trauma, but it is unknown if any such event even occurred.

Trauma cases are very common in veterinary practice. The range of severity of the trauma varies dramatically, from two pets playing roughly to gunshot wounds to being hit by a car (Figure 16–5). The treatment obviously depends on the type and location of the injury that results. For many mild cases, merely restricting activity for a period of hours or days allows the animal to recover fully. Medication designed to decrease pain and inflammation may also be used.

In many cases, standard first aid treatment is very helpful. The same principles that are used in humans also apply to pets. In cases where there is bleeding from open wounds, applying a clean pressure bandage is helpful. Even with serious lacerations, maintaining pressure on the site can often control the bleeding.

Unfortunately, the lacerations are not always in a location where this can be accomplished. Several months ago, I received a call from a panicked owner. She had been trimming her dog's nails and the dog tried to lick its foot at the same time the owner clipped a nail. Not only was the nail trimmed, but also a piece of tongue had been removed. The tongue, rich in blood vessels, was bleeding steadily. In addition, saliva made the volume look much larger. The dog did not cooperate with having direct pressure applied to its tongue. This dog was anesthetized and the tongue sutured.

Trauma can also be severe enough to fracture bones. Again, depending on location, splints can be applied to stabilize the leg. The splint may not be the final treatment, but it prevents further damage to the bone and soft tissue in the area. Without the splint,

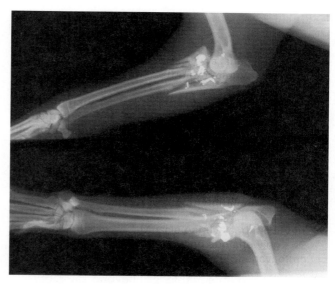

FIGURE 16–5 Radiograph of a dog with a fractured radius and ulna. The bone has been shattered by a bullet (seen as dense white regions on the radiograph).

the bone end might puncture through the skin, adding bacterial contamination to the bone. This can prevent normal healing if an infection is established in the bone.

Applying a pressure bandage is an important technique in first aid. Bandages can be used to stop bleeding and to protect open wounds from further damage and contamination. Adding a rigid support within the bandage can be used to support a leg with bone or joint damage (Figure 16–6). Many variations are possible in bandaging. The location, type, and extent of the injury influence the type of bandage that is used.

Certain principles exist for all types of bandages. For example, the bandage should be kept clean and dry. A bandage that becomes wet and contaminated

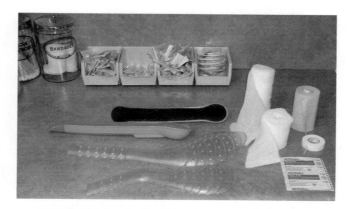

FIGURE 16–6 A selection of bandage material and splints.

increases the risk for infection. The bandage itself can damage the underlying skin if it is too tight. If the bandage is too tight, it may also hinder the circulation to the lower portion of the leg. This may cause the animal discomfort and increase the likelihood that it will chew at the bandage. If the foot or toes are exposed, they can be checked to determine if the bandage is too tight. If the foot begins to swell or the toes become cold, the bandage needs to be loosened immediately.

Nondisplaced fractures on the lower leg of a dog or cat can often be stabilized with a splint. In fracture repair, the goal is to immobilize the joint above and below the site of the fracture. For example, a radius fracture requires the splint to stabilize the carpus and the elbow. The basic steps in applying a splint are:

1. Stirrups: Strips of tape are applied to the sides of the legs and an equal length beyond the foot (Figure 16–7A).
2. Cast padding: A thin, cotton padding is applied evenly over the leg. This layer allows the splint to apply even pressure over the entire leg, preventing pressure sores under the bandage (Figure 16–7B).

3. Splint: A rigid support (often plastic) prevents the bandage from bending (Figure 16–7C).
4. Gauze: Wrapped snuggly around the splint, gauze provides good stabilization of the leg. This must be applied tight enough to keep the leg from moving and loose enough to allow normal circulation. The ends of the stirrups are turned onto the bandage. This ties them into the bandage to help keep it from sliding off the end of the leg.
5. Tape: This is applied to keep the entire bandage together (Figure 16–7D).

This basic technique is used in many bandages. This same technique, without the rigid splint, produces an excellent pressure bandage. On open wounds, a nonadherent pad is placed over the area to prevent the healing wound from sticking to the bandage. These bandages should be changed if discharge begins to soak through the padding or if the bandage appears to slide down the leg.

More severe trauma can also damage internal organs. At times this may require emergency surgery and treatment. Animals that have been hit by car may

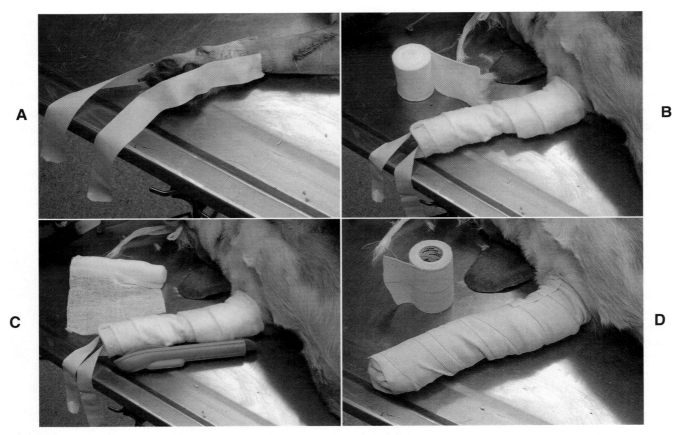

FIGURE 16–7 Applying a splint. *A.* Stirrups of tape are added on both sides of the leg. This helps to prevent the bandage from slipping. *B.* A layer of padding is added to the leg. *C.* At this stage a splint can be included in the bandage and secured with gauze. *D.* The stirrups are folded into the bandage and an elastic tape has been used to cover the bandage.

have ruptured a major organ, such as the spleen or liver, resulting in severe internal bleeding. This may require an exploratory surgery to find the site and correct the damage. Such trauma can also damage the lungs to the extent that air leaks from the tissue. The air becomes trapped between the lungs and the chest wall (a condition called **pneumothorax**). This air can be drawn out of the chest through a chest tube that enters through the skin and then penetrates through the muscles between the ribs. Suction is applied to the external end of the tube. In this way, air or fluid can be drained from the chest cavity (Figure 16–8). In more severe cases the lungs may need to be repaired surgically.

Not all damage is immediately evident following the trauma. If an animal has a full urinary bladder when a car strikes it, there is risk that the bladder may rupture. The tear may actually be small enough to allow the animal to urinate while small amounts leak into the abdomen. These animals may appear normal immediately following the trauma but become much sicker over the next 1 to 2 days. The urine causes a chemical irritation in the abdomen, and the waste products are reabsorbed. The end result is called **peritonitis,** an inflammation throughout the abdomen. The bladder must be surgically repaired in these cases.

Parturition can cause trauma to the mother. During a difficult delivery, tears may occur in the uterus, cervix, and vagina. A tear in the uterus results in a leakage of fluid, which often contains bacteria, into the abdomen. Again the result can be peritonitis. It is also possible that a major vessel is torn and the animal can die from internal bleeding.

Recently I worked on a cow that had prolapsed her uterus. At some point the uterus was torn. I was not able to tell if this occurred during the delivery or following the prolapse. I sutured the tear before correcting the prolapse and began treating this animal with an antibiotic because of the contamination of the internal organs. The cow was doing fine initially but was at a high risk for peritonitis.

Toxins are a common cause of illness in animals. If the toxin is known, a specific antidote or treatment can be administered. Difficulty arises when the toxin is unknown, making diagnosis and treatment difficult.

Animals may be exposed to toxins by ingestion, inhalation, or contact. With contact, the product may only be toxic in particular locations, such as the cornea or mucous membranes. Some of these toxins cause irritation to the surface of the skin or mucous membranes. However, other toxins are absorbed into the animal through the skin. The signs of the toxin are often related to the route of entry (for example, skin irritation with contact toxins, vomiting and diarrhea with ingested toxins, and breathing difficulties with inhaled toxins). However, many toxins affect other organ systems once they gain entry into the body.

Veterinarians are often presented with a sick pet whose owners are concerned that the animal may have been poisoned. The history is that the animal is free to roam and is now sick. I am often asked if I can do a blood test to see if the animal has been poisoned. Many tests are available to detect poisons in the blood or urine, but there are thousands of poisons and many cause identical clinical signs. To do proper testing, the veterinarian must have some idea of what the animal may have ingested.

In other situations owners discover what the animal has ingested. They may catch the animal in the act or discover the empty container. With toxins, the length of time that the animal is exposed is very important in determining the treatment. Initiating treatment before large amounts of the toxin have been absorbed can be lifesaving.

The list of products that can be toxic is huge. Many everyday products found around the house are potentially toxic to pets (Figure 16–9). Even products designed for pets can become toxic when the animal is exposed in large amounts or to a product designed for a different species (for example, a cat receiving medication designed only for a dog). Toxins can also be household cleaners, over-the-counter or prescription medications, plants, and certain animals. (For example, some toads have toxins in their skin that irritate the mouths of dogs that bite them.)

Problems can result when animals gain access to illicit drugs. In these situations the owners are often reluctant to admit the potential exposure, because of concern for them. A complete and accurate history is necessary for the veterinarian to be successful. As with

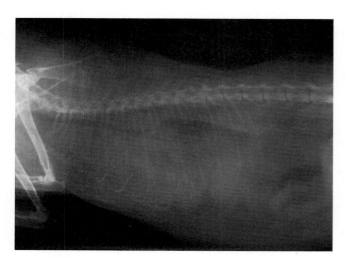

FIGURE 16–8 Radiograph of a cat with a large amount of fluid in its chest. Treatment might include a chest tube to drain the fluid from the chest.

FIGURE 16–9 Many products commonly found in households can be toxic to pets.

many toxins, rapid intervention in these cases is often necessary to save the animal.

Table 16–2 lists a number of plants that are capable of causing toxicity in animals. Geographic location is important in the discussion of toxic plants. Certain plants only grow in specific regions. Knowledge of the plants common to a given area is necessary. Not every part of these plants is toxic. For example, an apple is not toxic, but a large quantity of apple seeds is. A goat consuming cherry leaves from a live tree does not become sick. However, the wilted leaves from a fallen cherry tree can contain cyanide, which can be extremely toxic.

Proving exposure to toxic plants may also be difficult. In pastures with grazing animals, there are often large numbers of toxic plants. In general the animals do not consume large enough amounts to become sick. Problems are often worse in drought conditions. Many of these toxic plants are commonly found along stream banks. As dry weather decreases normal pasture grasses, grazing animals may eat more of the fresh green plants close to the stream. In these conditions the animals may consume enough to become ill.

Table 16–3 lists a number of common products and medications that can cause toxic effects in animals. From this list it is obvious that many common products can be damaging to companion animals. Prevention is so important in toxic diseases. These diseases are completely preventable by eliminating access of the animals to these products.

When an animal is exposed to a toxin, beginning treatment as soon as possible can often make the difference on success. The odds of success are much higher when the animal is first exposed than when clinical signs have already developed. The first goal of treatment is to minimize the amount absorbed. When the animal has skin exposure, bathing can be very effective at minimizing the exposure.

The rate at which ingested toxins are absorbed is quite variable. When the ingestion is detected quickly (less than 2 hours), inducing vomiting can be considered. When the animal vomits, much of the toxin can be eliminated from the animal, greatly decreasing the amount absorbed. Several different products are available to induce vomiting.

Vomiting should only be considered under the guidance of a professional. With certain toxins and conditions, vomiting can be hazardous to the animal. Animals exposed to caustic or irritating toxins should not be induced to vomit. When vomited, the toxin can cause further damage to the esophagus and mouth. When an animal is showing certain clinical signs, vomiting can also be dangerous. If the animal is seizing, is having difficulty breathing or swallowing, or is extremely weak, vomiting should not be induced.

Following vomiting, toxin may still remain in the intestinal tract. Activated charcoal is given orally to bind or adsorb the toxin still remaining. Weak chemical bonds attach the toxin to the activated charcoal. The charcoal is not absorbed from the intestines, so the toxin is effectively removed from the animal in the feces. Note that the feces of a treated animal will appear black from the charcoal treatment. Activated charcoal can bind a large number of different toxins (but not every toxin).

When possible, a specific antidote can be administered following exposure. For example, many mouse and rat poisons cause bleeding by inactivating vitamin K. Administration of vitamin K by injection or orally is a specific antidote. Antifreeze, ethylene glycol, only becomes toxic when the liver metabolizes it. An antidote can be administered that prevents this metabolism, allowing it to be excreted by the kidney without damage. Many toxins have no antidote available.

Once clinical signs of toxins have begun, treatment is directed to be supportive. Treatment varies considerably with each type of toxin. The treatment is directed at the specific organ systems affected. For example, if an animal is in respiratory distress, oxygen can be supplemented. If seizures occur as a result of the toxin, medications such as phenobarbital can be used to control them until the toxin is eliminated from the body. Intravenous fluids are valuable in many cases to help maintain blood pressure and increase urine production.

TABLE 16–2 Common Toxic Plants

Type	Plant	Clinical Signs
Trees	Oaks	Anorexia, weakness, icterus, dehydration, may cause death
	Cherry (wilted leaves)	Respiratory distress, death
	Apple (seeds)	Vomiting, salivation, respiratory distress
	Black walnut	Laminitis in horses bedded with shavings
	Red maple	Hemolytic anemia
Houseplants	Oleander	Vomiting, diarrhea, pain
	Mistletoe	Irritation of stomach and intestines, shock
	Dumbcane	Irritation of mouth and throat
	Caladium	Irritation of mouth and throat
	Philodendron	Irritation of mouth and throat
Flower garden	Foxglove	Irregular heartbeat
	Lily of the valley	Stimulation of heart, dizziness
	Daffodil	Vomiting, trembling
	Hyacinth	Diarrhea
	Iris	Respiratory distress
	Autumn crocus	Stomatitis
Vegetable garden	Tomato (leaves)	Vomiting, diarrhea
	Rhubarb (leaves)	Vomiting, weakness, cramps
	Onion	Hemolytic anemia
Ornamental	Mountain laurel, rhododendron, azalea	Salivation, pain, vomiting
	Japanese yew	Sudden death
	Wisteria	Vomiting, pain, diarrhea
	Holly	Diarrhea
	Juniper	Vomiting, diarrhea
Forests	Braken fern	Fever, anemia, respiratory distress
	Mayapple	Vomiting and diarrhea
	Jack in the pulpit	Irritation of mouth
	Dutchman's breeches	Salivation, diarrhea, incoordination
Wetlands	Water hemlock	Salivation, twitching, seizures
	Skunk cabbage	Irritation of mouth and throat
	Fox tail, horse tail	Incoordination, weakness
Fields	Nightshade	Diarrhea, weakness, convulsions, coma
	Jimsonweed	Restlessness, twitching, paralysis
	Poison hemlock	Weakness, recumbency
	Belladona, deadly nightshade	
	Poke weed	Pain, vomiting, diarrhea
	Milkweed	Convulsions, coma
	Buttercup	Stomach irritation, pain, diarrhea, twitching
	Curly dock	Respiratory distress, death, abortions
	Red berried elder	Vomiting, diarrhea
	Lambsquarter	Respiratory distress, death, abortions
Western dry range and grasslands	Halogeton, barilla	Low calcium, kidney damage
	Greasewood	Low calcium, kidney damage
	Crazyweed, locoweed, milkvetch	Irritable, incoordination
	Groundsel	Liver damage
	St. Johnswort	Skin becomes sensitive to sunlight, blisters, scabs
	Ergot (fungus on grasses)	Lameness, sloughing of skin
	Tansy ragwort	Liver damage
	Yellow star thistle	Involuntary chewing
	Johnson grass, sudan grass, sorghum	Cyanide toxicity, sudden death
	Bitterweed	Weakness, salivation, vomiting and diarrhea
Mountain	False hellebore	Salivation, convulsions, weakness, vomiting
	Larkspur	Affects nervous system, convulsions, collapse
	Monkshood	Vomiting, diarrhea, paralysis
	Lupine	Vomiting, pain, ataxia, muscle paralysis

TABLE 16–3 Common Toxins

Acetaminophen	Facial swelling, vomiting, anemia
Alcohol	Ataxia, depression, coma, death
Anticoagulant rodenticides	Bleeding disorder caused by inhibiting vitamin K
Chocolate	Vomiting, diarrhea, seizures, coma
Copper	Sheep very sensitive, weakness, anorexia, hemolytic anemia
Ethylene glycol (antifreeze)	Vomiting, ataxia, seizures, coma, death
Lead	May be found in old paint, used motor oil, batteries. Causes anemia, brain damage
Nicotine	Excitement, diarrhea, vomiting, depression, muscle weakness
Organophosphate insecticides	Blocks acetylcholinesterase in the nervous system. SLUD—salivation, lacrimation (tearing), urination, defecation, vomiting, diarrhea, respiratory difficulty
Zinc	Can occur after ingestion of pennies (after 1983) and galvanized metal, causes hemolytic anemia
Over the counter anti-inflammatory drugs (e.g. aspirin, ibuprofen, naproxen)	Vomiting, abdominal pain, renal failure

Many references are available to help veterinarians understand the principles of toxins and the treatments available. When unable to find information on a product, several hotlines are available with a large databank of information:

ASPCA National Animal Poison Control Center (Allied Agency of the University of Illinois College of Veterinary Medicine), 1-888-426-4435

Animal Poison Hotline (Sponsored by the North Shore Animal League America in New York City and PROSAR International Animal Poison Center), 1-888-232-8870

These services do have a fee to support their operation.

The I in the MAD TIN mnemonic covers an array of diseases. The I can stand for infectious, immune, iatrogenic, or **idiopathic**. Infectious and immune diseases have been covered to a large extent in previous chapters and will not be reviewed.

Iatrogenic diseases describe those conditions that develop as a result of treatment. Hyperadrenocorticism can be iatrogenic. Many pets have chronic allergy problems, which may require frequent treatment with cortisone. Certain autoimmune diseases may also require high dosages of cortisone. Treating a pet with cortisone for long periods or with high dosages can cause all the signs of naturally occurring hyperadrenocorticism. Pets are monitored for signs consistent with this disease when they are on such treatment.

When on high levels of cortisone, the animal produces less of its own. The medication provides the negative feedback, keeping natural production low. If the medication is removed suddenly, hypoadrenocorticism can be created. This again is an iatrogenic disease.

Surgery is another source of iatrogenic diseases. Surgical excision of the thyroid glands is a standard treatment for hyperthyroidism. The parathyroid glands are closely associated with the thyroid glands. Sometimes during surgery it is impossible to identify the parathyroid glands. Due to this problem, the parathyroid glands are often removed with the thyroid glands. When this occurs, the animal will have iatrogenic hypoparathyroidism. This problem must be anticipated and can be treated.

Idiopathic describes a condition in which current medical knowledge cannot explain an underlying cause. In horses a number of diseases can cause severe bleeding into the urine. This can result from cancer or from a tract that opens from an artery into the ureter. When no underlying cause is detected, the disease is called idiopathic renal hematuria (*heme* refers to blood; *uria* refers to urine).

Neoplasm and *nutrition* represent the N in MAD TIN. Many nutritional diseases have already been discussed. Nutritional diseases often overlap with certain metabolic diseases. The distinction between the two classes is not always clear. The important feature is that this class of disease is considered in the differential diagnosis list.

A neoplasm or tumor develops with growth of cells in an uncontrolled manner. When examined microscopically, a neoplasm has cells that have a similar appearance to the type of tissue from which it arises but usually no organized structure. (Chapter 1 includes a more detailed description of the cell structure in neoplasms.) Neoplasms have no useful function. In the case of endocrine tissues, the neoplasm may actively secrete hormone, but without the normal control or feedback.

Neoplasms are divided into two major classes, benign and malignant. As the cells of a neoplasm continue to divide without control, they may just expand in the local area or actually invade the surrounding tissue. Benign tumors only expand locally, without invading the surrounding tissue. In general, benign neoplasms are much slower growing than malignancies. Benign tumors generally have a well-defined border. *Encapsulated* describes this property where the edges of the tumor are clearly defined.

Malignant neoplasms are very destructive and invade the surrounding tissues (Figure 16–10). *Cancer* is generally used to refer to malignant tumors. (In some writings, cancer may be used to describe any tumor, benign or malignant.) The edges of the tumor are often not clearly defined. The cells of the tumor may extend beyond the edge that can be seen and felt.

Malignant neoplasms have the potential for spreading to other locations within the body, a process called **metastasis**. Tumor cells can be spread through the bloodstream or lymphatic system. Individual tumor cells invade these vessels, become detached, and then are carried to distant locations (Figure 16–11). With lymphatic spread, the new tumor is often detected in the regional lymph node that drains that area. When it spreads into the bloodstream, the secondary location may not be as predictable. The lungs, liver, and spleen do have a high incidence as the site of metastasis. The extensive blood supply in these organs increases their risk for developing one of these secondary tumors.

Benign tumors can also be quite damaging, even though they do not spread. A benign tumor growing on or under the skin is usually not damaging or harmful to an animal. However, if the benign tumor is located close to a vital area, the effect can be quite devastating. Benign brain tumors are an example. A benign brain tumor located in a vital area can cause severe signs due only to the local expanse of the growth.

The terminology used to describe tumors is descriptive of the originating tumor. The suffix added to the term helps to classify the tumor as benign or malignant. The suffix *–oma* is added to terms for benign tumors. The suffix *–sarcoma* is added to malignancies arising from connective tissue. The term *carcinoma* is added to malignant tumors arising from epithelial tissues. (Other terminology does exist for even more tumor types). Examples include:

Cell type	Benign	Malignant
Fat	Lipoma	Liposarcoma
Fibrous tissue	Fibroma	Fibrosarcoma
Bone	Osteoma	Osteosarcoma
Squamous epithelium	Squamous cell papilloma	Squamous cell carcinoma
Duct or gland	Adenoma	Adenocarcinoma

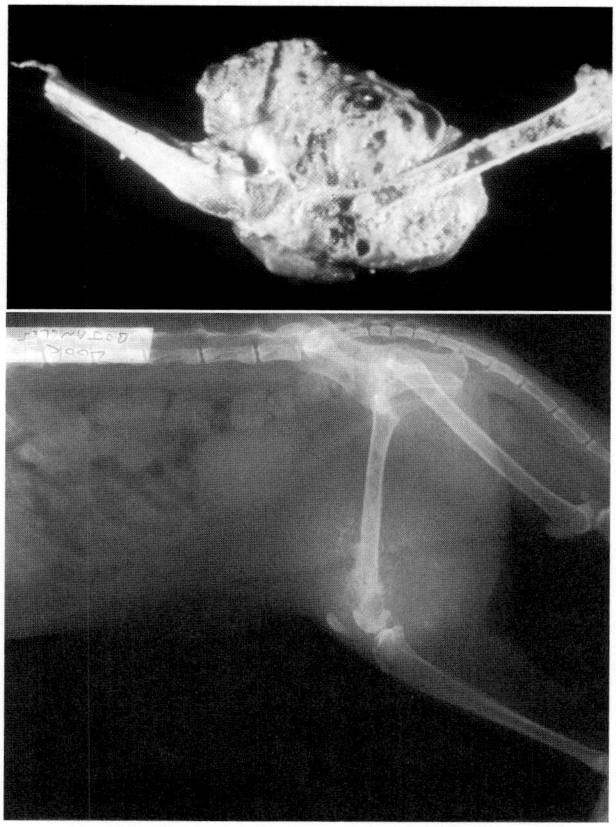

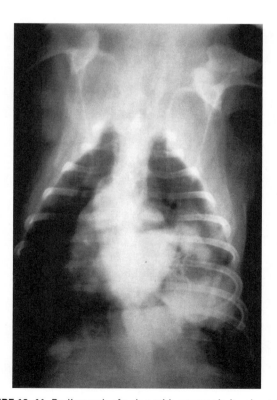

FIGURE 16–10 *A.* An osteosarcoma, a malignant bone tumor. *B.* Radiograph of an osteosarcoma.

FIGURE 16–11 Radiograph of a dog with a tumor in its chest. The tumor had spread, or metastasized, from another tumor. *(Photo by Anne E. Chauvet, DVM, University of Wisconsin School of Medicine.)*

SUMMARY

Classifying diseases into such categories as parasitic, infectious, metabolic, and toxic helps practitioners, owners, and students make sense of a vast field of study. Even so, some diseases may be classified in multiple categories or cause conditions that could be classified in other categories, therefore causing confusion. Nevertheless, a disease classification system gives veterinarians a tool in the study of their discipline.

REVIEW QUESTIONS

1. Define the following terms:
   schistosomus reflexus   peritonitis
   congenital   idiopathic
   hemophilia   neoplasm
   arthritis   metastasis
   pneumothorax

2. True or False: By law, veterinarians can give medical advice to humans because humans are animals.

3. True or False: Benign tumors can be damaging even though they do not spread to other parts of the body.

4. _____ describes all the processes that occur within an animal.

5. Diabetes can be classified as a _____ disorder.

6. Does the risk of tumors increase or decrease with age?

7. Give one possible antidote to injection of a rat or mouse poison that causes bleeding.

8. What disease category does hemophilia belong to?

9. Is antifreeze toxic if ingested by a dog?

10. Should vomiting be induced in all cases of toxin ingestion?

11. What does the suffix *–oma* mean?

12. Describe the technique mentioned in the text for stimulating urination in sheep.

13. Describe the appearance of a calf with schistosomus reflexus.

14. What kind of disease is arthritis?

15. How does oral administration of activated charcoal aid in the removal of toxins?

ACTIVITIES

Material needed to complete activities:
Telephone book

1. Attempt to identify the poisonous plants found in your own house and yard. Then visit <http://www.about-house-plants.com> to determine if these plants are potentially toxic to your pets.

2. Examine the various commercial products in your house for other potential toxins (for both animals and people). Do the labels of these products give guidance on how exposures should be treated? Do you have a poison control number readily available for emergencies? If not, use local telephone books to find the number. Post the poison control telephone number along with other emergency telephone numbers by your home telephone. Also, write directions to your home from the local emergency service provider and put that also by the telephone. This can be of great assistance to an individual already stressed by an emergency situation.

17

Zoonoses

OBJECTIVES

Upon completion of this chapter, you should be able to:

- List and describe several diseases common in domestic animals that are contagious to humans.
- Relate the academic material learned in this chapter to clinical practice.

KEY TERMS

rabies
visceral larva migrans
cutaneous larva
  migrans

toxoplasmosis
cat scratch fever
ringworm
RNA viruses

Q fever
pasteurization
anthrax
mad cow disease

scrapie
West Nile fever
brucellosis
tuberculosis

INTRODUCTION

All veterinarians have their own reasons for entering the profession. Many veterinarians enjoy science and medicine, working with animals, and the technical challenges of the profession. Once in the profession, veterinarians also undertake responsibilities to people around animals. Many diseases in animals are transferable to humans. The veterinarian has responsibility in minimizing these diseases and protecting the people exposed to these animals.

A DAY IN THE LIFE

What Are Friends For?

Veterinary school was extremely challenging. The amount of material covered was just tremendous. In the final year, students often put in long, hard hours in clinical rotations. During that time, I made some of the most important friendships of my life. The support of friends was so important during those times when veterinary school seemed overwhelming. In addition to working hard, we did find time to have fun and enjoy life. We could find things to laugh about, even if they were directed at me.

In Chapter 15 we discussed basic principles in preventing diseases. In a hospital setting, such as the barns at my veterinary school, contagious diseases are of particular concern. *Salmonella*, a bacterium that causes severe diarrhea, is one of these diseases. There were times in school when horses that were presented for other conditions (such as surgery) would develop diarrhea. To diagnose *Salmonella*, feces from the animal were cultured. It generally took several days for the results to be available. If the disease was confirmed, the horse was transferred to the isolation ward.

While waiting for the culture, precautions were taken to minimize the risk of spreading *Salmonella* to other horses in neighboring stalls. An area outside the stall door was marked with tape. Within this area, disposable plastic boots, a protective gown, and gloves were worn. When leaving the area, shoes were sanitized in a disinfectant footbath. As a student, I did not look forward to this type of case because it increased the time commitment in an already busy day. But the situation was quite common, and everyone knew to follow procedure.

Salmonella is also of concern because humans can contract the disease. The clinical signs of human and animal presentation are similar: high fever and diarrhea. Many humans with salmonellosis need to be hospitalized for supportive treatment and possibly antibiotics. My senior year, I developed signs consistent with *Salmonella* infection, although it easily could have been the flu. I was taking large animal pathology, a rotation in which autopsies are performed on horses and cows. It was quite possible that during that rotation I could have been exposed to the organism.

When I developed symptoms, I left the postmortem room and returned to my dormitory. I felt miserable, weak, and feverish. I also experienced severe

FIGURE 17–1 Humans can succumb to many animal diseases. Dr. Lawhead's classmates hoped to prevent the spread of disease in the dormitory.

diarrhea. Late that first night of my illness, I once again had to leave my room to go to the bathroom. I felt so weak and had no desire to get out of bed. But even in my state of misery, I did laugh when I opened my door (Figure 17–1). My supportive friends had taped off my room, complete with disposable boots and a protective gown! I had become the biohazard of the dormitory. Fortunately, I made a quick recovery, with no additional evidence of *Salmonella*.

I was quite fortunate in that case, but diseases that affect humans can be devastating. Last summer I was presented with a young kitten that had become ataxic. The kitten was taking high steps with its hind legs and was having trouble staying upright. The kitten was not vicious but was highly anxious and agitated.

This kitten had been found as a stray almost 2 months earlier. At that point, the kitten did have a couple of small wounds on its hind legs. The owner also mentioned to me that the kitten had bitten her earlier that day when she was trying to pick it up. The history made me nervous. The case was classic for **rabies**. Rabies is of particular concern because it can cause a fatal infection in humans. The kitten continued to worsen and was euthanized. The animal was submitted for testing and did come back positive for rabies. The owner then had the concern that a rabid animal had bitten her.

continued

Yesterday I visited a farm to examine a cow that was unable to rise. The cow had calved 3 months earlier, which made milk fever an unlikely cause. When I examined the cow, she had a drooped ear and her eye on the same side seemed sunken in the socket. I noticed that she was unable to blink with that eye. The farmer did mention that she was holding her head to the side the previous day, but he had not become concerned. This cow likely had a brain infection called listeriosis. The outlook on this cow was not good, but I had more important topics to discuss. I needed to talk about the risk to the farm family.

ZOONOTIC DISEASES

OBJECTIVE

■ *List and Describe Several Diseases Common in Domestic Animals That Are Contagious to Humans*

Zoonotic diseases can be transmitted from animals to humans. The transmission may occur through direct contact with the animal, through a vector (such as fleas or ticks), or through food contamination. The type and frequency of zoonotic diseases varies tremendously between geographic areas. In industrialized countries the number of people with direct contact to food animals is very low. For example, fewer than 2% of the population in the United States works directly with farm animals. The number of households with pets has increased dramatically over the last century. This has increased the significance of pets as a source of human diseases.

As with any infectious disease, the higher the number of infected animals, the higher the risk of spread. In addition, the behavior of the owners and the type of interaction with the animals directly influences the risks. Proper sanitation and hygiene can greatly reduce the risk of many zoonotic diseases. Children are at higher risk for many zoonoses. Children are more likely to have close physical contact with their pets and often have worse hygiene than adults. Young children are also more likely to place contaminated items in their mouths and are less likely to wash hands frequently.

Visceral larva migrans (VLM) is one such disease that puts children at a higher risk when they place things in their mouths. (*Viscera* is a term used to describe the internal organs.) The larvae of the roundworms found in dogs and cats cause VLM. *Toxocara canis* is the most common cause of this condition.

Infected dogs pass eggs in the feces (Figure 17–2). Depending on the amount of moisture and the outside temperature, the eggs develop larvae in approximately two weeks. At this point the eggs become potentially infective to other animals. Ingesting these

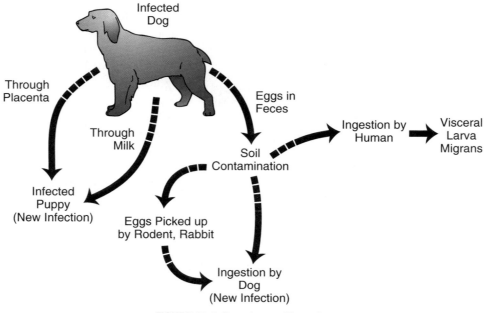

FIGURE 17–2 Roundworm life cycle.

eggs or larvae can infect other dogs. The larvae then migrate through the liver and lungs of the dog. Eventually they are swallowed, and adults develop in the intestines. Puppies can also be infected before birth as the larvae migrate through the mother's body, cross the placenta, and enter the pup. Larvae can enter the pup through the mother's milk as the third means by which new infections can occur.

The adult roundworms are capable of passing large quantities of eggs in the feces of the dog. Over time the feces disappear with exposure to sun and rain. The eggs remain in the ground and develop into an infectious stage. It is this contamination of the ground that puts children at risk. If toys are used to play in the dirt and then placed in the child's mouth, the larvae can be ingested.

Once the larvae are ingested, they migrate through the human. Humans are not a true host, and therefore adults do not develop. The larvae can cause damage as they migrate through the liver, lungs, and eyes. The clinical signs often include fever, coughing, wheezing, an enlarged liver, and abdominal pain. The signs are not very specific and can be misdiagnosed. The disease is generally self-limiting, and children recover with supportive treatment. Deaths have been reported when the larvae damage the heart or central nervous system.

Cutaneous larva migrans (CLM) is caused by hookworms (especially *Ancylostomum braziliense*) in dogs, which have a life cycle similar to that of roundworms. Hookworm larvae that develop from eggs are also capable of infecting other dogs by penetrating the skin. The larvae then migrate through the body and eventually develop into adults within the intestinal tract.

When these infective larvae penetrate the skin of humans, they migrate within the layers of the skin. This usually causes a raised red and itchy path within the skin. The disease in humans has also been called creeping eruption. Fortunately this disease is generally mild and self-limiting (that is, it resolves without treatment).

Maintaining good sanitation with household pets can prevent VLM and CLM. Immediately removing feces from the environment eliminates the eggs before they become infective. This is very practical for owners and their own pets that are confined to a yard. Problems arise in public areas, such as parks, where multiple dogs are free to roam and contaminate the environment. Parents should be aware of this potential risk. Control must also revolve around training children to keep nonfood items out of their mouths and to wash their hands frequently. Veterinarians also play a significant role by routinely examining dogs for parasites and administering proper medication to eliminate infections.

Another parasitic disease in animals that is of concern to humans is **toxoplasmosis**. *Toxoplasma gondii* is a protozoal parasite that infects cats. Cats are the definitive host, meaning that they are necessary to complete the life cycle. Infected cats pass oocysts (a type of egg) in the feces, which are then capable of infecting others (Figure 17–3). If other animals, such as cattle, pigs, and sheep, ingest the oocysts, the protozoa forms cysts in the muscle tissue. These hosts generally do not become sick from the infection and do not pass oocysts in the feces. The muscle tissue of these animals is a possible source of infection if consumed by other carnivores.

Healthy humans who happen to consume *Toxoplasma* cysts often do not show any signs or only experience vague, flulike symptoms (fever, chills, muscle aches). This disease is of much greater concern in pregnant women and immunocompromised individuals, such as those infected with human immunodeficiency virus (HIV), those being treated for cancer, or those who have undergone organ transplantations and may not have a normal immune system. In these people a life-threatening infection can occur. *Toxoplasma* organisms can invade the brain or lungs in these individuals, causing severe damage.

If women become exposed to the organism during their pregnancy, there is a chance that the fetus will also become infected. This is a risk primarily in women who have had no prior exposure to *Toxoplasma*. When a pregnant woman develops **toxoplasmosis,** miscarriage, stillbirth, or birth defects may result. The birth defects generally involve the brain and eyes. The result may only be minor visual defects but may also be blindness and severe mental retardation.

Cats are important in the spread of the disease because they are the definitive host, responsible for maintaining the life cycle. Direct contact with cats is a very unusual means of spreading the disease. Cats in general only shed the eggs periodically, and the eggs are not infectious when they first leave the cat. The real risks are from fecal contamination that has been present for several days and from consuming improperly prepared meat from an infected animal.

The emphasis on control revolves around immunosuppressed individuals and women who may become pregnant. Several critical control factors should be followed by at-risk individuals. These include the following:

■ Litter boxes should be cleaned daily. This way, feces are removed before any *Toxoplasma* eggs become infective. It is generally recommended that pregnant women have someone else, such as their spouse, clean the litter box in these situations. (This recommendation is greatly appreciated by women.)

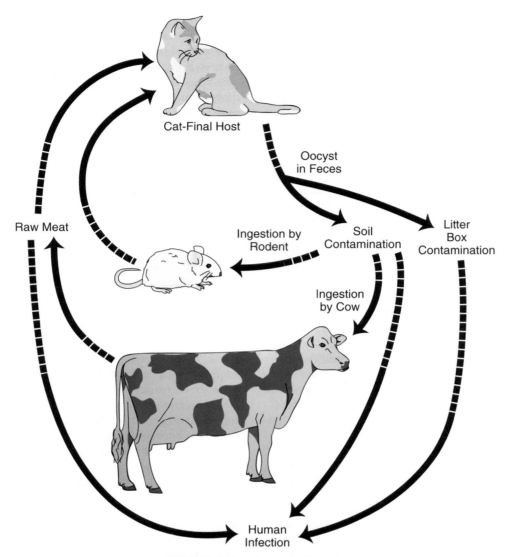

FIGURE 17–3 Toxoplasmosis life cycle.

- Cats should be discouraged from being hunters. Even rodents can be infected with *Toxoplasma* organisms and can spread it to cats.
- Sandboxes should be kept covered when not in use. Cats find the loose sand appealing as a litter box.
- When gardening or working in soil, gloves should be worn.
- Meats should be adequately cooked before consumption. Temperatures should reach at least 66°C (150°F).
- Hands should be washed thoroughly after handling raw meats and after contamination with soil.

Bite wounds from pets are a common occurrence. This is not exactly a disease transferred from the pet, but it is a condition in humans that results from direct contact with them. Many bite wounds are minor, but some can be so serious that they require suturing, antibiotics, and even hospitalization. Many alarming cases have been reported in which a dog or pack of dogs has even killed humans.

Many groups are at a higher risk for animal bites. Obviously veterinarians and animal control officers are placed in situations that increase the risk. Professions, in which the individual must enter another's home are at higher risk as well. These professions include letter carriers, meter readers, and any home service profession. Many dogs feel threatened as strangers enter their property, and they become very defensive.

Children may represent the individuals at highest risk. Often children do not understand proper animal handling and lack fear of pets. Young children may handle pets roughly, and in response to pain the animals bite. It is alarming that children are much more likely to be bitten in the face. Their smaller size and their approach toward the head of animals increase this risk. Head bites can be quite hazardous and often produce long-lasting scars.

Not every bite wound can be prevented. However, some basic principles can help to limit incidences. Young children should only have contact with pets when supervised. Children should be taught at a young age how to properly interact with pets. When entering a house with a barking dog, the owner should be present to keep the animal under control. Never try to strike at a dog in attempt to scare it; this often increases the likelihood that the animal will try to bite.

Bite wounds are particularly alarming when the animal doing the biting escapes. Bite wounds spread rabies, a disease that can infect all mammals. It may also be possible to spread the disease with contamination of deep scratches and mucous membranes from saliva or cerebrospinal fluid.

Rabies is caused by a virus that infects the central nervous system. Wildlife has the highest incidence of rabies. Interaction with wildlife occasionally brings the disease to our domestic animals. In my home region (central Pennsylvania), skunks, raccoons, and foxes have all been detected with rabies. In my career I have seen rabies affect cats, horses, and cows, all resulting in potential human exposure.

When a rabid animal bites another animal, the rabies virus is introduced into the body. The virus enters nerve fibers and migrates toward the central nervous system. Eventually the virus reaches the brain, where it establishes infection. The incubation period for rabies is extremely variable. Because the virus must migrate through the nerves, a bite around the head results in clinical disease much more quickly than a bite on the hind foot. The incubation period can range from a couple weeks to many months.

The virus then moves down the nerve fibers to the salivary glands, where the virus can be shed in the saliva. The virus can be present in the saliva for days before the animal shows clinical signs. Therefore the clinical appearance of the animal doing the biting cannot be a reliable assurance that it is not rabid.

The clinical signs of rabies can be extremely variable, but all involve the central nervous system. Initially the animal may become restless and have a change in behavior. The animal may then become quite agitated and aggressive, snapping and biting at inanimate objects. Other animals may show signs of being in a stupor, almost nonresponsive. Many neurologic signs, such as ataxia, paralysis, and seizures, commonly occur with rabies. The paralysis may involve the muscles that control swallowing. The common description of a rabid animal frothing at the mouth is due to the inability to swallow. Rabies has

also been called hydrophobia (fear of water). The animal is not truly afraid of the water but is unable to drink properly. Eventually rabies is fatal.

Rabies cannot be diagnosed in a living animal. The brain must be tested to confirm the diagnosis. Tests are available to detect the rabies virus within the brain tissue. When a human or pet is bitten by a suspicious animal, the animal must be available to test. The diagnosis cannot be made based only on the clinical signs. No treatment is available for an animal or human with the clinical disease.

Prevention is essential in the control of rabies. Many vaccines are available to help prevent rabies. In animals the vaccine must be given before the animal is exposed. Many states and localities have laws mandating the vaccination of pets. These laws have been developed to protect the humans in the communities. In Pennsylvania, dogs and cats must be 12 weeks old for their first rabies vaccination. A booster is required in one year. Many vaccines are currently approved for a booster every three years thereafter.

Exposure to cats also puts humans at risk for **cat scratch fever** or cat scratch disease. Classically the disease develops after a bite or scratch from a cat. Any contact with a cat increases the potential for developing this disease. Often the disease begins with a sore developing at the site of the scratch. Fever and an enlargement of the lymph nodes, typically the ones responsible for draining that region, follow this. Usually patients feel tired and weak in response to the disease. Other signs include headache, weight loss, swollen conjunctiva (surrounding the eye), sore throat, and enlarged spleen.

Cat scratch fever is generally self-limiting, and patients recover within two to four months. In some patients the signs can last much longer. The disease is believed to be caused by a Gram-negative bacillus (*Bartonella henselae*). Antibiotic therapy has not been shown to speed the recovery of infected individuals. This organism generally does not cause clinical disease in the cat that carries it. Kittens are much more likely to carry the disease.

Ringworm is a fungal infection in the skin of animals. The disease can infect any domestic species, and all may be a potential source of infections for humans. In animals, ringworm causes hair loss and scaling (Figure 17–4). Often the regions of hair loss are circular, but they can be of any shape. The regions start small and spread outward. In humans, ringworm also causes hair loss on the scalp. On regions of the skin without hair, the first sign is often a raised red lump. The redness then spreads outward and the center heals. This spreading reddened ring provided the name *ringworm*, even though no worm is involved.

The disease is not highly contagious, but close contact with infected animals increases the risk. Prepubescent children have a higher risk of developing

FIGURE 17–4 Gray, crusty patches of hair loss are classic signs of ringworm.

clinical signs. Avoiding contact with infected animals and thorough washing of contact areas are important control measures in preventing the disease in humans. When clinical signs are visible in an animal, avoiding contact is straightforward. Cats quite often are carriers of the ringworm spores without clinical signs. In households in which humans develop ringworm infections, the pets need to be evaluated even if clinical signs are not evident.

Horses are susceptible to a group of **RNA viruses** that cause an inflammation of the brain. These diseases are Eastern, Western, and Venezuelan equine encephalomyelitis. In horses these diseases often begin with a high fever, stiffness, anorexia, and often colic. As the disease progresses, the horses begin to show neurologic signs. These may include a severe depression, ataxia, sleepiness, head pressing, circling, and constant walking. Some horses may become very agitated and even aggressive (making rabies an important differential diagnosis). These viruses are often carried in birds and are transmitted among animals by a mosquito.

In humans the signs may be quite similar. Fever, headache, confusion, coma, seizures, and death are all possible. Humans develop the disease from mosquitoes as well and are not a threat to other humans. Currently there is no specific cure, and treatment is supportive for an infected individual.

Vaccines are available to prevent the disease in horses. For human protection, control revolves around the mosquito. Eliminating potential breeding areas (stagnant wet areas) and insecticide treatment help to lower the incidence of the mosquito vector. Large wet areas such as marshes and ponds can be a breeding ground for mosquitoes. But mosquitoes can also use smaller collections of water, such as birdbaths and used tires. Monitoring the disease in animals can be useful in alerting humans in that region to watch for potential signs.

Food animals represent a potential source of infectious diseases to humans through dairy and meat products. Listeriosis, or circling disease, in cattle, sheep, and goats is a potential threat to humans. The Gram-positive bacteria (*Listeria monocytogenes*) may be found in harvested feed.

The organism penetrates the roof of the mouth, entering the nerve fibers. The bacteria migrate to the brain, where they cause nerve damage. Many animals show signs of paralysis on one side of the face. This may cause a drooping ear and eyelid and a loss of the ability to blink. Often these animals hold their head to one side and when walking always turn one direction (hence the term *circling disease*). When this occurs, it is nearly impossible to force the animal to turn the opposite direction. Some animals continually walk, whereas others may enter a corner and press forward against a gate or wall. The organism may also cause abortion in cattle.

Infected animals may shed the organism in their milk. Humans consuming unpasteurized dairy products have the potential of becoming infected. In humans the organism also infects the brain. Headaches, fever, nausea, vomiting, coma, and death are all possible signs. Consuming only pasteurized dairy products is essential in preventing the disease in humans. Proper sanitation when handling animals and aborted fetuses is also essential.

Q Fever, caused by *Coxiella burnetii*, results in abortions in sheep and goats. Many infected animals do not show clinical signs. The disease is of much more importance because of the potential threat to humans. In humans, Q fever has a sudden onset of flulike symptoms. Humans may show fevers, headaches, chills, and weakness. Some humans develop chronic signs due to damage to the heart. Only a very small percentage of cases result in death. Raw milk is the primary source of infection. **Pasteurization** is effective in destroying the organism.

Foodborne pathogens are a major concern in food animal medicine. A number of bacteria that are normal intestinal inhabitants and many pathogens of animals can cause serious illness in humans. The goal of animal producers, veterinarians, meat processors, and the human medical profession is to ideally eliminate or, more realistically minimize, the number of cases of foodborne illness.

The major human pathogens include the bacteria *Salmonella, Escherichia coli* (*E. coli* O157:H7 is the strain most commonly seen currently), and *Campylobacter* and the protozoa *Cryptosporidia*. The bacteria all may cause clinical diseases in farm animals, and some, such as *E. coli*, may be part of the normal flora. Primarily these organisms cause intestinal signs, such as diarrhea and anorexia, in the infected animals. *Cryptosporidia* is a protozoa found in food animals that causes severe chronic diarrhea. This organism is a potential threat to contaminate water supplies. Large numbers of all these organisms may be passed in the feces. It is important to recognize that humans are a potential reservoir of these organisms as well.

In recent years, many cases of food poisoning (often *E. coli* O157:H7) have resulted from improperly cooked food. Several of these cases have occurred at fast food restaurants, resulting in large numbers of people being exposed and becoming ill. In humans, vomiting, diarrhea, and fever are often the primary signs. Children, the elderly, and immunosuppressed patients face the highest risk. The disease often becomes so severe that kidney failure results. Many deaths have been reported.

Many food sources may become contaminated with pathogenic bacteria. These organisms can survive on the food source outside the animal. The list of potential contaminated foods is actually quite long and includes beef, pork, dairy products, lamb, poultry, eggs, raw vegetables, apple cider, fish, shellfish, mushrooms, and water. In this chapter the discussion emphasizes the risks associated with meat and dairy products. Many of these principles apply to these other foods as well.

As food is delivered from the farm to the dinner table, there are many potential sources of contamination. The entire process begins with the producer on the farm and then is transferred to the slaughter or processing plant. From this point the food is shipped to a retailer, purchased by the consumer, stored, and eventually prepared and served.

Due to the large number of organisms that may be shed in the feces, the farm does represent a major potential in the contamination of the product. Animals shipped to slaughter typically have some degree of fecal contamination on their hide and hair coat. Producers strive to maintain a healthy herd, which maximizes their productivity. However, not all disease can be prevented, and many healthy animals may be shedding a pathogen. Maintaining good sanitation and minimizing stress of the shipped animals helps to keep contamination to a minimum.

At the slaughter facility, the animals must be skinned and the meat processed. Contamination on the hide can potentially spread to the carcass. Removing the abdominal and thoracic organs also provides a potential source of contamination. Every step, including grinding and cutting, provides a step where the meat is handled and potentially contaminated. Strict

guidelines are in place that emphasize the cleanliness of the facility, the equipment, and the personnel. Inspectors are present to ensure that proper procedures are carried out and that the animals being slaughtered are not showing evidence of disease.

Meat and dairy products are then shipped to a retailer, where proper storage is essential. All the steps involved are useful in minimizing the number of bacteria on the food. It is not realistic to think that all bacteria have been eliminated. Maintaining the food at proper refrigerated temperatures keeps the organisms from dividing.

Likewise, once purchased by the consumer, proper storage is essential. Meats should be thawed in the refrigerator to keep the bacteria from rapidly dividing. An alternative is to rapidly thaw the meat using a microwave. The surface of meats thawed at room temperature can allow bacteria to flourish before the center of the meat is thawed. Meat can become contaminated in the household as well. Proper sanitation in the home is a key step in maintaining a healthy product.

The ultimate responsibility in food safety lies at home. Many precautions are taken at every step until the food is delivered to the consumer. The final step is proper preparation of the food. Meats must be cooked to adequate temperatures to ensure that any existing bacteria have been killed. Red meats need to be cooked to an internal temperature of at least 160°F (poultry should be up to 180°). A meat thermometer is useful to confirm the internal temperatures.

Several diseases have recently reached popular notoriety. Articles are common on **anthrax, mad cow disease, scrapie,** and **West Nile fever.** Their effects on human health make the public very concerned about these diseases.

Bacillus anthracis causes anthrax. This is a spore-forming bacteria that can infect humans as well as animals. Anthrax may be seen in cows, sheep, goats, and horses. The most consistent sign of anthrax in animals is sudden death. Animals may show fever, anorexia, and bloody urine or diarrhea. Just as in humans, the spores can be introduced into an animal by ingestion, by inhalation, or through a wound in the skin. The organism exists throughout the world. Within the United States, certain states have a higher incidence, including South Dakota, Arkansas, Louisiana, California, and Missouri.

Often outbreaks occur during the warm months and may follow heavy rainfall. Heavy downpours may bring anthrax spores to the surface. The most common form in animals is the intestinal form, after contaminated grasses are ingested. The bacteria can then penetrate the lining of the intestinal tract and begin to replicate.

The organism does not survive for long periods within a dead animal. However, these animals often have a bloody discharge (urine, feces, or saliva) that contaminates the environment. It is the contamination of the external environment and the hair coat of the animal that provides the highest risk for humans. Individuals in professions handling animals or animal products are at the highest risk for developing the disease. These occupations include veterinarians and any worker that processes hide, hair, or wool.

In humans the cutaneous, or skin, form of the disease causes a severe sore that becomes very swollen. The inhalation form of the disease causes a severe pneumonia that may be fatal. The intestinal form is the least common and causes fevers, cramping, and diarrhea. The intestinal form may also be fatal. The disease can be treated successfully with antibiotics if detected early. The difficulty is that the disease progresses rapidly and treatment must begin early.

West Nile fever was first detected in 1999 in the United States. This disease is typically found in the Middle East, Africa, and parts of Europe and may infect many animals, including humans, horses, dogs, cats, and several species of birds. In many animals the disease is mild and goes undetected. In a small percentage of cases, more serious illness may result. The West Nile virus can cause inflammation within the central nervous system. The signs may be quite variable but often include ataxia, paralysis, and weakness and may be fatal. The disease is often fatal in birds. Dying birds may provide a clue that the disease is present in an area. Public health officials often monitor the death of birds to detect the presence of the virus.

The virus is spread between animals by mosquitoes. A mosquito that bites an infected animal can then transfer the virus to humans. In humans the disease typically produces flulike symptoms such as fever, muscle aches, headaches, and vomiting. The disease can be fatal in many cases.

Mosquito control is a very important part of managing the spread of West Nile fever. Insecticide spraying and minimizing breeding grounds is essential. Early detection of the disease in dead birds helps to identify the spread of the disease to new geographic areas. This provides health officials with information on where to emphasize control measures.

Mad cow disease and scrapie have become household names. These two diseases and the human disease Creutzfeldt-Jakob disease (CJD) are classified as transmissible spongiform encephalopathies (TSE). TSE is descriptive of the damage caused to brain tissue. Encephalopathy is a disease of the brain. *Spongiform* describes the holes that form in the brain tissue, making it appear spongelike. These diseases are transmissible; that is, they can spread from one infected animal to another.

These diseases are not caused by any of the classic infective organisms described in the Chapter 14. A prion, a poorly understood infectious agent, causes this group of diseases. All other infectious organisms contain protein and nucleic acid. A prion is an abnormally shaped protein molecule. No nucleic acid has been identified within a prion. Prions have an extremely long incubation period, typically measured in years, whereas most bacteria and viruses have an incubation period in the range of days to weeks.

These diseases typically produce a gradual onset of neurologic signs. Mad cow disease, or bovine spongiform encephalopathy (BSE), often begins with a change in the animal's behavior. Infected cattle often become very agitated and aggressive. Small noises or sensations often stimulate a violent reaction from the cow. The disease is eventually fatal.

Scrapie is a similar disease found in sheep and goats. The clinical signs are quite like BSE. The disease often begins with behavioral changes and weight loss. The animal becomes restless and nervous. Some animals develop an incessant itching, causing a sore area on the body. As the disease progresses, the animal worsens, with seizures or severe weight loss and anorexia. There is no cure for scrapie.

CJD has been known in humans for many years. In recent years a relationship has been found between mad cow disease and a new variant Creutzfeldt-Jakob disease (nvCJD). It is believed that the prion responsible for mad cow disease is also the causative agent of nvCJD. It is suspected that ingesting beef products from infected animals is the source of this infection in humans.

Mad cow disease has primarily been a problem in the United Kingdom. The United States has maintained an aggressive surveillance program since 1990. At no point has BSE been detected in the United States. Other control measures have been implemented in an attempt to prevent its occurrence. There is currently a ban against any import of ruminants from countries at high risk for BSE. Mad cow disease may be transmitted through feeding of meat and bone meal from other ruminants. As a result, this practice has also been banned. The U.S. Department of Agriculture and the Food and Drug Administration are taking a very active role in trying to prevent the occurrence of BSE in the United States.

CLINICAL PRACTICE

OBJECTIVE

■ *Relate the Academic Material Learned in This Chapter to Clinical Practice*

This chapter discusses a variety of pathogens that may be shared between humans and animals. Many more zoonoses exist. This discussion helps to illustrate the veterinarian's role in protecting the health of humans. The specifics of each disease vary, but the principles tend to be the same.

Veterinarians work with owners and producers in attempt to keep all animals healthy. Many of the principles discussed throughout this text are once again used. Proper nutrition, minimizing stress, and vaccinating for important diseases helps to protect the health of the animals. When all this fails and animals do become sick, veterinarians also work to keep the health of the humans protected. Educating the client is an important step in this process.

The cow infected with listeriosis that was discussed at the beginning of this chapter offers an excellent example of the need for client education (Figure 17–5). This particular farm family consumed raw milk from their own cows. They had done it for many years and had never suffered an illness. This cow presented a threat to the health of the family. There is a real potential that this cow or another in the herd was shedding *Listeria* organisms in the milk, which could infect a member of the family. I emphasized this risk to them and mentioned that many other organisms, such as *Salmonella*, might also be present in the milk. I had to make a strong recommendation that only pasteurized milk be consumed.

Pasteurization was developed to minimize the risk of pathogens in dairy products. Flash pasteurization is a process in which milk is quickly heated to 71°C and maintained at that temperature for 15 seconds. The milk is then quickly cooled. This process was initially used to kill common pathogens, such as *Brucella* and *Mycobacterium tuberculosis*.

Brucellosis is a reproductive disease that causes abortions in cattle. Humans can develop the disease through contact with infected fetuses or by ingestion of raw milk. In humans, brucellosis is also called undulant fever or Bang's disease. In humans the signs are much like a severe flu, with intermittent and irregular fevers (that is, the fevers undulate up and down).

Tuberculosis is a disease that typically causes a debilitating respiratory infection in humans. Humans often have fever, weight loss, coughing, and chest pain when suffering from tuberculosis. The disease is very difficult to treat, often requiring long-term antibiotic therapy. The disease is still quite common in many parts of the world, although the incidence is low within the United States. The causative organism of tuberculosis, *Mycobacterium tuberculosis*, can be shed in the milk of infected cows. Farmers may also develop the disease with aerosol transmission from the cow. Pasteurization of milk and frequent testing of cattle have helped to decrease the incidence within the United States.

Currently brucellosis and tuberculosis occur infrequently, thanks to stringent testing and control measures. The pasteurization process is still effective on the large numbers of pathogens that still exist. Pasteurization is not a complete sterilization of milk. Some heat-resistant organisms can still survive, which is the reason that milk has a limited shelf life even when stored in refrigeration. Pasteurization is used because it has little effect on the taste of the product.

Rabies is not treatable once clinical signs have begun. When a rabid animal bites a person, the disease is still preventable. The first step is an injection of antibodies at the site of the wound. The goal is that any virus present will be bound with the specific rabies antibodies. The person then undergoes a series of five vaccinations to raise the level of active immunity. The vaccines are extremely effective as long as exposure is known.

People in certain professions, such as veterinarians, wildlife officers, and animal control officers, are at a higher risk than the general population for becoming exposed to a rabid animal. Because of this risk, many elect to undergo rabies vaccination before any exposure. Just as in animals, the rabies vaccine requires a series of vaccinations and then periodic boosters. If a rabid animal bites a vaccinated individual, the person is immediately given boosters but no injection of antibodies. A vaccinated individual will quickly raise their own level of immunity with the booster shot.

As a profession, veterinarians strive to maintain a healthy animal population. In addition, educating the public on proper safety and health concerns helps to maintain the health of the humans involved with the animals. Monitoring the health of animals, conducting routine vaccination programs, controlling parasites, and engaging in sound sanitation practices all help to protect the general public.

FIGURE 17–5 A cow infected with listeriosis; she was unable to rise.

SUMMARY

Recent occurrences of anthrax in humans have caused nationwide concern even though anthrax outbreaks in the human population are extremely rare. Veterinarians and human medical doctors alike are trained to identify more commonly occurring zoonotic diseases. Students of veterinary science should familiarize themselves with the causes, symptoms, treatments, and prevention of such disease conditions.

REVIEW QUESTIONS

1. Define the following terms:

   rabies
   visceral larva migrans
   cutaneous larva
      migrans
   toxoplasmosis
   cat scratch fever
   ringworm
   RNA viruses

   Q fever
   pasteurization
   anthrax
   mad cow disease
   scrapie
   West Nile fever
   brucellosis
   tuberculosis

2. In order to prevent food poisoning, red meat should be cooked to an internal temperature of _____.

3. What causes salmonellosis?

4. Name the disease associated with roundworm larvae invasion.

5. What is another name for creeping eruption?

6. What animal serves as the definitive host of *Toxoplasma gondii*?

7. Is rabies is a viral or bacterial disease?

8. What is another name for rabies?

9. What disease is caused by the Gram-negative *Bartonella henselae*?

10. What shape does a ringworm infection take?

11. What is another name for listeriosis?

12. What problems can arise for pregnant women who contract toxoplasmosis?

13. What treatments are available for a person who has presented with clinical signs of rabies?

14. What group of viruses, mentioned in the text, cause brain inflammation?

15. What is the best defense for prevention of *E. coli* infection?

ACTIVITIES

1. Visit <http://www.fda.gov> to take a quiz about the safety of your own kitchen. Key search words are kitchen and safety.

2. Learn how to prepare and handle foods safely at <http://home.earthlink.net>. Key search words are food and safety.

3. Find a detailed table listing storage times for frozen and refrigerated foods at <http://msucares.com>. Key search words are food storage, frozen, and refrigerated. Also look for a detailed discussion of proper food handling, storage, and cooking.

4. Visit <http://www.cdc.gov> to learn more about rabies. Key search word is rabies. Information is available on the disease, its spread, and its occurrence. Maps at the site show the occurrence of rabies across the United States.

18

Diagnosis of Disease

OBJECTIVES

Upon completion of this chapter, you should be able to:

- List the major methods used to diagnose disease and cite examples of disease diagnosis with each testing method.
- Discuss the clinical significance of disease diagnosis.

KEY TERMS

signalment
borborygmi
ophthalmoscope

packed cell volume
chemistry panel

complete blood cell
count (CBC)

serology

INTRODUCTION

Throughout this text, you have been introduced to a variety of diagnostic tests. In general these tests have been discussed in reference to a specific disease or type of disease. The challenge arises when a case first presents. Every clinician must develop a systematic approach to reaching a diagnosis. There is no perfect system, but common principles do exist.

A DAY IN THE LIFE

A Road I Traveled…

Graduating veterinarians have a variety of choices on how to direct their careers. Opportunities exist with pharmaceutical companies, research, consulting, teaching, and advanced training with internships and residencies. The majority of veterinarians enter private practice. They take on a role very similar to the family physician in human medicine. This is the road I chose to travel.

The transition from veterinary school to private practice can be quite challenging. I joined a practice that at the time had two other veterinarians. I entered practice with a tremendous amount of knowledge but lacked experience. Fortunately, my employers were extremely helpful and guided me as I began my career. New graduates bring the latest information and techniques to their new job. Their employers possess tremendous experience that plays a critical role in the profession.

I was out of school less than a year when I visited a farm to see a cow that had trapped a foot beneath a metal plate. The cow was no longer trapped when I arrived. The injury was severe, though; she had nearly amputated one claw of her front foot. It was obvious that the damage was too severe to save this toe. I thoroughly cleaned the area and proceeded to finish the amputation. I controlled the bleeding and bandaged the foot.

As I sanitized my equipment, I explained the necessary bandage care. I also warned the producer that the cow's life span was likely to be limited. The stress of bearing weight on only one toe would eventually take its toll. Most cows in this condition only last one lactation.

We continued to visit, and the farmer began asking about veterinary school. He inquired, "How do they get all of these different cases, so that you can learn how to do these different things?" He was pleased that the new vet could handle such an unusual situation. I told him that my only experience came from a book! I had never witnessed an amputation of a claw but had been taught about the procedure. Veterinary school had provided me with all the necessary knowledge to handle this situation. My practice experience was improving with time.

One reward of my profession comes from systematically working a case, making a diagnosis, and successfully treating the condition. This process can be quite challenging, but in the end it is also quite rewarding. There are many facets to the entire process. It takes communication with the owner, examination of the animal, and a combination of knowledge and experience to be successful.

A major difference existed between my veterinary school experience and that of private practice. In veterinary school there were often multiple veterinarians working on each case, and the results of testing were often available within hours. For certain tests, such as radiographs, the results were available quickly and in addition had been interpreted by a specialist.

In private practice I do not have the luxury of waiting on immediate results before deciding on a treatment. When I visit a farm and examine a sick cow, I would love to know what her blood calcium level is. Unfortunately, I do not have an on-the-farm test available, so I must rely on my clinical findings. I must decide if the treatment is warranted and then proceed. Often, administering the calcium proves the diagnosis. After receiving calcium, the afflicted animal often responds quickly, thus showing that the calcium was actually needed.

Over the years I have seen many examples of the same cases, and to this day I continue to see new variations of the same diseases. My education is continuing, even after 15 years of practice. Through differing experiences, I see the variation that can occur with any disease.

This chapter discusses the process followed for making a diagnosis. A tremendous number of diseases exist. Furthermore, many diagnostic tools are available. Not every possibility can be discussed. The goal is to show the organized approach that helps in reaching the diagnosis. As you proceed through this chapter, think about the following case that was seen recently at our office. Then consider what procedure would help diagnose the disease.

Merzee, a 10-year-old female dachshund, presents to your office for the first time. The owners are concerned because Merzee has been lying around more than normal and is not eating well. The owner thinks that Merzee has been losing weight.

DISEASE DIAGNOSIS

OBJECTIVE

■ *List the Major Methods Used to Diagnose Disease and Cite Examples of Disease Diagnosis with Each Testing Method*

All cases must begin with a thorough history. The range of questions possible in a good history is extensive. History proves extremely valuable because of the information that it provides. It is also important to recognize that the history does not increase the cost, is not invasive, and has no risk associated with it. It is obvious that in veterinary medicine, the owner's perception is crucial to guiding the process of making a diagnosis. A thorough knowledge of veterinary medicine helps to plan for the questions asked in the history. Certain clues that present with one may lead to further questions.

Veterinarians must also adapt their questioning based on the species of animal. For example, asking what type of work a horse is doing (for example, pleasure, racing, or jumping) helps determine a diagnosis in a case of lameness. On the other hand, that question has no relevance in a case of a vomiting cat. In this case, inquiring whether the cat has access to the outdoors is more important. Cats free to roam have a higher likelihood of ingesting a toxin than a closely observed, strictly indoor cat.

There are basic principles that every history follows. **Signalment** is the basic description of the animal. The signalment includes the name or number of the animal, age, breed, sex (including spay/neuter status), reproductive status (for example, whether the animal could be pregnant, time since last delivery), and use or activity (such as indoor/outdoor cats and dogs, training status of a horse). These simple questions are the first step in developing the decision-making process. The obvious differences in male and female anatomy lead to other important considerations. Certain breeds stand at much higher risks for certain diseases.

The next step is to discuss the chief complaint, the reason that the animal is being presented for examination. Once this point is established, the questions can provide further details on the history of the problems. Common questions revolve around the length of the problem, the course of the disease (such as sudden or gradual onset), and how the disease has progressed. Many times one sign develops early and then progresses into further problems.

It is always important to inquire about home remedies. At times these remedies may be masking subtle signs. For example, an animal being treated with aspirin may not experience the degree of pain felt earlier. The question is also important because the treatment may not be benefiting the animal and may even be worsening the signs. Aspirin again offers an example, where the treatment for a painful condition may result in a dog beginning to vomit.

It is important for the veterinarian to control the flow of the history. Owners have their own perception of what is important and may not recognize other valuable points of information. For example, a case in which a dog is eating aggressively but losing weight requires questions about diarrhea and water consumption. A dog with chronic parasitism may lose weight even though it is eating well. The history of diarrhea may help point to that diagnosis. A diabetic dog often shows the same signs but has an excessive thirst.

It is important to ask very specific questions as well. In one familiar case the veterinarian mistakenly asked if the owners had ever seen diarrhea from their cat. They replied that they had not. Later the veterinarian came to find that the cat never used the litter box in the house and they had never seen a bowel movement. The practitioner assumed that the cat did not have diarrhea. Had the veterinarian asked if they could describe the cat's bowel movements, the owners would likely have replied that they did not know. The veterinarian discovered the cat had diarrhea after it was hospitalized. Although the veterinarian assumed that the diarrhea began in the hospital, the cat's entire history became clearer with further discussion, and the possibility of previous diarrhea was noted.

The next step in diagnosis is the physical examination. During the physical examination (also called the physical or the exam), all the organ systems need to be evaluated (Table 18–1). In many diseases, more than one system may be affected. Detecting the problems in all systems is important in determining the underlying cause. Not every possible point is evaluated on every physical examination. For example, a complete neurologic examination requires testing of all the cranial nerves and the peripheral reflexes. This is not necessary in the kitten that presents for sneezing, sore eyes, and a fever. Emphasis is placed on the major problem during the examination.

Veterinarians develop a system to perform an organized physical. By systematically and repeatedly performing each step in the physical, important systems are not missed. Most veterinarians perform the steps of physical exams in a very specific order. Often during this time, the clients are discussing problems and asking questions. By maintaining a routine, veterinarians are sure to perform all steps involved in a physical.

The first step in the physical is to take a general observation of the animal. Many questions can be answered during the first contact with the animal. Is there any lameness or weakness? Are the eyes sunken, as if dehydrated? Is the breathing normal? Is there a normal awareness to its environment? Is there any distention of the abdomen? Is there fecal contamination

TABLE 18–1 The Physical Examination

General appearance	Awareness/attitude
	Body condition
	Movement: lameness, ataxia
Vital signs	Temperature
	Pulse
	Respiration
	Capillary refill time, color of mucous membranes
	Evaluation of hydration status
Cardiovascular system	Heart: sounds, rhythm
	Pulses: strength, regularity
	Distension of blood vessels (e.g. jugular veins)
	Swellings of extremities or dependent areas (e.g. under the jaw, lowest part of chest)
Respiratory system	Respiratory sounds: evaluated over lungs, trachea and upper airways
	Evaluate respiratory difficulty: open mouth breathing, laboring, does difficulty occur on inhalation, exhalation, or both?
	Can a cough be caused by squeezing on the trachea?
Digestive system	Manure: amount, color, odor, consistency
	Abdominal palpation: abnormal masses, pain, splinting of the abdomen
	Rectal examination
	Mouth examination: normal teeth, foreign bodies
Musculoskeletal system	Normal movement of head, neck, and legs
	Evidence of swelling
	Symmetry between legs
Nervous system	Awareness
	Coordination
	Eyes: structure, reflexes
	Cranial nerves and reflexes
	Peripheral nerves and reflexes
Skin and hair coat	General appearance
	Hair loss
	Sores, rashes
Mammary system	Swellings
	Milk characteristics
Lymphatic system	Lymph nodes: shape, size, pain, symmetry
Urinary system	Abdominal palpation
	External structures
Reproductive system	External genitalia
	Rectal examination (horses, cattle)

on the hair coat that may indicate diarrhea? Does the hair coat appear healthy?

Next most veterinarians take the animal's vital signs. This includes the temperature, pulse, and respiration rate. The pulse is evaluated not only for beats per minute but also for strength and rhythm. In cattle the rate of rumen contractions is also considered a vital sign. This is a very important monitor of the cow's health.

A normal animal's gums or mucous membranes should be pink. An anemic animal has much paler membranes. When pressed by a finger, the gums become much whiter. When the finger is released, the gum should quickly return to its pink color. The elapsed time is called capillary refill time (CRT) (Figure 18–1). In a healthy animal the CRT is less than 1 or

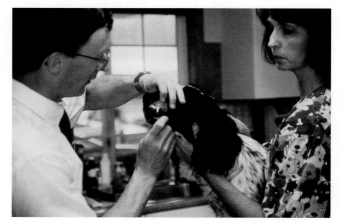

FIGURE 18–1 The dog's mucous membranes are checked for color and capillary refill time. *(Courtesy Patricia Curley.)*

two seconds. In an animal whose blood is not being distributed effectively through the capillaries (such as occurs with animals in shock), the CRT can be much slower. Students can evaluate their own CRT in the fingernail bed. The tissue under the fingernails should be pink. When the nail is pressed, the blood is forced out of this tissue, creating a white appearance. When the nail is released, the blood should quickly return to the nail.

The stethoscope proves a valuable tool in the physical examination. The heart sounds are evaluated to determine if the beat is regular and if there are any murmurs present (Figure 18–2). Review Chapter 4 for an understanding of the normal heart sounds and murmurs. A pulse point may be palpated at the same time the heart is evaluated. It is important to confirm that a pulse is created with each heart contraction. The respiratory system is also evaluated with the stethoscope. Listening to the chest evaluates the lungs for any evidence of disease. Review Chapter 5 for an understanding of lung sounds.

The stethoscope can also be used to listen to the normal sounds of the gastrointestinal tract. In ruminants the activity of the rumen allows for an important evaluation of the health of the animal. A displaced abomasum, a disease discussed earlier in the text, fills with gas and moves upward, out of normal position. To diagnose this, the stethoscope is placed over the side of the cow, toward the end of the rib cage. While listening, the body wall is struck rapidly with a finger. Because the stomach is distended with gas and also contains fluid, the procedure creates a pinging noise (the ping has been described as sounding like a coin dropped in a well). The presence of a ping is diagnostic of gas distending an organ that also contains liquid.

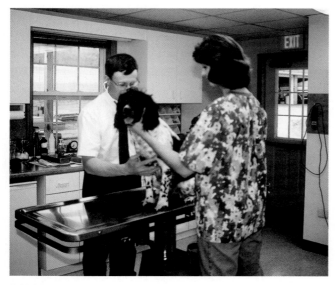

FIGURE 18–2 A stethoscope is used to listen to heart and lung sounds. *(Courtesy Patricia Curley.)*

The size and location of the ping helps to distinguish what structure is distended.

In horses it is also essential to evaluate the normal sounds of the intestinal tract. **Borborygmi** (singular is borborygmus) describes the normal noise made within the intestinal tract as gas and fluid move through the tract. (This is the noise in humans that is commonly described as stomach growling.) This type of noise should be audible in all regions of the horse's abdomen. A lack of noise could represent a problem that has limited the activity of the intestines in that region. Examples include an impaction of feces or a twisted loop of intestine.

In small animals, palpating the abdomen is extremely important. In large animals, only the outer edges of the abdomen can be felt. In small animals, several organs can be identified if the animal is relaxed. In many dogs and cats the kidneys, urinary bladder, spleen, and caudal edge of the liver may be felt. These can be evaluated for normal size, shape, and sensitivity. The palpation of the abdomen is the most important aspect in evaluating the urinary system in small animals. Other organs may become palpable if there is an abnormality. Normally the intestines cannot be identified. The same structure with a large foreign body or intussusception may be felt.

A rectal examination evaluates the size and shape of the prostate gland in male dogs. In addition, the internal bony structure of the pelvis may be felt. The character of the feces in the colon is also detected during this procedure. A rectal exam can be a very important diagnostic tool in evaluating a dog straining to defecate. An older male dog that has been having difficulty defecating could have constipation with dry, hard stool or a prostate gland that has enlarged to the extent that the feces has difficulty moving through the narrow opening that remains.

In horses and cattle the rectal examination may provide even more information. In these animals the veterinarian may insert an entire arm and palpate many structures in the caudal regions of the abdomen. This is most routinely done to evaluate the reproductive tract. In addition, regions of the intestinal tract, rumen, pelvis, kidney, and lymph nodes may be felt. Just as in small animal medicine, much can be learned about the fecal characteristics during this procedure.

Evaluation of the nervous and musculoskeletal system begins with observation of the animal. The gait or movement of the animal can easily show lameness, weakness, or neurologic problems affecting the motor nerves. The interaction of the animal with its surroundings can also provide clues as to whether the animal can hear and see.

A complete neurologic examination evaluates all the cranial nerves and reflexes of the body. More often, enough tests are performed to localize a problem. For example, a dog with hind limb paralysis due to intervertebral disk disease would undergo extensive examination of the peripheral reflexes. The animal would be evaluated for awareness and to interpret the prominent cranial nerve functions (for example, whether the animal can see). The neurologic examination may then be able to localize the problem to the spinal cord and rule out problems in the brain.

Examination of the eye may be considered in the neurologic examination. The reflexes of the pupils and the ability to see are an essential part of the exam. In addition, the optic nerve is the only portion of the nervous system that is actually visible. An **ophthalmoscope** is the instrument used to observe the structures in the interior of the eye, such as the optic nerve, retina, and retinal blood vessels (Figure 18–3).

The veterinarian must rely on visual observation and palpation of the musculoskeletal system. Musculoskeletal problems may be seen and felt as a swelling. Often lameness may direct attention to a particular limb. During the exam, joints are bent and manipulated to detect any limitations or unusual movements. At times the movement may create an unusual noise or feel. A damaged joint may develop a clicking or popping feel as the joint is bent. The animal also gives feedback if the procedure is painful. With pain, the animal may attempt to bite or kick the veterinarian. Even well-mannered animals often develop tense muscles in the region when a procedure causes pain.

The lymphatic system can be examined by palpating lymph nodes that are located under the skin. These structures can be felt in many animals even when they are normal. The lymph nodes are evaluated for enlargement or pain. An enlargement may give evidence

of infection in that region or could possibly be an indication of cancer.

When examining lactating animals, the mammary system is checked. The mammary glands can be palpated to evaluate for any swelling or heat. The milk is also examined for any signs of abnormal secretion. Mastitis typically causes flakes or clots in the milk. These clots are formed by large numbers of white blood cells (WBCs) that enter the mammary gland due to inflammation.

A case is evaluated following the history and physical examination. The specific approach taken by each veterinarian varies. One method is called the problem-oriented approach. In this method the problems are listed. Next, a list of differential diagnoses is developed. Ideally, all the problems should be covered by one disease. Unfortunately, this is not always possible. There are times where an animal has two diseases occurring simultaneously.

The diagnoses are ranked with the most likely causes first. A diagnostic plan is then developed to confirm or rule out the diseases. A large list of potential tests is available. These tests include blood work, urinalysis, radiology, biopsy, and pathology. These tests have been discussed with individual diseases throughout the text.

Wide ranges of blood tests are available. **Packed cell volume**, complete blood cell count, and a **chemistry panel** are among the most commonly used blood tests (Figure 18–4). The packed cell volume is a rapid test that provides the percent of the blood composed of red blood cells. In this test, blood is placed in a

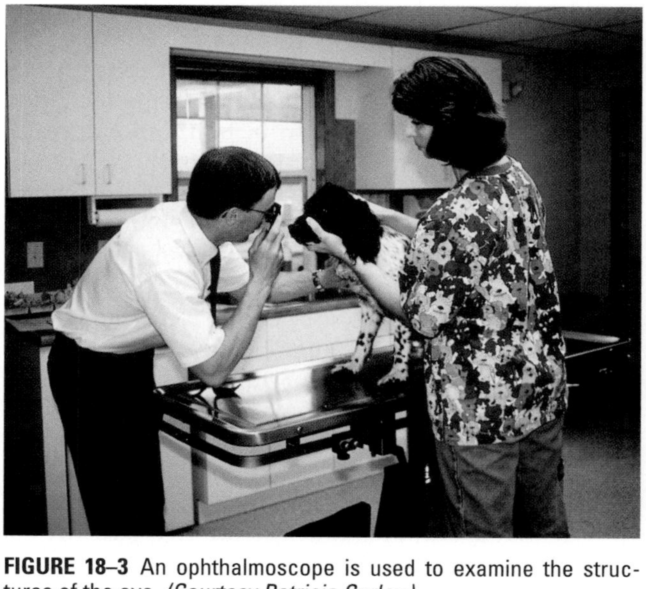

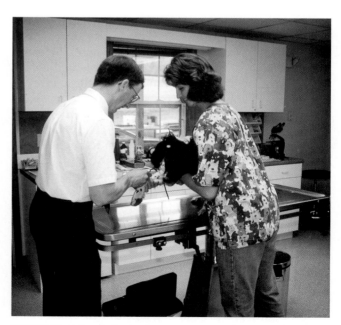

FIGURE 18–3 An ophthalmoscope is used to examine the structures of the eye. *(Courtesy Patricia Curley.)*

FIGURE 18–4 A vein in the front leg is a common site to obtain a blood sample. *(Courtesy Patricia Curley.)*

small capillary tube and spun in a centrifuge. The red blood cells settle to the bottom of the tube and plasma rises to the top. A simple measurement then allows the packed cell volume to be established. The plasma from the tube can be placed on a refractometer to measure the protein. This test helps in monitoring patients for changes. The test requires only a small amount of blood and is completed within a few minutes.

A **complete blood cell count (CBC)** requires a larger volume of blood. This test also evaluates the red blood cell (RBC) count (often a packed cell volume is a part of this testing). In addition, the size of the RBCs and the amount of hemoglobin is measured. This can provide a more complete evaluation of the anemic patient. The number of platelets present is also evaluated.

The complete blood cell count also measures the total white blood cell (WBC) count and provides a breakdown of the types of cells present. This aids in evaluating infectious and inflammatory diseases. The role of the individual white blood cells was discussed more completely in Chapters 4 and 11.

A classic example of WBC changes occurs in a bacterial infection. Typically the total white blood cell count, and primarily neutrophils, increases rapidly. This change occurs rapidly as mature neutrophils are released from the bone marrow. As the infection continues, the bone marrow releases more immature neutrophils (band cells) trying to maximize the infection fighting cells available.

A tremendous number of neutrophils can be lost from the bloodstream during a serious infection. A high band cell count combined with a low mature neutrophil count can be an indication that the bone marrow is rapidly releasing immature cells but is unable to keep numbers at adequate levels. This can occur in an animal with a widespread, overwhelming infection. Monitoring the changes that occur in white blood cell counts over the course of infection can be very informative.

Numerous individual tests can be run on blood. Most laboratories offer a collection of tests measuring components that are often evaluated. The collection of tests often has a specific name, such as a chemistry profile or screen. The efficiency of routinely performing the same group of tests allows this profile to be offered at a lower price than if all the tests were run individually. The specific tests included vary among laboratories.

The chemistry profile often evaluates blood sugar, electrolytes, protein, liver and pancreatic enzymes, bilirubin, and nitrogen-containing wastes. Blood sugar obviously evaluates the endocrine function of the pancreas but also can be altered with diet and stress. Cats are especially prone to having an elevated blood sugar due only to the stress of being at the veterinarian's office. The electrolytes are essential to evaluate the hydration status of the animal and also evaluate the hormones that regulate these levels. For example, a diagnosis of hypoadrenocorticism might be suspected from the results of a chemistry profile with elevated potassium and low sodium levels.

The protein level in the blood may also be elevated when an animal is dehydrated. Other factors also influence the total protein in the blood. Protein divides into albumin, which is produced in the liver, and globulin, the protein of antibodies. Albumin levels can be decreased due to low production in an animal with a diseased liver. In addition, excessive loss from parasites or a disease of the kidney or intestines may also result in low levels. When albumin levels are too low, the osmotic pressure of the blood may become so low that fluid leaks from the blood vessels and into the surrounding tissues. If the globulin fraction of the protein is elevated, it may be a result of a chronic infection. In response to the long-term infection, the body has had the opportunity to produce a large amount of antibody.

The liver is a metabolically active organ. It plays a number of roles and is able to carry out these functions due to the variety of enzymes contained in cells. A small amount of these enzymes leak into the bloodstream on a regular basis. Two examples of liver enzymes are alanine aminotransferase (ALT) and alkaline phosphatase (alk phos). ALT is found only in the liver cells, and therefore an increase provides direct evidence of damage to liver cells. Alkaline phosphatase is found in other locations than just the liver, so this must be interpreted. Within the liver, alk phos is found in the cells of the bile ducts. A large increase in alk phos can represent a blockage of the bile ducts. Alkaline phosphatase also increases in response to corticosteroids, so an increase can be found in hyperadrenocorticism.

Bilirubin is also measured in many chemistry profiles. Bilirubin is produced as red blood cells are destroyed. The liver is responsible for clearing the blood of this pigment. An elevation in bilirubin can result when the liver is not functioning properly or in cases where RBCs are being destroyed too quickly. Combining a complete blood cell count and the bilirubin measure can help to detect if RBC destruction is occurring.

The exocrine portion of the pancreas produces the enzymes lipase and amylase. In much the same manner as the liver enzymes, any inflammation in the pancreas causes an elevation of these enzymes in the blood. Pancreatitis, an inflammation of the pancreas, is a common cause of vomiting in dogs. This is especially common in overweight dogs that are fed a meal high in fat.

Creatinine and urea nitrogen are two measures of nitrogen-containing wastes. These two compounds are cleared from the bloodstream by the kidneys. An elevation can be a result of kidney disease or dehydration. In dehydration the protein levels are usually increased as well. In diseases such a kidney failure, other features (for example, elevated phosphorus and anemia) often help to rule out dehydration.

The complete blood cell count and chemistry profile provide a tremendous amount of information. This discussion is not designed to explain all the possible changes that occur. The goal is for the student to understand the thought process designed in choosing such tests. These two simple tests provide an evaluation of a wide range of organ systems. Understanding the physiology and function of these systems is essential to interpreting the results. A large portion of this text has emphasized this information in an attempt to provide an understanding of the disease conditions that may result.

A tremendous number of tests may be performed on blood. Hormone levels and the normal feedback control of these hormones can be evaluated. The speed at which blood clots, as well as the individual clotting factors, may also be measured. Toxins can also be detected within the blood, but it generally requires some suspicion of the specific toxin involved. Many of these tests are quite expensive, and most toxins require that a specific test be requested.

In addition to evaluating the function of the kidneys through blood testing, urinalysis may provide information as well. Chapter 6 has a detailed description of urinalysis. Often urinalysis is used in conjunction with information derived from the chemistry profile. For instance, consider a dog with a chemistry profile showing a moderate elevation in both urea nitrogen and creatinine. To determine if this is a result of dehydration or kidney disease, a urine sample is collected. Typically, if the animal is dehydrated, the specific gravity of the urine should be very high, as the kidneys attempt to conserve the body's water. In kidney disease it is much more likely that the specific gravity will be much lower.

Radiographs have been shown throughout this text as a means to look within the body (Figure 18–5). Evaluating bone often provides the clearest examples of how beneficial a radiograph can be. A complete fracture of a bone is an example of a diagnosis that is definitively confirmed with a radiograph. Not all radiographs provide such a concise diagnosis.

Radiographs provide valuable information because the x-rays penetrate tissues of different densities at different levels. The five densities found in animals include air, fat, soft tissues, bone, and mineral (see Chapter 3). On a radiograph these would range from dark to light, respectively. Changes from the normal pattern can be interpreted.

Consider a chest radiograph. The thorax is clearly outlined with the bones of the vertebrae, ribs, and ster-

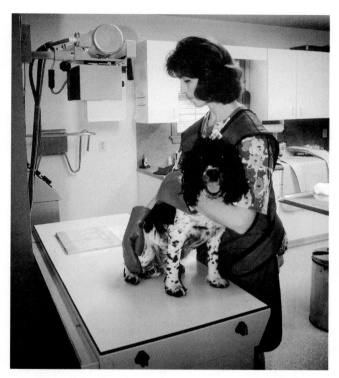

FIGURE 18–5 Wearing protective apron and gloves, a veterinary assistant prepares to radiograph a dog. *(Courtesy Patricia Curley.)*

num (these appear light on the radiograph). The air-filled lung tissue appears dark. The heart and surrounding vessels appear as a medium density. The lung field should fill the entire chest cavity. If fluid is present in the thorax, the lungs are compressed and the fluid can be seen to outline the lower portion of the lung lobes. The detection of fluid in the thorax may at times be quite obvious. However, the radiograph does not define the type of fluid. The fluid might be blood, pus, or lymph. Although the radiograph detected the problem, further diagnostic tests must be used to finalize the diagnosis.

Knowledge of anatomy proves key in radiology. The normal positioning of structures must be evaluated, but also the normal size in relationship to other structures is interpreted. The chest radiograph again provides an example. The size of the heart on the radiograph can be interpreted in relationship to the size of the chest. This can be used to detect an abnormal enlargement of the heart.

In many situations, radiographs show clues that must be combined with the results from other tests. In the discussion of chemistry profiles it was mentioned that amylase is elevated with pancreatitis. Radiographs are often taken of these cases to rule out intestinal foreign bodies and tumors. In pancreatitis, inflammation surrounds the pancreas, which lies behind the stomach and liver. On a radiograph this appears as a loss of detail. The distinction between adjoining structures becomes less obvious.

The differences in density of normal structures may not be adequate in making the diagnosis. Certain conditions may require the use of a contrast medium to help outline certain structures. Many examples have already been shown. Administering barium orally can help to detect obstructions in the intestinal tract. Dye may be injected into the vertebral canal to confirm a diagnosis of intervertebral disk disease. Certain bladder stones and tumors can be shown more clearly with contrast radiographs. In this situation a gas, a dye, or both can be used to fill the bladder.

Another commonly used technique is submission of tissue samples or biopsies to a laboratory for evaluation. It is common to surgically remove a tumor and submit the entire structure (Figure 18–6). In this excisional biopsy the goal is to remove the entire tumor. The purpose of the biopsy is to determine the type of tumor and whether normal tissue is present at the margins of the specimen. If tumor cells are present at the margins of the specimen, there is a much greater chance that the cancer will recur.

Determining the type of tumor also provides the information necessary to decide on further treatment. Various tumors behave differently. If the diagnosis is a benign tumor and there are no tumor cells present in the surgical margins, no further treatment is usually required. With malignant tumors the biopsy can provide information on future outlook. Some malignancies invade the local tissues, whereas others are likely to spread to more distant locations. A decision can then be made regarding whether additional treatment may be necessary. Radiation or chemotherapy can be used to kill cancer cells that remain in the body.

There are situations in which the entire tumor cannot be removed or more information is needed prior to surgery. In these situations a small portion of the tumor may be submitted. This sample may be taken surgically; that is, a region of the tumor is removed. Another technique is to aspirate cells from the region with a needle and syringe. The needle is inserted into the tumor, and suction is applied with the syringe. Cells gathered in the needle can then be forced onto a slide and that sample submitted. This technique is less invasive than surgery but collects a much smaller sample. There are times when the aspirate does not provide enough diagnostic information. Within a tumor there can be normal cells as well, and by collecting a relatively small sample, the diagnosis may not be possible. In addition, the structure of the tissue is disrupted and the cells are spread over a slide. Other tools are available that allow a small piece of tissue to be collected. The sample is still small but maintains the normal structure of the tissue being sampled.

The same techniques can be used to diagnose any diseased tissues, such as those from animals with autoimmune and infectious diseases. In the section on radiology, disease conditions in which fluid is present in the thorax were discussed. Often a final diagnosis cannot be made from the radiograph. The fluid can be aspirated from the chest cavity and submitted for analysis. The contents of the fluid and an analysis of the cells from within the fluid are often used for the final diagnosis.

With all these processes, the pathologist must have a tremendous knowledge of cells and tissues. The pathologist must first be able to recognize what normal structures are present and interpret all the abnormalities. Pathologists also perform autopsies, or postmortems, on animals that have succumbed to a disease. Autopsies have a tremendous role in herd health. With an individual animal, an autopsy may help to explain the cause of death and help the owner cope with the loss. In herd situations the autopsy may provide the information necessary to prevent a large outbreak within a herd. In such a situation the loss of an animal may provide valuable information that prevents a much larger loss.

The pathologist is aided by a complete history. With this information the pathologist may direct attention to the most likely affected regions. The pathol-

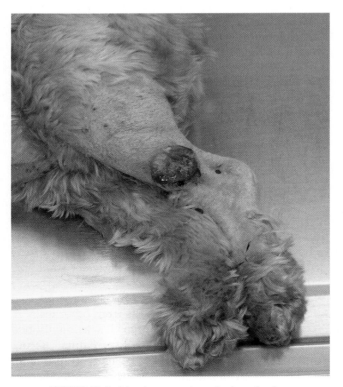

FIGURE 18–6 A benign growth on the leg of a dog.

ogist has the option to do biopsies on any tissue. If a specific region is suspected, extra samples may be collected. For example, the brain and spinal cord cover a large area. If a specific disease is suspected based on the clinical signs, additional samples may be harvested from the central nervous system. In addition to sampling the tissues and organs for biopsy, samples may be taken to isolate infectious organisms or test for toxins.

Bacterial and fungal culturing were discussed in Chapter 14. Fluid from infected regions or swabs of infected tissue may provide the organism that allows for its identification. These samples can be taken from a living animal or during an autopsy. In the living animal the fluid may be milk, urine, blood, swabs from mucous membranes, or fluids aspirated from diseased regions. Many veterinarians perform bacterial and fungal cultures at their office. For a more detailed identification or for isolation of viruses, samples are often submitted to a diagnostic laboratory.

In certain cases it is difficult to isolate an infectious organism. **Serology** measures the presence of antibodies against a specific organism. It is essential to understand the humoral immune response to interpret the results of a serologic test (see Chapter 11). A positive result shows that the animal has been exposed to the organism but does not prove that it was the cause of the disease. A negative result can occur if the test is taken before the animal has the opportunity to increase antibodies to a detectable level.

To confirm a diagnosis with serology, the animal is tested twice, with usually 3 to 4 weeks between samples. If the level of antibody or titer changes four times, it is proof that the animal was exposed and responded to the organism. The change can either be an increase or decrease depending on the timing of the testing.

A tremendous number of tests are available. Typically one or more tests are run that are most likely to confirm or eliminate the diagnoses on the differential list. When the results are available, the problem list is adjusted to account for any new findings. Combining all the information available from the history, physical examination, and subsequent testing, the differential list is reevaluated.

Several possibilities exist. Ideally a diagnosis is confirmed with the information and treatment then begins. For example, a 15-year-old cat presents with an increased thirst, increased urination, poor appetite, weight loss, and vomiting. The blood chemistry shows elevated creatinine, urea nitrogen, and phosphorus. In addition, blood sugar and thyroid hormone levels are normal. The CBC shows a normal WBC count but an anemia. Urinalysis shows extremely diluted urine, even though the animal appears dehydrated. The col-

lection of information confirms the diagnosis of renal failure, and appropriate treatment options may begin.

After the initial testing, one of the diagnoses may become more likely, but further testing is necessary to confirm that fact. Consider a dog that presents for weakness, vomiting, and diarrhea. The dog had seemingly been very normal until the signs began today. The list of differentials for a dog with vomiting and diarrhea is quite long. The chemistry profile shows a mild elevation in urea nitrogen and creatinine, elevated potassium, and low sodium. The result is very suggestive of hypoadrenocorticism but does not completely confirm the diagnosis. With this high level of suspicion, a more specific test may be run to confirm the diagnosis.

Another option that is less favorable occurs when the initial testing offers no clues toward furthering the diagnosis. One solution is to run additional tests. For example, when a chemistry profile and CBC offer no guidance, radiographs may provide more information. At times all the options available to the veterinarian do not provide the answer. These cases may then be referred to a specialist. The specialist has taken advanced training in a particular field (such as surgery, medicine, cancer, or ophthalmology). The advanced training of specialists and the availability of more sophisticated equipment often provide the diagnosis. Many of the advanced technologies that are available in human medicine are also available at some of the university veterinary hospitals and advanced referral centers. It is beyond the scope of this text to discuss these tools, but the student may be familiar with the names of the equipment (for example, computed tomography [CT] scans, magnetic resonance imaging [MRI], ultrasound machines, and endoscopy). At this time the economics of such technology prevents their availability at most local veterinary clinics.

A final diagnostic tool comes only with the years of training and the subsequent experience that every veterinarian gains. Each experience helps to train the veterinarian and show the variations that occur with each disease. In many situations, treatment must be started before all the tests are completed. The veterinarian relies on experience to begin treatment, often directing it at the one or two most likely diseases. An animal responding to a specific treatment can be one more hint confirming a diagnosis.

There are many times when the results of a test will come back normal, not supporting the results that are anticipated. It is tempting to report this as not providing any helpful information. Even normal results provide valuable information. Normal values help to rule out many diseases. It is the total information provided by normal and abnormal testing that allows the veterinarian to make a final decision.

CLINICAL PRACTICE

OBJECTIVE

■ *Discuss the Clinical Significance of Disease Diagnosis*

The process of diagnosis may seem very confusing at this point. The huge number of different tests available may seem overwhelming. The goal of this chapter is to show the different types of tests available and how the thought process proceeds. This section uses the sample case from the introduction to summarize the diagnostic procedure.

Case Summary: Merzee, a 10-year-old female dachshund, presents for the first time. The owners are concerned because Merzee has been lying around more than normal and is not eating well. The owner thinks that Merzee has been losing weight.

This is the first visit of this dog to the office, so a complete history is necessary.

History: *Owners' responses in italic.*

1. How long has she had the problem? *It may have been going on for a couple weeks but has gotten much worse over the last 2 days.*
2. Has there been any vomiting or diarrhea? *Not really, although she did vomit once 3 or 4 days ago.*
3. Is she up-to-date on her vaccinations? *Yes.*
4. Has she been spayed? *No.*
5. Has she been in heat recently? *Maybe, but we're not sure. She acted a little funny a few weeks ago.*
6. Has there been any change in her water consumption? *Now that you mention it, she may be drinking more water recently.*
7. Is she eating anything? *She has been eating a little bit but hasn't eaten anything today.*
8. What is her normal diet? *She eats anything we eat.*
9. Does she eat any dog food? *Oh my, no, she doesn't like dog food.*
10. Does Merzee run free at all? (Evaluates the likelihood that she could have been traumatized or ingested a toxin.) *She never leaves the house unless she is on a leash.*

Physical Examination:

Temperature: 102.8°F (slightly elevated)
Pulse: 150 beats per minute (seems normal for an excited dog)
Respiration rate: panting (again, very excited)
General appearance: mildly obese, walks well with no limping, very aware of its surroundings
Mucous membranes: good pink color, slightly tacky (mild dehydration?), with capillary refill time < 1 second

Cardiovascular system: regular heart rate, no murmurs heard, good pulses
Respiratory system: normal lungs sounds, no difficulty breathing
Digestive system: abdomen very tense and somewhat distended, difficult to palpate, normal bowel movement with no parasites found
Urinary system: unable to palpate kidneys or bladder
Neurologic system: no abnormalities found
Musculoskeletal system: no abnormalities found
Skin: no abnormalities found
Mammary system: no abnormalities found
Lymphatic system: no abnormalities found

Problem List:

Anorexia
Weight loss
Increased thirst
Mild fever

Differential Diagnoses:

Infected uterus (consistent with the dog's age, fever, possible history of recent heat; increased thirst, anorexia, and weight loss common)
Kidney failure (increased thirst, anorexia, weight loss, age are consistent; fever not typical, but dog was very excited)
Diabetes mellitus (increased thirst and weight loss are consistent; initially appetite is increased; usually no fever unless there is a secondary infection)

Diagnostic Plan:

Blood chemistry—evaluates kidney function, liver function, electrolytes, and blood sugar
Complete blood cell count—evaluates for infection, along with red blood cell count
Abdominal radiographs—evaluate for an enlarged uterus (Figure 18–7).
Urinalysis—evaluates for kidney disease

Results: The group of blood tests has helped to illustrate the cause of Merzee's illness. The chemistry profile shows a slight increase in urea nitrogen (Table 18–2). This is consistent with mild dehydration. The CBC shows elevated WBCs, neutrophils, and band cells. These changes, along with the fever, are consistent with the body's response to an infection. The abdominal radiographs show a region in the lower portion of the abdomen with a solid density. Intestines are generally present in this location, so there is evidence that some structure is displacing them.

It is not unusual to proceed with treatment without a guarantee of the diagnosis. Merzee's case offers

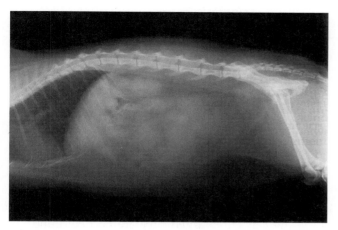

FIGURE 18–7 A radiograph of a dog with a uterine infection. The uterus is quite large and filled with pus. The intestines are displaced from their normal position.

a perfect example. The history of an older (not spayed) female dog, recently in heat, combined with the blood test results and the radiographs makes the diagnosis of uterine infection extremely likely. Often the radiograph does not definitively prove that it is the uterus that is distended, but the combination of diagnostic clues allows treatment to begin.

Merzee did have an infected uterus, which was removed surgically. Fortunately, she responded well. An infected uterus can be a life-threatening condition. The infection can become overwhelming, even entering the bloodstream. The kidneys can also be damaged with this condition, leading to kidney failure. This condition is one reason that routine spaying of young healthy dogs is highly recommended.

TABLE 18–2 Merzee's Blood Results

Test	Result	Reference Range	Units
Glucose	84	65–120	mg/dl
Urea nitrogen	18	6–24	mg/dl
Creatinine	1.2	0.4–1.4	mg/dl
Sodium	148	140–151	mEq/L
Potassium	4.6	3.4–5.4	mEq/L
Chloride	110	105–120	mEq/L
Bilirubin	0.1	0.0–0.4	mg/dl
Alkaline phosphatase	119	10–150	units/liter
Alanine aminotransferase (ALT)	12	10–70	units/liter
Total protein	7	5.2–7.2	g/dl
Albumin	2.6	2.4–4.3	g/dl
Globulin	4.4	0.9–4.0	g/dl
Calcium	10.2	7.9–12.0	mg/dl
Phosphorus	3.7	2.1–6.8	mg/dl
Amylase	702	400–1400	units/liter
White blood cell count	86800	6000–17000	cells/μl
Red blood cell count ($\times$1000000)	5.52	5.5–9.00	cells/μl
Packed cell volume	37.5	37.0–54.0	%
Mature neutrophils	55600	3000–11800	cells/μl
Bands (immature neutrophils)	20800	0–300	cells/μl
Lymphocytes	5200	1500–5000	cells/μl
Monocytes	2600	0–800	cells/μl
Eosinophils	0	0–800	cells/μl
Basophils	0	0–100	cells/μl

SUMMARY

Veterinarians begin examination of animals by establishing a history. The history will give the practitioner clues to disease diagnosis. After taking the history, the veterinarian follows a step-by-step process designed to eliminate causes and eventually lead to a diagnosis. By routinely moving through all examinations in this same orderly manner, veterinarians can confidently come to diagnostic conclusions.

REVIEW QUESTIONS

1. Define the following terms:

   signalment
   borborygmi
   ophthalmoscope
   packed cell volume

   chemistry panel
   complete blood cell
      count (CBC)
   serology

2. True or False: Establishment of home remedies previously given is irrelevant when diagnosing a disease condition.

3. True or False: The problem-oriented approach is one means veterinarians use to arrive at a diagnosis.

4. The third step in diagnosing a disease is performing a _____.

5. _____ is produced as red blood cells are destroyed.

6. To confirm a diagnosis with serology, the animal in question is usually tested _____ with several weeks between samples.

7. If an animal is dehydrated, will the specific gravity of urine be high or low?

8. Can stress interfere with blood test results in cats?

9. Can a stethoscope be used to evaluate intestinal sounds?

10. What is the first step a veterinarian must take in diagnosing a disease?

11. After establishment of a history, what is the second step in diagnosing a disease?

12. Describe the mucous membrane of an anemic animal.

13. How can an enlarged prostate be determined in a dog?

14. List two problems that a lack of noise in the intestinal tract could represent.

15. List specific information found in the signalment.

ACTIVITIES

In this case the instructor will serve as the moderator to provide information about the animal. The pertinent history has been provided. You may ask additional history questions. The instructor may answer with relevant information or may say "I don't know." Unfortunately, in the real world the person bringing the animal to the veterinarian does not always know the answers.

Step 1: The initial history and signalment.

It is 12:30 A.M. on December 27 and the phone rings. When you answer, a very upset man is telling you that Buffy the cocker spaniel has been vomiting for the last 5 hours. He is very concerned that she has made a mess all over the floor, and by the way, she seems to be getting weak. He wants to bring her in immediately. As you crawl out of bed, you tell him that you will meet him at the office in 15 minutes.

A young man and woman bring Buffy into the office. Buffy is a 60-pound spayed female cocker spaniel. She is six years old. As you look over the records, you see that she has been well vaccinated and is up to date. The couple tells you that they are dog-sitting Buffy for the actual owners.

Step 2: Develop a history from the sample questions provided. Your questions may not be identical to other students. Your questions about topics not cov-ered can be answered with "I don't know" because it is not this couple's dog.

Step 3: Ask about specific findings that concerned you on physical examination. Office examination is $28.00.

The initial impression of the dog as it enters the room: Dog walks without lameness, is aware of surrounding, has a slight arch to its back as if having abdominal or back pain. The dog is obese.

Temperature: 102.6°F
Pulse: 185, strong and regular
Respiration: panting (very anxious)
Mucous membranes: pink, slightly tacky or dry
Capillary refill time: less than 1 second
Cardiovascular system: regular heart rate, no murmurs heard, good pulses
Respiratory system: normal lungs sounds, no difficulty breathing
Digestive system: abdomen very tense; Buffy acts like she's in a lot of pain when front portion of the abdomen is being palpated; normal bowel movement, no parasites found
Urinary system: kidneys feel normal in size and shape; bladder small, no stones palpable
Neurologic system: no abnormalities found

Musculoskeletal system: no abnormalities found

Skin: oily but appears healthy; skin turgor slightly decreased

Mammary system: no abnormalities found

Lymphatic system: no abnormalities found

Step 4: Establish a problem list.

Step 5: Prepare a list of differential diagnoses. You may not have a specific name for the disease, but you should suggest likely causes. Use the MAD TIN mnemonic to aid in the process.

Step 6: Choose from the list of diagnostic tests that follow. Remember, economics are very important. Run only tests that are necessary. The owners will be very upset if they find that unnecessary tests have been run. Choose a test or group of tests that need to be run immediately. Once the results are available, further testing may be selected. After selection of tests, contact your instructor for results.

Packed cell volume and total protein	$4.00
Complete blood cell count	$16.00
Platelet count	$11.00
Chemistry profile	$24.00
Urinalysis	$8.00
Chest radiographs	$53.00
Abdominal radiographs	$53.00
Myelogram	$250.00
Barium series	$130.00
Adrenocorticotropic hormone (ACTH) stimulation test	$70.00
Dexamethasone suppression test	$60.00
Exploratory surgery and biopsy of stomach and intestines	$500.00

Step 7: Based on the results of the testing, physical, and history, choose a course of action from the following. The instructor will share the correct course of action.

Buffy likely has an intestinal foreign body and requires immediate surgery.

Buffy is in the final stages of kidney failure and immediate intravenous fluid therapy is necessary.

Buffy likely has a stomach tumor and should be treated with chemotherapy.

Buffy likely has pancreatitis and should be treated medically to control the vomiting and correct the dehydration.

Buffy should be referred to the university hospital for a more thorough evaluation.

UNIT IV

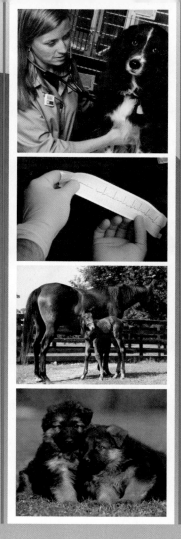

Surgery

19

Principles of Surgery

OBJECTIVES

Upon completion of this chapter, you should be able to:

- *Explain the clinical significance of the basic principles of successful surgery.*
- *Explain the clinical significance of healing of lacerations by first and second intention.*
- *Explain the clinical significance of common considerations in veterinary surgeries.*

KEY TERMS

aseptic technique
disinfectants
sterilization
autoclave
antiseptic

granulation tissue
first intention healing
golden period
second intention
 healing

proud flesh
débridement
hematoma
seroma
abscess

dehiscence
intestinal anastomosis
gastric dilation–
 volvulus syndrome
necrotic

INTRODUCTION

Surgery remains a challenging field in veterinary medicine. In addition to requiring knowledge of anatomy, physiology, and disease conditions, surgery requires delicate manual skills. Much can be taught in a classroom setting about the techniques and principles of surgery. Eventually the final learning process must come with experience.

A Chance to Cut Is a Chance to Cure...

This adage is supposed to represent the mindset of a true surgeon, someone who is anxious to operate on any animal in an attempt to save its life. Realistically, surgeons evaluate each case individually and decide if surgery is the best option. Surgery is one of the most visible skills that a veterinarian possesses.

When I attended veterinary school, the first 2 years provided little exposure to live animals. There were a few exercises on topics such as physical examination, restraint, and foot trimming, but there was very little responsibility for the animal's well-being by the veterinary student.

During third year we finally had the opportunity to have responsibility for the livelihood of a living animal. Our class was divided into groups of three. Each group had a dog to care for by performing physical examinations to ensure that the animal was healthy for anesthesia and surgery. Finally the big day arrived—our first experience at performing surgery. Within the group, one student performed the anesthesia, one acted as the primary surgeon, and the third stood as assistant surgeon.

This was truly an exciting time. With anesthesia the animal is sedated to a point that it does not outwardly respond to pain and is immobile. The concern in the back of everyone's mind was the realization that administering too much would cause the dog to die. The surgeons had their own responsibilities. For the first time, we were going to cut into a living animal and repair a wound. After all the long hours of bookwork, we had finally reached the real world! We were sure nervous!

As students, we had a tremendous amount to learn. Our movements were inefficient; we were slow and ever so cautious. As in any surgery, there was bleeding. It was quite scary the first time the student surgeon cut through an artery and we all saw blood pulsing across the table. To control the bleeding, we had to hold pressure on the bleeding vessel with a gauze sponge, clamp it with an instrument, and then tie it off with suture. It seemed so simple in theory, until the first time that blood was spraying us.

The same process that was so scary then is one that I have now done thousands of times. I am much more comfortable as a surgeon after many years of experience. But even with experience there are times when surgery can make my heart race. Surgery has provided me with some of the greatest and worst moments of my career.

As a new veterinarian, my confidence was tested at times. To this day there are still cases that do not work out the way I plan. As a recent graduate, though, the failures made me question my own skills.

Surgery to correct a displaced abomasum was discussed in Chapter 2. I became very comfortable performing this surgery during my first year out of school. The abomasum can become displaced either on the left or the right. In a left displacement the stomach is out of its normal position and needs to be moved to the anatomically correct position. When the abomasum displaces to the right, the stomach can actually twist on itself (often referred to as torsion). This is a much more serious condition and becomes life threatening if not treated quickly. In torsion the contents cannot escape, and the blood supply to the stomach can be compromised.

It was a hot August afternoon when I pulled into the farm lane. The owner was not there that day, but the hired man showed me which cow was sick. I detected a right-sided abomasal torsion on a very sick cow. The cow was dehydrated, with a very rapid heart rate and weak pulses. I told the hired man that the outlook was not good. We had the option of trying surgery or selling her as a cull cow.

The hired man called the owner of the farm and I then explained the same options to him. The owner elected to try surgery; the cow was a valuable animal if she survived and was worth very little in her present condition. With the hired man's assistance, I moved the cow into position to perform surgery. Everything was going well until I maneuvered into the abdomen and attempted to correct the position of the stomach.

The torsion of the stomach was severe. The stomach was extremely distended and had become discolored from the poor circulation. These stomachs not only contain a large amount of trapped gas, but they have a large volume of liquid as well. I knew exactly what to do; I just did not know if I could accomplish the task. I reached into the abdominal incision as deep as I could go and placed the flat of my hand and arm against the stomach. I lifted and pushed as hard as I could, but the stomach would not come back into position.

Between the heat of the day and my nerves, the sweat was rolling down my face. Things were not going well, and I was beginning to doubt myself. I had corrected displaced abomasums before, but none as large

continued

as this. Then disaster struck. With the poor circulation, the severe distention, and my pressing as hard as I could, the stomach tore. I now had stomach contents pouring into the abdomen.

I felt horrible as I realized how serious the condition had become. This cow was definitely going to have peritonitis, and I still had not corrected the underlying problem. I discussed the situation with the farm hand and then closed the incision. I had to admit defeat; the cow was going to die. I was crushed, and any confidence I had developed was destroyed. At that moment, surgery was not a fun part of my job.

Looking back on that moment, I realize that there are cows that I will not be able to save. To this day I still encounter abomasal torsions that are equally bad, and the same problem can occur. Fortunately, I have enough experience that I do not question my skills. I have to help the farmer make an informed decision, and together we realize that the outcome may not be positive. The discouragement is still very real, though, when an animal dies in surgery. The loss may not shake my confidence, but it is still a blow to my ego that I was unable to save the patient.

Surgery can also provide the experiences that renew my enthusiasm. Toby, a 6-year-old Doberman pinscher, presented with wounds to his nose and mouth. It was obvious that Toby had been shot with a high-power rifle. The bullet had entered through the top of his nose, penetrated the roof of his mouth, and fractured his mandible, leaving a gaping wound under his jaw. The bullet then continued, leaving a laceration on his shoulder.

Toby was an exceptionally nice dog. I discussed with the owners how severe the wounds were. I had to raise the options of referral to a specialist or euthanasia. The owners just felt that we had to give Toby a chance, but they did not have money for a specialist. We agreed that I would do the best that I could. I started intravenous fluids and anesthetized him. I then attempted to put the pieces back together. Unfortunately, there were teeth, pieces of bone, and gum missing. I repaired the mandible with a piece of stainless steel wire and sutured the gum together as much as possible.

Between the contamination and the extent of the injuries, my initial repair failed. The mucous membranes did not heal over the bone, and the bone was not healing. I once again offered referral, but the reality was that it was going to be too much money. I told them that I would look into options that I might be able to perform.

I called a friend, Dr. David Sweet, a veterinary surgeon. I explained the situation to him, and we discussed what options I had. Finally we developed a plan. I called

the owners and explained that I was willing to make one last attempt. I emphasized that this was going to be a learning experience for me. I had never attempted anything like this, and there were no guarantees. I made one last offer to refer Toby to a specialist. The owners agreed to allow me to try. I reached an agreement with the owners that I would donate my professional time, but there would still be substantial charges for materials, anesthesia, and antibiotics.

Dr. Sweet and I had broken the repair into three stages. I first stabilized the fractured mandible with a set of stainless steel pins driven into the bone (Figure 19–1). The pins were joined together with another rod on the outside of the jaw. I then collected cancellous bone from the humerus to pack into the defect still present in the mandible. This bone graft was going to provide mineral and bone cells to help heal the defect. The final step was to move a flap of mucosa from the inside of the cheek to cover the exposed bone. The flap remained attached to the cheek, maintaining its blood supply.

I was exhausted after the 5 hours of surgery. I had been tense the entire time. I had done extensive reading and preparation before the surgery. Dr. Sweet had talked to me many times about the various procedures. At the end of surgery, I just was not sure how Toby was going to do. Fortunately, this case turned into a success. The bone healed, and the flap of mucosa remained intact.

Toby was a sweet dog, and I was tremendously pleased with the outcome. The owners were very appreciative. This surgery provided me with one of my greatest accomplishments. Although I had not performed any of these procedures before, I was successful. Surgery requires knowledge of anatomy and physiology. The actual procedures require proper care to prevent infection and proper handling of tissues to allow them to heal.

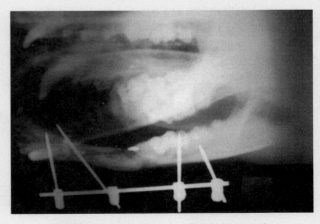

FIGURE 19–1 Radiograph of Toby's jaw repair.

PRINCIPLES OF SURGERY

OBJECTIVE

■ *Explain the Clinical Significance of the Basic Principles of Successful Surgery*

The skin and mucous membranes provide a barrier to invading pathogens. Surgery disrupts this barrier and provides a potential site for bacteria to invade. Animals' normal bacterial flora, surgeons, and the environment all are potential sources of bacteria. Special care must be taken to minimize the exposure of the patient to bacteria. **Aseptic technique** describes the general practices used to minimize the risk of infection that may occur following surgery.

Disinfectants are used to thoroughly clean the facilities within the operating room. Disinfectants are used on inanimate objects and not on the patient. Disinfectants are too harsh to be used on skin or mucous membranes, but they kill a majority of the pathogens present on the equipment. By thoroughly cleaning the operating facility, the exposure of the animal to bacteria is lessened. In addition, it is ideal to minimize the flow of people through the room. This helps to reduce the spread of bacteria through the air.

Every surgical instrument is a potential source of contamination. Initially the instruments are thoroughly cleaned of any debris, such as tissue or blood. The surgical instruments are then sterilized prior to surgery. **Sterilization** is a procedure in which all microorganisms are destroyed. Pressurized steam and chemicals are two means of sterilization.

An **autoclave** is used to sterilize instruments with pressurized steam (Figure 19–2). Water boils at 220°F. Adding additional heat may make the water boil

faster, releasing more steam, but it does not in itself increase the temperature. Water within the autoclave is brought to a boil. The autoclave has a tight seal, which allows pressure to increase. By increasing the pressure within the autoclave, higher temperatures may be reached.

Instruments are generally packaged within a special wrap that allows the steam to penetrate. These surgical packs can then be handled on the outside without contaminating the surgical instruments. These packs are placed in the autoclave and maintained at 250° to 275°F for 15 to 30 minutes. At these temperatures, all microorganisms, including spores of bacteria or fungi, are destroyed.

Because sterilized equipment has no different appearance, special heat-sensitive markers are used to identify surgical equipment that has been sterilized. A common example is a special tape used to close the wraps around the instruments (Figure 19–3). Initially this tape is light brown with faint lines. Following sterilization, these faint lines become a very dark brown. This identifies the surgical equipment that has been processed through the autoclave. The tape does not guarantee that the instruments within the pack are sterile, but it does ensure that adequate temperatures have been reached on the exterior of the pack.

A collection of surgical instruments necessary for a given procedure is often put together in a pack. The specific instruments vary with the preference of the surgeon. Good quality surgical instruments are made of stainless steel. The stainless steel is produced with a hardness that allows good durability, especially for instruments with a cutting edge. The metal in the instrument is also treated to minimize corrosion. Without this treatment, the steam sterilization of an autoclave could result in severe corrosion.

The complete list of surgical instruments available is extensive. Many instruments have been designed for specific surgeries or procedures. The following

FIGURE 19–2 An autoclave provides high-pressure steam heat to sterilize surgical instruments.

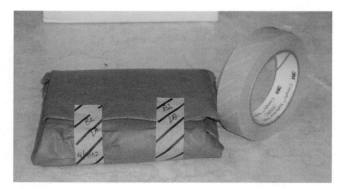

FIGURE 19–3 A surgery pack—a collection of instruments used for one procedure. A roll of indicator tape is next to the pack. Note that the unused tape has light-colored lines. The surgical pack has been autoclaved, and the indicator lines have turned dark.

list includes many commonly used instruments (Figure 19–4).

1. Scalpel: Most surgeries begin with an incision into the tissues. The scalpel or surgical knife makes that incision. The actual blade is designed for a single use, is disposable, and is attached to a scalpel handle. The handle is the portion that is reused. Different size and shapes of scalpel blades are available.

2. Needle holder: This instrument holds the needle used for suturing. Many needle holders have a serrated tungsten carbide insert in the jaws that holds the needles. This material is very hard and lasts a long time. As this insert becomes smooth, it becomes difficult to hold the needle. The insert can be replaced rather than the entire instrument. Most needle holders have a ratchet lock, so that the needle can be clamped and held without maintaining grip pressure on the tool.

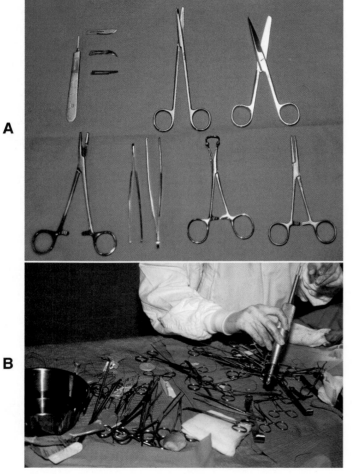

FIGURE 19–4 *A.* Surgery instruments. Top, left to right: scalpel handle and blades, Metzenbaum scissors, Mayo scissors. Bottom, left to right: needle holders, thumb forceps, towel clamp, Hemostat. *B.* Orthopedic surgery. A wide selection of instruments are used during surgery.

3. Scissors: For surgery, several different styles of scissors are available. Curved scissors are often more maneuverable during the procedure, whereas straight-blade scissors more readily cut through tougher tissues. Surgical packs commonly have Metzenbaum dissecting scissors, used for delicate tissues, and heavier Mayo scissors for heavier tissues.

4. Tissue forceps: Many tissue forceps have the appearance of a pair of tweezers. The tip of tissue forceps varies depending on the usage. The tip may have a serrated end for handling delicate tissues or interlocking teeth for a secure grip. Another type of tissue forceps has a ratchet lock for securely holding tissues. In general the ratchet lock type should not be used on delicate tissues.

5. Hemostatic forceps: These forceps are generally called hemostats. These instruments have a ratchet lock and are used to clamp blood vessels. Once clamped, the vessel can be ligated (or tied off) to prevent bleeding. These come in different sizes depending on the type of procedure being performed.

6. Retractors: These tools are used to hold tissues and expose the surgical area, so that the surgeon has a better view. Retractors can be held by an assistant or they can be self-retaining.

7. Towel forceps or towel clamps: Surgical drapes are used to cover the animal, except for the region of the incision. This prevents contamination of the surgeon and instruments by the surrounding areas on the dog and surgical table. Towel clamps are used to clamp this material to the animal.

8. Spay hook: The spay hook is used to bring the uterus through a relatively small incision.

As stated, stainless steel surgical instruments can withstand the high temperatures of an autoclave. However, such heat may damage other surgical instruments. These instruments may be sterilized by chemical means. Ethylene oxide is a gas used for chemical sterilization. The gas is quite toxic and requires special chambers to expose the instruments to the gas without contaminating the environment.

The surgeon also represents a source of bacterial contamination. A large number of bacteria are present with normal skin. Hair and cells from the outer skin layers are constantly being shed and often carry bacteria. In addition, air exhaled from the surgeon can be contaminated with pathogens. Proper preparation of the surgeon minimizes contamination risk.

The surgeon wears a head cover and surgical mask. The head cover is designed to cover all hair. Surgeons with facial hair wear a head cover that protects these regions as well. The surgical mask covers the mouth and nose of the surgeon. These facemasks are designed to block the spray of saliva that may occur while talking.

Surgeons then wash their hands and forearms with a surgical scrub. An **antiseptic** soap is used to thoroughly clean the skin. The scrubbing procedure not only removes dirt and oil from the skin, but also greatly decreases the number of bacteria present. The scrubbing procedure needs to be very thorough, covering all areas of the hands. This includes cleaning under the fingernails. Generally a scrub brush is used along with the antiseptic soap.

Contact time is extremely important with the use of antiseptics. Antiseptics do not immediately kill bacteria and require prolonged contact with the microorganism. Proper technique dictates that the skin be scrubbed all over the hands and forearms down to the elbows. Then an organized scrubbing is performed. This allows the antiseptic to be in contact with all areas while the scrubbing continues. The standard rule is for the surgical scrub to last 5 minutes. This provides adequate time for the antiseptic to greatly reduce the number of bacteria.

During the scrubbing, the hands are held higher than the elbows. The principle is to hold the cleanest part the highest. In this manner, if the elbows have a higher bacterial count, water does not run from the elbows to contaminate the hands. Using a sterile towel, the hands are then dried, followed by the arms.

The surgeon then puts on a sterile surgical gown with the aid of an assistant. Sterile surgical gloves are then placed on the hands, with the cuffs of the gloves sealing over the sleeves of the surgical gown (Figure 19–5). While not actively operating, the hands are held in front of the surgeon (still above the elbows). This prevents any accidental contact of the hands with a nonsterile item, such as the surgical table.

The question may arise as to why such a thorough scrubbing is necessary when gloves are being worn over the hands. The gloves that are used are sterile and are ideal for handling the surgical instruments and tissues. Unfortunately, the gloves are easily damaged with sharp tools or even tissues, such as a tooth or bone fragment. A small puncture in the glove does not result in severe contamination when the hands have been thoroughly scrubbed with an antiseptic soap.

The animal has to be prepared for surgery as well. The hair is clipped from the surgical region. Hair needs to be clipped in a large enough area that the surgical incision can be enlarged if it becomes necessary during the procedure. The skin is then scrubbed with an antiseptic soap. Just as with the surgeon, the scrubbing removes dirt and oils and greatly reduces the bacterial population. Contact time is equally important for the animal preparation (Figure 19–6). During scrubbing, the central region over the anticipated incision is scrubbed first, working in larger circles to the outside of the clipped area. This prevents dirt and hair from the peripheral region from contaminating the central region.

Following all the preparation of the surgeon and the animal, the actual surgery may begin. The preparation described previously is for a clean surgery. Examples are a spay or castration, where healthy skin is entered and healthy tissue is being removed. Not all surgeries are this clean. Surgeries involving the mouth, the intestines, and contaminated wounds may have varying degrees of contamination. Obviously the greater the contamination, the greater the risk of infection following the procedure.

Conditions surrounding the surgeon also vary dramatically. The entire procedure for animal and surgeon preparation minimizes the risk of infection following the surgery. The photograph of the surgery

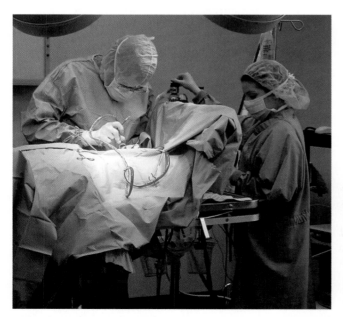

FIGURE 19–5 The sterile surgical environment. Dr. David Sweet is performing a surgery, while Ann Zackim, Certified Veterinary Technician, monitors anesthesia. *(Courtesy Dr. David Sweet and Ann Zackim.)*

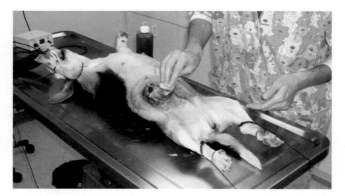

FIGURE 19–6 A veterinary assistant prepares a dog for surgery. *(Courtesy Patricia Curley.)*

on a cow in Addendum B does not show all these precautions. Although some veterinary clinics have a large-animal surgery facility, I do my cattle surgeries in the barn. I still do a thorough job of scrubbing the animal and myself but I do not take the precautions of using a cap, mask, and sterile gown. These surgeries are kept as clean as possible, but they do have the increased risk of contamination from the environment. A gust of wind blowing dust or a cat jumping onto the surgical field is possible when operating in barn conditions.

LACERATION HEALING

OBJECTIVE

■ *Explain the Clinical Significance of Healing of Lacerations By First and Second Intention*

Surgery creates a wound. Understanding the healing process is essential for both surgery and traumatic wounds. Skin will be used as an example of how tissues heal. Although the healing is divided into phases, the steps actually overlap in time.

The inflammation phase begins immediately following the trauma or the incision. Bleeding is actually beneficial in flushing the wound of contamination. Instantly the blood vessels constrict, slowing the flow of blood. The blood begins to clot, protecting the animal from excessive blood loss. The clot fills the wound and seals it. As the clot dries, a scab is formed. The scab helps to protect the damaged area and allows for healing to occur beneath it.

The blood vessels eventually dilate, bringing more white blood cells to the area. These white blood cells help to destroy damaged tissue and invading bacteria. The damaged tissue leaks plasma into the wound area. The dilated vessels increase the heat in the region and produce a reddened appearance. The leaking plasma adds to the swelling, which applies pressure to the nerve ends, resulting in pain. Heat, redness, swelling, and pain are the classic signs of inflammation that defines this stage.

The healing process then enters the repair phase. This process has actually begun during the same time as the inflammatory phase. In this step, connective tissue cells enter the damaged area and begin to form fibrous connective tissue. In addition, capillaries begin to grow into this newly forming connective tissue. In an open wound the combination of capillaries and connective tissue takes on a fleshy red appearance. This healing tissue is called **granulation tissue**.

Granulation tissue has an essential role in allowing the healing process to continue. This tissue obviously fills the gap between the wound edges and sets up a barrier to infection. The rich supply of capillaries delivers white blood cells to the area to engulf any invading bacteria. The granulation tissue provides a surface for the epithelial cells to bridge across the wound.

As the granulation tissue forms, epithelial cells from the wound edges begin to move across the wound. These cells initially form a single layer beneath the scab, until the two edges connect. The cells then continue to replicate and the layer thickens. Early removal of the scab can damage the newly forming epithelial layer and slow healing. Therefore parents everywhere are right to tell children not to pick at scabs!

The connective tissue within the healing wound actually allows for the wound to decrease in size over time. This wound contraction occurs as the healing tissue becomes more organized and actually shrinks. In a large wound the contraction helps to bring the skin edges closer together, decreasing the time required for the epithelial cells to reach across the wound. In certain situations the contraction may fail before the wound edges have met. If the tension on the skin edges is too high, or if too much motion exists, the wound may not be able to bridge the gap. It is also possible for contraction to cause restricted motion, especially if the wound is across a joint. A developing scar is not very elastic and may restrict the ability to move.

During the first 4 to 6 days, there is not a dramatic increase in the strength of the wound. During this time any further trauma could easily reopen the wound. Over the next 1 to 2 weeks, the wound increases in strength rapidly as the connective tissue is formed and becomes better organized. Even following this time, the wound continues to increase in strength, a process that can continue for years.

Whenever possible the wound edges in a surgical incision are brought together with sutures. The skin edges are held close together to speed the healing process. In a well-closed incision the epithelial cells can cross the defect within 1 to 2 days. **First intention healing** describes a wound with the edges closely apposed. The healing process is generally quite rapid and successful.

Following an injury, bacteria begin to multiply within a wound. The damaged tissue, blood, and plasma provide an excellent medium on which the bacteria can grow. The **golden period** describes the first 6 to 8 hours, when bacterial numbers are still at a moderate level, allowing the wound to be closed. During this period the laceration can be closed to allow for first intention healing.

Beyond the golden period (more than 6 to 8 hours), the bacterial contamination may be too high to consider suturing the wound and allow first intention healing. Grossly contaminated wounds may also be left open to heal. **Second intention healing** describes the healing that occurs in wounds where granulation tissue must fill the gap between the skin edges and subsequently allow the epithelial cells to grow. The size of these wounds can vary from a small bite wound to a very large region of skin lost in a burn.

Second intention healing effectively heals large wounds. Horses are prone to a condition called **proud flesh**. Proud flesh appears as an overgrowth of granulation tissue that prevents epithelial cells from growing across the wound. This most commonly occurs on wounds of the lower legs. In these regions there is little underlying tissue and the skin has relatively high tension as it heals.

To prevent proud flesh, it is ideal to suture lacerations on horses to allow for first intention healing. Unfortunately, this is not always possible (for example, the wound is too large or too contaminated). Limiting the movement of the area is beneficial if second intention healing is necessary. This is generally accomplished by securely bandaging the lower leg.

If proud flesh does occur, it must be removed before the epithelium can grow across the wound. Granulation tissue has little nerve supply, so it can be manually cut back to skin level with little sedation or anesthesia. Granulation tissue is rich in blood supply, so this process does result in significant bleeding. Bandaging following this procedure is necessary. Topical products are available that may chemically reduce the proud flesh. The decision regarding what method is used depends on the preference of the veterinarian, the size of the wound, and the degree of proud flesh.

Several basic principles are essential in minimizing healing time and ensuring success in a surgical incision. As already emphasized, asepsis is extremely valuable. All of the preparation of the patient and surgeon aid in this process. During surgery, gentle handling of the tissues helps to minimize the extent of the inflammation that occurs. Excessive swelling can slow the healing process. When making surgical incisions, surgeons consider the blood supply to the region. Incision should be made at locations that do not disrupt the blood supply. Damaging the blood supply can greatly slow the healing.

When suturing traumatic wounds, dead tissue and contaminants within the wound or on the skin margins should be removed. **Débridement** describes the process in which the damaged and contaminated tissue is removed. Furthermore, the body's immune system must remove any of this tissue remaining before healing can occur. Within hours of sustaining a wound, the skin edges begin to deteriorate and dry. These edges are trimmed to provide a fresh edge with blood oozing from the capillaries. These edges can then be successfully sutured. Care must be taken so that excessive tension is not applied to the tissue when it is being closed. The tension increases the likelihood that the suture material will cut through the tissues. This can be a problem in larger traumatic wounds or following the removal of a large tumor.

Wounds caused by a high-velocity projectile (such as a bullet from a high-power rifle) have unique difficulties. The bullet not only creates a hole, but the shockwave from the high velocity damages tissue around the defect. It is important to débride the wound aggressively. The bullet also contaminates the wound, dragging dirt and hair into the wound.

In any wound there can be bleeding from arteries and veins. Before a wound is closed, the bleeding should be well controlled. Oozing of blood from capillary beds may not be a problem, but any significant bleeding may separate tissues and provide bacteria a pocket for growth. Both of these factors slow healing.

Dead space, an important concept to consider in surgery, is actually a potential space that is present because of separation between tissues. Consider a surgery in which a large tumor is removed from the subcutaneous tissue. If the skin is closed over the area where the tumor was present, a large pocket exists. Initially it appears normal, but the skin is not attached to the underlying connective tissue. Fluid accumulates in this space from blood or plasma that leaks from the tissues damaged in the surgery.

Hematoma describes an accumulation of blood in the dead space. A **seroma** has fluid that is similar to serum with only a small number of red blood cells. The fluid in a seroma may be straw colored (basically no red blood cells) or a light red (with a few red blood cells). An **abscess** is a third fluid accumulation in the dead space. In an abscess the fluid contains bacteria, white blood cells, and dead tissue.

This collection of fluid increases the tension on the skin incision and prevents the skin from adhering to the underlying tissue. The fluid needs to be removed before complete healing can occur. If only small amounts are present, the body will reabsorb the fluid following the inflammation phase. If greater volumes accumulate, the veterinarian may need to remove the fluid through a needle or a small surgical incision. If left untreated, the increasing pressure may cause the wound to break and allow the fluid to leak from the pocket.

Two major methods prevent this type of fluid accumulation. In smaller pockets of dead space, the tissues may be sutured together. For example, the edges of subcutaneous tissue may be sutured. This effectively minimizes the size of the pocket. The second technique places a drain in the region. For example, a piece of soft latex tubing is placed in the dead space and brought out through the skin (Figure 19–7). The tubing maintains an opening in the skin that allows any accumulating fluid to leak from the pocket. The fluid does not drain through the center of the tubing.

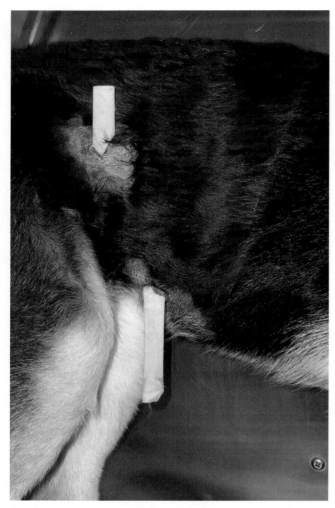

FIGURE 19–7 A large fluid accumulation in the flank of a dog has been surgically drained. The Penrose drain protruding from the holes maintains an opening to allow any additional fluid to drain to the exterior.

The purpose of the tubing is to maintain an opening in the skin for fluid to have an escape.

Dehiscence, breaking of wound edges, may occur along the entire length of the suture line or a smaller region. Many factors may contribute to this failure to heal. As already mentioned, having too much tension on the incision may result in the sutures cutting through the tissue edges. Improperly placing the sutures too close to the skin edge or tying them too tightly may also result in the dehiscence. Infection at the incision may also cause it to break open. Finally, animals licking or scratching their own wounds may cause enough trauma to produce dehiscence.

The type of suture material used depends on the strength required and the type of tissue being sutured. Suture material is divided into two major classes, absorbable and nonabsorbable. In general (exceptions do exist), absorbable suture material is used within the body. Over time the body destroys the material, and eventually it will be eliminated completely. The goal is for the material to maintain strength long enough for the incision to heal adequately. Surgical gut or catgut is a commonly used absorbable suture material. The material is not made from cats, but from the connective tissue found in sheep or cattle intestines. Catgut is primarily made of collagen. Many other synthetic absorbable materials are available. The various suture materials each have different advantages, such as less tissue reaction, greater knot security, or greater and longer strength. Selection of material varies with the preference of the surgeon and the type of tissue being sutured.

Nonabsorbable suture material is not destroyed by the body and is commonly used to close the skin incision. Because the animal does not destroy these materials, these sutures are later removed. Nylon, polypropylene, silk, and stainless steel are all examples of nonabsorbable suture materials. Although nonabsorbable materials are most commonly used to close the skin, there are times where they are used within the body. In certain instances the long-lasting strength of the nonabsorbable materials is warranted. The concern of burying such a material within the body would be if the animal reacted to the material. If this occurred, a second surgery would be required to remove the sutures.

The needles used with suture material come in a variety of shapes and sizes as well. The needles may be straight or curved, often being judged on the portion of a circle that they form (such as $\frac{3}{8}$ circle). Needles may also be a tapered point or cutting. In cross-section the taper-point needle would appear as a circle, whereas the cutting needle would appear as a triangle. Taper-point needles are used on delicate tissues that allow easy passage of the needle. Dense tissues, such as the skin, generally require the use of a cutting needle. The edges of the triangle help to cut through the tissue, allowing the needle to pass more readily.

Suture material must also be selected based on the strength required to hold the incision. In addition to selecting the type of material as discussed previously, suture materials are selected on their diameter (Figure 19–8). A scale has been developed that sizes the suture material from the smallest diameter of 10-0 to the largest diameter of 7. This United States Pharmacopeia (USP) size allows for the surgeon to easily communicate the suture selection. For instance, I often use a 3-0 suture to close the skin of dogs and a 2 or 3 for the skin of cattle. The selection is based on the fact that the tension on the incision line of the cow is much greater than that of a dog. The smallest sizes require the use of magnification to see the actual material. These sutures are obviously used in the finest surgeries.

A number of suture patterns can be used to close an incision. A commonly used pattern is called simple interrupted (Figure 19–9). In this pattern the suture passes straight through the two sides of the incision, is

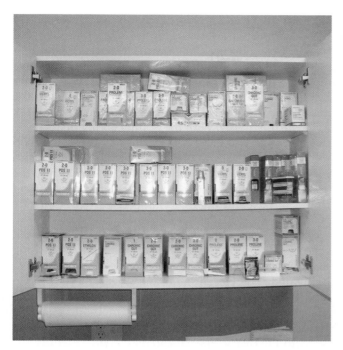

FIGURE 19–8 A variety of suture materials in different sizes and types are available for surgery.

a number of passes through the tissue until the opposite end is reached. The suture is tied at the opposite end and then cut. This pattern contains one long piece of suture material, with a knot at each end. The simple continuous pattern is much quicker to perform than the simple interrupted pattern. The disadvantage of the simple continuous pattern is that if one knot or strand of suture fails, the entire suture line is no longer secure.

Another commonly used suture pattern is called a mattress suture. A horizontal mattress suture is one example of this pattern. In this suture the needle passes straight through the two edges of the incision. The needle is then moved farther down the incision and passed back across the edges. The two ends are then tied, which keeps the knot on one side of the incision. The advantage of this pattern is that it increases the area over which the suture applies pressure. A mattress suture is often used in incisions with high tension. This makes it less likely (relative to simple interrupted) to cut through the edges of the incision. Mattress sutures can also be performed in a continuous manner.

Tying knots is an essential skill for surgeons. Multiple square knots are tied to secure the suture. The number of knots placed depends on the type of suture material used. Some suture materials have poor knot security and require a higher number of knots. Granny knots and half hitches are much less secure and are not recommended. It is critical to not excessively tighten the sutures. The suture should be applied to simply appose the edges, not crush the tissue. Overtightening makes it uncomfortable for the animal and increases

tied, and the ends cut. The simple interrupted pattern apposes the edges of the incision neatly together. An entire row of individual sutures is used to close the incision. A simple continuous pattern begins with a similar suture and is tied. With the continuous pattern, the end is not cut. The suture pattern then continues with

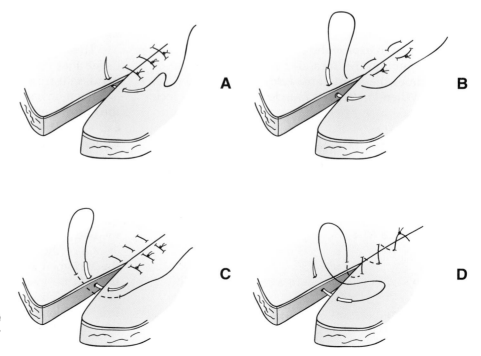

FIGURE 19–9 Suture patterns. *A.* Simple interrupted. *B.* Horizontal mattress. *C.* Vertical mattress. *D.* Simple continuous.

the likelihood that the suture will cut through the tissues. This can lead to the animal removing the suture too early or the wound dehiscing.

The stack of square knots used in tying suture materials is actually a series of overhand knots. When the tissues are under tension, a single overhand knot may pull apart before the second knot is applied. To prevent this, a surgeon's throw is used. A surgeon's throw takes a second pass of the suture within the first overhand knot. The remainder of the square knots are then tied as usual.

SURGICAL CONSIDERATIONS

OBJECTIVE

■ *Explain the Clinical Significance of Common Considerations in Veterinary Surgeries*

It is useful to realize that there is not just one method of performing a given surgery. Many variations are possible. The differences may involve the approach, the order of the steps performed, or the specifics of the technical methods used. The following descriptions are only one possibility of how procedures may be performed.

One of the most commonly performed abdominal surgeries in pets is the ovariohysterectomy, or spay. In this procedure the animal is anesthetized and secured to the surgery table. The abdomen is clipped and scrubbed in preparation for the surgery. Most commonly, spaying is performed through a ventral midline incision. *Ventral* refers to the lowest portion of the abdomen (which becomes the highest point when the animal is on its back). The midline is used because the connective tissue from both sides of the abdominal wall meets at this point, the linea alba (Figure 19–10). The incision is made through the skin and subcutaneous tissues to expose the linea alba. An incision can be made through the linea alba with very little bleeding. At times the incision may deviate from the exact midline. When this occurs, muscle is incised. The incision can still be easily closed, but it is much more likely that bleeding will occur from the muscle tissue. This bleeding can obscure the view of the surgery field.

Special care must be used when incising the linea alba to prevent lacerating any abdominal organs. Once the linea alba is open, the surgeon has access to the abdomen. The surgeon then uses a finger or spay hook to locate the uterus. With the animal laying on its back, the uterus is generally found close to the spinal column and surrounding muscles. With the ventral midline approach the uterus is found deep within the abdomen. Once located, a horn of the uterus is brought to the surface and the ovary associated with that side is identified.

The ovarian artery supplies blood to the ovary and the cranial portion of the uterus. This artery branching from the abdominal aorta close to the kidney needs to be ligated. The closely associated ovarian vein is ligated at the same time. One technique of ligating major vessels is called the three-clamp method. The first clamp (hemostatic forceps) is placed immediately beneath the ovary. Special care is taken to identify the entire ovary and avoid clamping it with the hemostat. If a portion of the ovary is left within the animal, it may still show estrus activity. Two more clamps are placed across this vessel, each slightly farther from the ovary.

A suture is placed around the ovarian vessels at the level of the deepest clamp. As this clamp is taken off the vessels, the suture is tightened into the groove crushed by the clamp. Depending on the size of the vessels, a second ligating suture may be used. The vessels are then cut between the two remaining clamps. The stump of the vessels is carefully released, while the veterinarian observes for any evidence of bleeding. If this suture slips off these vessels, life-threatening bleeding may occur. The same procedure is then performed on the opposite ovary.

The broad ligament is a thin sheet of connective tissue that runs the entire length of the uterus, suspending it from the dorsal body wall. This thin sheet is ligated on each side. This frees the uterus, allowing the entire length of the uterine horns and body to be exposed. The cervix can be palpated to identify its location. The uterine body and the uterine vessels running on opposite sides are then ligated. A three-clamp method can be performed for this procedure as well. After the ligation, the uterine body is incised and the uterine stump carefully released. In a spay the ovaries, uterine horn, and a majority of the uterine body are removed.

Once the entire uterus and ovaries are removed, the abdomen is examined for evidence of bleeding. If there is no bleeding, the abdominal incision can be closed. An absorbable suture on a taper-point needle is used to close the linea alba, generally in a simple interrupted or continuous pattern. The subcutaneous

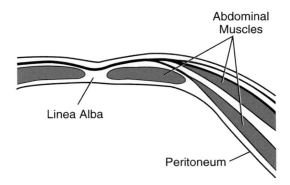

FIGURE 19–10 The structure of the linea alba. During abdominal surgeries this is a common site of incision.

tissue is also sutured with an absorbable suture to minimize dead space. The skin is then closed using a nonabsorbable suture material on a cutting needle.

Spaying becomes a very routine procedure for most veterinarians. The benefits of the procedure have already been discussed in Chapter 8. It is important to recognize that even though it is routine, the procedure is a major abdominal surgery. It does require general anesthesia and ligation and incision of major vessels. Uncontrolled bleeding is a possible consequence. Obese animals make the procedure much more difficult because of the fat deposits in the connective tissue associated with the uterus and ovaries. Large amounts of fat in these tissues make the vessels more difficult to identify and make the tissues more difficult to handle. This is one reason that young, healthy, and active (and therefore lean) animals are ideal surgical candidates. Another potential consequence of the procedure may occur if the ureters become entrapped in one of the ligating sutures. This obstructs the flow of urine from the kidney, which may permanently damage the kidney associated with that ureter.

A ventral midline approach is commonly used to explore the abdomen of a horse as well. Table 7–4 in Chapter 7 lists a number of causes of colic in horses. Depending on the severity, many of these causes may require surgical correction. In abdominal surgery a horse is anesthetized, and placed on its back, and secured. A large incision is made along the linea alba, allowing the surgeon to explore the abdomen. By feel and observation, the entire intestinal tract is examined for obstructions and disease conditions. A thorough knowledge of anatomy is critical to interpret the positioning and location of the structure of the intestinal tract.

One possible finding is a loop of intestine or colon that has twisted upon itself. This intestinal torsion cuts off blood supply to this section of the intestinal tract.

The region takes on a dark blue or even black appearance. The loop can be properly positioned and if caught early may improve in coloration. The tissues may also be too badly damaged and will not recover following the correction. In these situations the diseased region must be resected or removed.

When removing a length of intestinal tract, the incisions must be made at a region where the tissue is completely healthy. The blood supply to the damaged region must be ligated to prevent bleeding. The intestines are clamped with special forceps that do not crush the tissue but do prevent spillage of the intestinal contents. The damaged section of the intestines is removed. The newly created ends must then be sutured together to create a tight seal. **Intestinal anastomosis** describes the procedure in which two regions of intestine are joined. This can be done manually or with a special tool that allows the placement of stainless steel staples. The stainless steel does not cause tissue reaction, does not support the growth of bacteria, and can be done quite quickly.

This type of colic is severe, and many complications may result following the surgery. The risk of peritonitis is significant following intestinal anastomosis. If too large a region is affected, the procedure may not be practical. These horses are under tremendous physical stress during such a procedure, and secondary infections such as salmonellosis are quite possible.

A condition called **gastric dilation–volvulus syndrome** (GDV), or gastric torsion, is a surgical emergency in dogs. Classically this occurs in large, deep-chested breeds. Often these dogs have consumed a large meal or large volumes of water. Many times these dogs have been physically active following the meal. The combination of all these factors allows for the accumulation of gas and the rotation of the stomach (Figure 19–11).

Gastric Dilatation Volvulus

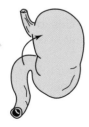
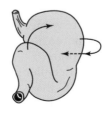
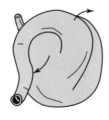
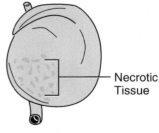

Pyloric antrum is displaced aborally

Pylorus crosses midline, passes underneath distended oral part of stomach

Fundus moves ventrally and becomes located in ventral abdomen

Gastric dilatation displaces greater curvature ventrally

Necrotic Tissue

FIGURE 19–11 The formation of gastric torsion.

These dogs develop signs of abdominal distention and frequent attempts at vomiting. Dogs with GDV are quite uncomfortable, and signs worsen quickly. Just as in the horse with intestinal torsion, GDV can compromise the circulation to the stomach. These dogs often develop signs of shock. Diagnosis is made based on the clinical signs and abdominal radiographs. The abdominal radiographs show a stomach severely distended with gas (Figure 19–12). The volvulus or twist of the stomach gives the appearance that the gas is divided into compartments.

Immediate treatment is essential for the success of these cases. Intravenous therapy is begun to reverse the effects of shock. Occasionally it is possible to pass a tube down the esophagus to relieve the pressure from the stomach. More commonly, if the stomach is twisted, the tube is not able to pass into the stomach. Surgery is then required. One approach is to do a ventral midline incision in the cranial region of the abdomen.

The stomach is easily visible and can then be drained to make correction easier. The stomach is then put into its normal position and evaluated. GDV can result in irreversible damage to the stomach and spleen, due to the lack of circulation. **Necrotic** (dead) regions of either organ need to be removed prior to finishing the surgery. This is acceptable as long as limited regions of the stomach are affected. Once the surgeon is comfortable with the appearance of the organs, the stomach is attached to the abdominal wall. This procedure is performed to prevent recurrence. The abdomen is then closed in a similar fashion to the procedure used in the spay.

Following the surgery, complications may still occur. Arrhythmias may occur in the heart rhythm.

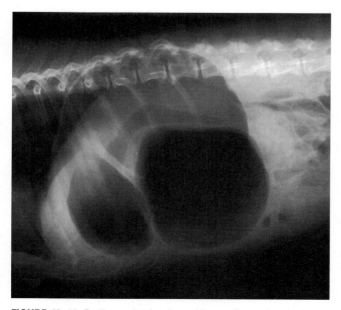

FIGURE 19–12 Radiograph of a dog with gastric torsion. Note the large gas-distended stomach.

The shock can persist, and further regions of the stomach may deteriorate. The first few days following the surgery remain critical for the survival of the patient.

Another abdominal surgery performed in companion animals is the removal of bladder stones. The approach is once again a ventral midline approach in the caudal abdomen. This allows for exposure of the urinary bladder. The urinary bladder is brought through the abdominal incision for easy access. Gauze sponges are packed around the bladder to prevent spillage of urine into the abdomen. The bladder is incised in a region away from the ureters. The incision is extended to allow for the urinary stone to be removed.

Once the stones are removed, the bladder incision must be closed. This can be performed with a two-layer closure. It is essential that the suture line is closed securely to prevent the leakage of urine into the abdomen. Continuous mattress sutures are often used. The suture line used will invert the edges of the incision into the bladder itself. A second layer of similar sutures is then performed. By inverting the edges, the size of the bladder is made somewhat smaller. The bladder has such an ability to distend that it is usually not a significant problem. The midline abdominal incision is closed in the standard fashion.

Tumors within the skin and subcutaneous tissues present a common reason for surgery in small animal practice (Figure 19–13A). Once removed, the specimens are often submitted for biopsy evaluation. The biopsy helps to predict the nature of the tumor (benign or malignant) but also evaluates whether tumor cells are present at the margins of the specimen. The benign tumors are often well encapsulated, and the cells do not extend beyond the visible margin of the tumor. With malignancies, the margins are often not so well defined.

Because the surgeon cannot see the actual boundary of the tumor, a wide margin of apparently healthy tissue is removed in addition to the tumor (Figure 19–13B). This is often quite easy when the tumor is located in the skin in a region where the skin is loosely attached (such as the side of the abdomen or chest) (Figure 19–13C, D). This procedure becomes more difficult when the tumor is located low on a leg. Aggressive removal of tissue in this region may prevent primary closure of the incision. In this situation the surgical wound may need to heal by second intention.

Another challenge occurs if the tumor invades tissues deeper than the subcutaneous tissues. The surgery may become more difficult to perform if underlying muscles must be removed. In addition, this often creates a larger amount of dead space. I recently performed a biopsy on a large tumor on the side of a dog. The biopsy showed that the tumor was a malignancy arising from nerve tissue. The tumor was not freely moveable and seemed attached to the underlying muscle tissue. I did not feel comfortable with performing the surgical excision, and I referred the case

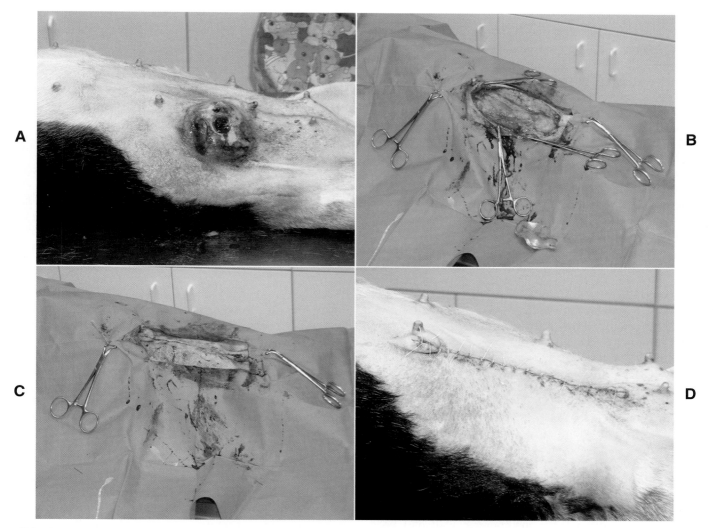

FIGURE 19–13 Surgical process. *A.* A large tumor on the ventral abdomen of a dog. The dog has been prepared for surgery. *B.* The tumor has been removed with a wide excision. The removal of the tumor has created a very large potential dead space. *C.* The tissue underlying the skin is closed with an absorbable suture. This process greatly decreases the amount of dead space. *D.* The skin layer has been closed with a line of single interrupted sutures.

to a local veterinary surgical specialist. She performed an aggressive surgery, removing a large section of skin, subcutaneous tissue, and muscle tissue from the side of the abdomen and thorax. The biopsy showed no evidence of tumor cells at the margins of the specimen. The incision was well healed at the time of the suture removal.

Surgery in the chest cavity has distinct challenges not present in abdominal surgery. Animals breathe by expanding the chest cavity, which subsequently expands the lungs. The expanding chest cavity lowers the pressure around the lungs, allowing them to expand. Once the chest wall is opened surgically, the animal is no longer able to breathe on its own. Expansion of the chest wall does not change the pressure around the lung tissues because air can enter through the incision. Therefore the anesthetist must breathe for the animal. An endotracheal tube is

inserted into the trachea. This tube has a balloon near the end that inflates, creating a seal within the trachea. Positive pressure can be applied to the oxygen entering the tube, forcing the lungs to expand.

The anesthetist must breathe regularly for the animal but must also coordinate the breaths with the surgeon. The surgeon has the added challenge when operating within the chest cavity of working around the expanding lungs and beating heart. Special care must be taken to avoid perforating the delicate lung tissues with a sharp instrument or needle.

Surgery on lung tissue also adds the challenge of developing an airtight seal on the incision. Although many structures are easily closed to prevent leakage of fluid, leakage of air is a common problem. Even the smallest suture material can create a hole large enough to allow a small amount of air to leak. With time the body will heal such a small hole in healthy tissue.

While the body is healing, a chest tube is used to allow air to be removed from the chest cavity.

The chest tube is placed with an end in the chest cavity, passing out through the chest wall with the other end accessible outside the body. The tube is put in place before the incision is closed. As the surgeon prepares to make the final closure of the incision, the anesthetist expands the lungs fully, forcing as much air as possible out of the thorax. The incision is fully closed. Suction can be applied (continuously or periodically) to the chest tube to draw out any new air accumulating within the thorax. The amount of air removed over time allows the surgeon to evaluate the progress of the patient. When air is no longer accumulating, the chest tube can be removed.

The chest cavity may be entered with an incision between the ribs or by splitting the sternum. Entering the chest between the ribs requires only an incision through the muscle layer between the ribs. The drawback of this procedure is that access is limited within the chest cavity. If the problem is in both the right and left lung fields, the surgeon may not be able to access the diseased tissue. A ventral midline incision provides a more complete access to the thorax but does require special tools necessary to split or saw the bones of the sternum.

The chest cavity can also be entered through the diaphragm in an abdominal incision. This becomes necessary with a diaphragmatic hernia. With this condition, a defect in the diaphragm allows abdominal organs to enter the thorax. A diaphragmatic hernia can occur with trauma, or the animal can be born with the condition. When the abdominal wall is opened, the patient is in the same condition as if the thoracic wall had been opened. The animal can no longer breathe for itself.

Once the abdomen is opened, the defect in the diaphragm needs to be identified. The abdominal contents must be carefully removed from the chest cavity. The defect in the diaphragm must then be sutured. It can be quite challenging to properly handle and tie suture material deep within the abdomen. The lungs are fully expanded while the final sutures close the diaphragm. The surgeon must evaluate whether a chest tube is required. The abdomen is then closed in a standard fashion.

All the surgeries mentioned to this point have been soft tissue surgeries. *Orthopedic surgery* describes those surgeries dealing with bones. Orthopedic sur-

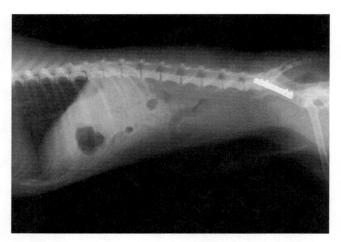

FIGURE 19–14 Radiograph of a pelvis repaired with a bone plate and screws.

gery requires another set of instruments and equipment necessary for repair of bones and related structures (such as joints and intervertebral disks).

Several methods are available for repairing fractures. The specific repair depends on the type and location of the fracture. I handle basic fracture repair with an intramedullary pin. The intramedullary pin provides excellent strength for preventing compression. This type of repair may not prevent rotation of the bone fragments.

The surgery to repair Toby's jaw, which was fractured by the bullet, used an external support. The stainless steel pins were driven through the bone of the jaw and then connected by another stainless steel rod. This type of repair holds the bone in the proper orientation and prevents compression. This repair was very effective because it allowed the bone graft to be placed within the defect.

This repair could also have been performed with a bone plate. In this procedure a stainless steel plate is securely fastened to the bone with screws (Figure 19–14). This too provides a rigid support that prevents rotation and compression. A large inventory of plates and screws must be kept on hand to allow for proper selection for the bone in question. The wide size range of animals that the veterinary surgeon treats makes the size range of bones very large. Consider the difference in the size of bones in a young kitten compared with those in a mature Newfoundland.

SUMMARY

Successful surgery occurs as a result of both extensive study and practical experience. Surgeons must follow precise aseptic techniques to prevent infection during surgery. Moreover, they must learn to match the tool to the surgical need. Understanding the wound-healing process of the first and second intentions gives surgeons the technical background to appropriately repair wounds.

REVIEW QUESTIONS

1. Define any 10 of the following terms:

aseptic technique	proud flesh
disinfectants	débridement
sterilization	hematoma
autoclave	seroma
antiseptic	abscess
granulation tissue	dehiscence
golden period	intestinal anastomosis
first intention healing	gastric dilation–
second intention	volvulus syndrome
healing	necrotic

2. True or False: Disinfectants can safely clean surgeons' hands prior to surgery.

3. How high does the temperature rise in an autoclave when instruments are being sterilized?

4. What is a scalpel?

5. Name one toxic gas mentioned in the text that can be used to chemically sterilize surgical instruments.

6. How long does the golden period last?

7. Does gastric torsion occur most typically in larger deep-chested dogs or toy-type canines?

8. What species mentioned in the text are especially prone to proud flesh?

9. Why should surgeons hold their hands higher than their elbows during scrubbing?

10. How can dead space be prevented in surgery?

11. What are hemostatic forceps used for in surgery?

12. What type of approach is used in the removal of bladder stones?

13. What is the primary reason to perform biopsies?

14. List the two main types of suture material.

15. List two means of sterilization.

ACTIVITIES

Materials needed for completion of activities:
   hemostatic forceps
   two pieces of rubber tubing
   a block of wood
   thumbtacks
   regular thread, yarn, string, or dental floss
   needles

Exercise: Visit one of the following websites: <http://www.bumc.bu.edu> or <http://cal.nbc.upenn.edu>.

1. Using a hemostatic forceps, practice performing an instrument tie. The practice board shown on the website can easily be replicated with two pieces of rubber tubing, a block of wood and a few thumbtacks. Regular thread, yarn, string, or dental floss can be used as the suture material. Examine the knots to ensure that each knot is a square knot.

   To simulate suturing, a needle can be added to the thread and the rubber tubing can be sutured. Create a simple interrupted pattern by passing the needle through each piece of tubing and then tying the suture with a series of four throws (basically two full square knots). The sutures should be tied only tight enough to appose the two pieces of tubing. Do not crush the tubing. In a real incision this could lead to dehiscence. Attempt to space the sutures evenly and close enough together that the tubing does not gap between the sutures. Practice the other suture patterns shown in Figure 19–9.

2. Students desiring a greater challenge can attempt to learn the one-hand or two-hand ties, shown on this website: <http://www.vetmed.iastate.edu>. Key search words are surgery and manual. When done efficiently, knots can be tied very quickly with this method. These techniques are also a great show of a surgeon's dexterity. The instrument tie is used much more commonly in veterinary medicine. The instrument tie does not require as long of ends to be cut and wasted. When suturing, the needle holder is in the surgeon's hand and is conveniently used to tie the suture. To do a hand tie the instrument must be repositioned or set down.

Addendum A: Decision Making in Veterinary Science

OBJECTIVE

Upon completion of this addendum, you should be able to:

- *Consider necessary factors in making informed decisions in the veterinary care of production animals and contrast the decision-making process in companion animals to production animals.*

INTRODUCTION

Every day veterinarians are required to make important decisions with their clients. Ultimately the client makes the final decision. The veterinarian plays a critical role in guiding the client, helping them to make an informed decision. The process varies dramatically for situations dealing with production animals and companion animals.

A DAY IN THE LIFE

Some Days You Laugh, Some Days You Cry...

As a veterinarian, my personal experiences influence my decision-making process. I base my judgement on the successes and failures of the past. Several years ago, I was presented with a 20-year-old cat that had a large tumor growing on its right shoulder. The tumor had been present for several months but had recently grown rapidly. The skin on the surface of the tumor had deteriorated and was now oozing blood. The margins of the tumor were not well defined and it invaded the deeper tissues.

I did not have a biopsy of the tumor, but I presumed that it was malignant. It had many classic characteristics of a malignancy: growing rapidly, not encapsulated, invading the deep tissues, and ulcerated on the surface. I discussed my concerns with the owners. The picture was bleak. The only option that I had for treatment was to amputate the leg. The tumor was too invasive to salvage a functional limb. My concern was that our patient was very old and had a cancer.

I explained that I see very few cats make it to 20 years of age. This was going to be a major surgery and there was always the possibility that other disease conditions could be developing. My fear was that we were going to spend the money on surgery and the result may not be favorable.

I discussed four options with the owners. The first option was not acceptable to any of us. Leaving the tumor untreated was just not practical with the way the surface was oozing. Secondary infection was likely to occur. A second option was to do a blood chemistry profile to evaluate the cat's overall health. My fear was that at 20 years of age, it was quite possible that the cat could have been developing kidney failure, hyperthyroidism, or a long list of age-related diseases. If the tests came back normal, we could proceed with the surgery. A third option was to just take the chance and do the surgery. With this option, we needed to assume the risk that the surgery could make other underlying conditions worse or that, realistically, the cat might fail to survive surgery.

The fourth option is always difficult to discuss. I felt that I had to mention euthanasia as an option. I wanted the clients to understand that I was making no guarantees about the success of any treatment option we chose. They had to realize that they might be spending money on a surgery where the outcome was going to be poor. I did not even know if the cat would be able

to walk after the surgery. Young cats do well following amputation, but this cat was already 20 years old.

I am often asked what I would do in the client's position. My answer is always based on the medical facts involving the case. I cannot put myself into the position of having 20 years of emotional attachment to the animal. I honestly felt in this case that euthanasia was the option that I would have chosen. I felt that the risks outweighed the benefits.

I had an open and frank discussion with the owners and in the end they elected to have surgery. They wanted to give the cat a chance, no matter what, so they did not have the blood tests performed. I was nervous about the surgery, but excited to give the cat a chance. The surgery went very well, and the cat recovered better than I had expected. Ten days later the cat returned for suture removal and the owners were quite pleased. The cat had adapted to three legs without any trouble and continued to jump onto a window sill, where it could enjoy the view.

I kept in touch with the owners, to see how the cat was doing. It continued to do well for almost 2 years after the surgery. I made a house call at that point, when the cat's general health had deteriorated severely. It was no longer eating and had lost a significant amount of weight. The time had finally come. The owners tearfully elected to put the cat to sleep. As sad as the moment was, we all rejoiced that the cat had survived 2 more years of enjoyable life following surgery.

This decision worked well for these owners. The risks of the procedure were high, but the gamble proved successful. However, everyone would not have made this decision. Clients must put an economic value on the emotional attachment to the pet.

In food animal medicine, the decision-making process differs greatly. Every day the client and I make decisions that we hope reward the client financially. Last week I visited a young Amish farmer who had recently begun to milk cows. He purchased cattle over the last several months from three different sources. He called because a number of cattle were not eating well.

The diagnosis was quite simple in this case. The sick cattle ran high fevers, breathed heavily, and coughed. When I listened to their lungs, it was obvious that the animals had contracted pneumonia.

continued

His biosecurity failed. The weather varied, with very cold mornings and hot afternoons. The barn was not well ventilated and likely became quite warm and humid during several afternoons. Moreover, the cattle from all three sources commingled within the one barn. The farmer did not investigate animal history before purchasing any of the cattle, but he was told that they were vaccinated. Upon investigation, the producer discovered that the vaccination history was doubtful and at most the last group had received one shot of a killed vaccine. Those animals never received a booster shot.

The treatment options were simple. First, we began treating the cattle with antibiotics and medications to relieve the fevers. The major decision concerned whether to vaccinate the remaining herd. I had to weigh the considerations. The first choice was to quickly vaccinate the remainder of the herd in the hope that they had been adequately vaccinated and a booster would provide a quick memory response. The second option was to not vaccinate. If the animals were already incubating the disease, vaccination would add to the stress of the animal and might actually make the outbreak worse. If they were never vaccinated, this single shot of killed vaccine would not provide adequate protection anyway.

Potential costs were associated with both options. By vaccinating, the farmer had the cost associated with purchasing the vaccine, the short-term loss of productivity that often occurs immediately following vaccine, and the potential loss associated with making the disease worse. By not vaccinating, the farmer did not have to spend any additional cash. However, if this option failed, a respiratory outbreak throughout the entire herd would cost even more, with medications, loss of productivity in the ill animals, and potentially death of animals.

We discussed the options and elected to vaccinate the remainder of the herd. A week later, our decision seemingly proved correct. No new cases developed in the remainder of the herd. The disease had been restricted to a group of seven animals that were purchased from one location. It would seem that these animals were not properly vaccinated prior to purchase or were exposed to an organism to which they did not have immunity. Veterinarians are responsible for helping farmers to make economic decisions that can affect their financial success. Often veterinarians must rely on their experience to guide the decision-making process.

One of the most difficult decisions in veterinary medicine involves euthanasia. I have talked to many people who had a desire to become a veterinarian but could not come to grips with putting an animal to sleep. Euthanasia is a difficult part of the job, often very sad and never easy. However, it is a crucial part of my work and often relieves an animal of the pain associated with a serious illness.

Over the years, I have put many animals to sleep. To this day, times exist when the task can be a tremendous struggle. Just a few months ago, I had the unpleasant experience of euthanizing one of my partner's dogs. Ali was a sweet 12-year-old golden retriever that had developed a cancer of the bone marrow, called leukemia. I had known Ali since she was a pup, and she often visited the clinic. Every time I saw her, she would roll over slightly, wagging her tail. She had trained me to pet her belly every time we met. After Ali was diagnosed, the veterinarians at our clinic aggressively treated her condition, but we knew that we were losing ground.

One evening I received the call that the time had come. We had all been through the process many times with many people, but this was going to be tough. I was going to euthanize a dog that I loved, for close friends. I felt their pain and knew my own as I scratched Ali's belly before we began. I tried to be strong for the family but could not talk much because there was a huge lump in my throat. I inserted a needle into a vein on her front leg and then steadily injected the medication. The medication sent Ali into a deep state of anesthesia and then reached a point of overdose, causing her heart to stop beating. Thankfully the process went very smoothly. It was a tough evening and a somber next day at work as well. We all felt the loss.

OBJECTIVE

■ *Consider Necessary Factors in Making Informed Decisions in the Veterinary Care of Production Animals and Contrast the Decision-Making Process in Companion Animals to Production Animals*

Primarily the decisions made within farm settings revolve around the economic return. Farmers raise animals as a source of income. Their desires are to maximize the health and productivity of these animals. In return, the profitability of the farms is maximized.

Numerous questions arise once an animal becomes ill. Often multiple treatment options exist. The decisions must be based on an evaluation of the cost, the likely success, and the cost of the treatment if there is a failure.

At times the calculations on the cost of treatment can be straightforward. Consider a dairy cow that has developed a severe case of foot rot. In this infectious disease, bacteria invade the skin between the claws of the hoof. If caught early, topical treatment with copper sulfate may relieve the condition. If the entire foot begins to swell, injectable antibiotics are often warranted. The disease can cause a severe lameness, which greatly hinders the productivity of the cow.

Many foot rot treatments are available. For simplicity, this discussion will evaluate two. The options are to use an injectable penicillin (cost of medication per day = $3.60) or ceftiofur (cost of medication per day = $9.15). Assuming that both medications have a similar effectiveness, it would appear initially that penicillin would be the better option. For a standard three-day treatment, the cost for penicillin would be $10.80 and for ceftiofur $27.45. Further analysis shows that ceftiofur does not penetrate into the milk and does not require the milk from the treated cow to be discarded. Penicillin requires that the milk be discarded for at least two days following the last treatment to prevent antibiotic residue in the milk. Therefore milk will have to be discarded for at least five days (three of treatment and an additional two). The value of the milk must then be calculated.

The value of the milk becomes the variable that influences the final decision. If the dairy cow was in her dry period and not producing milk, the decision would be easy—use penicillin. Intuitively it would seem that a cow producing 100 pounds of milk per day would be treated with ceftiofur. The farmer then raises the question: At what level of production does it become economical to use ceftiofur?

Two variables must be considered in this decision: the amount of value milk being produced. In general, farmers are paid for the number of hundredweights (cwt), or number of 100-pound units, of milk they produce. At the time this text was written, milk prices had dropped to almost $13.00 per hundredweight. Compare the two treatment options as follows:

Total cost of ceftiofur: $27.45
Total cost of penicillin: $10.80
Difference: $16.65

Ceftiofur becomes the treatment of choice whenever the value of the milk produced in the 5-day period is greater than $16.65.

$$\$16.65 \div \$13.00/\text{cwt} = 1.28 \text{ cwt or } 128 \text{ lb}$$
$$128 \text{ lb in 5 days} = 25.6 \text{ lb/day}$$

These calculations show that 25.6 pounds of milk per day is the financial breakeven point. At that milk weight, the cost of each treatment is identical. When milk drops below that level, penicillin becomes more economical. At any higher level of production, ceftiofur becomes the better option.

These calculations are very straightforward for a given course of treatment and a given milk price. Variations arise and there can be differences in the length of treatment, level of production, and the value of milk. Neither the farmer nor the veterinarian is anxious to perform such calculations with each change. Spreadsheet programs on computers provide an excellent means of accounting for these variations. Once the program is established, any of the variables can be changed easily. The computer then quickly performs the calculations.

This demonstration assumed an equal success with each treatment. Often success is one of the major variables that make the decision-making process difficult. Situations arise in which one treatment has a higher success rate than another. If the more successful treatment happens to be the cheaper option, the decision is easy. Generally the more successful treatment also has a higher associated cost.

One method of calculation used to aid in such processes is called decision tree analysis. This method uses the percentage of success and the cost of the two treatments to analyze which procedure will provide the most benefit overall. The decision tree analysis does not attempt to predict the outcome of an individual case. With this analysis method, the more cases that are treated, the more likely the results will follow the prediction.

This method uses probabilities to make the prediction. A probability is the chance or likelihood that an event will occur. For example, when flipping a coin, there is a 50% chance that it will show heads. In a small series of flips, the coin may consistently show tails. The results do not make the probability incorrect. The more coin tosses that are performed, the more likely it will be that the number of heads and tails will be similar. During the series of coin tosses, there will be streaks of multiple head tosses, as well as tails.

When making decisions about treatment, probabilities are helpful. However, when dealing with an individual, these probabilities can become meaningless. For example, the author's practice uses approximately 6000 doses of bovine respiratory vaccine every fall. Typically, one or two cows will develop an anaphylactic reaction as a result of the vaccination. This makes the probability of a reaction approximately 1 in 3000. When a large herd uses the vaccine, the risk is very manageable. Even with 300 cows, the chance is not great of having a cow react. If a client has only one cow and it happens to be the one that reacts, the overall risk was no higher. Unfortunately, the client will have the impression that the vaccine is extremely risky. The client's perception may be that the risk is 100%.

Decision tree analysis uses this concept of probabilities to predict which treatment method is going to be most economically beneficial when used in large numbers of cases. Decision trees can become very complicated as the options and branches increase. A simple example is provided to illustrate the concept.

The veterinarian is called to a farm that is having a herd mastitis problem. The veterinarian cultures a number of cows and determines that the problem is consistently an infection with *Staphylococcus aureus*. The veterinarian detects this organism in 20 cows that are averaging 60 pounds of milk. The farmer must get the mastitis problem under control or risk losing the current market to sell milk. This organism is very difficult to treat and is contagious among cows.

The veterinarian offers two different treatment protocols to the farmer. Option one involves a standard treatment with antibiotics for three days and has an average success rate of 30%. The second treatment protocol uses a series of three treatments with a new, more expensive antibiotic. This raises the success to almost 50%. The farmer asks the veterinarian which option he or she should choose.

The veterinarian calculates the cost of the two treatments. Treatment one with medication and discarded milk costs $48.40. Treatment two costs a total of $96.50. Due to the severity of the problem, the cows will be cured and saved in the herd, or culled if the treatment fails. The value of the milking cow is placed at $1500.00. The value of a cull cow is averaging $400.00.

These numbers are inserted into a simple decision tree. The tree has two initial branches for the two different treatment protocols. Each branch then has another two branches: a successful treatment and a treatment failure. The probability of success is placed on each branch, along with the value of the outcome.

The value of each branch is then calculated, with the cost of treatment being subtracted from the value

of the animal. This number is then multiplied by the success of the treatment. The values for the success and failure of each treatment are added together to calculate the final value for each treatment protocol. The numbers for each treatment protocol can then be compared to evaluate which has the best outlook for financial success for the farmer.

Referring to Figure A–1, the calculations for the two treatments are:

$$(\text{Value of successful treatment} - \text{Cost of treatment}) \times \text{Percent of success} +$$
$$(\text{Value of treatment failure} - \text{Cost of treatment}) \times \text{Percent of failure}$$

Treatment One $(1500 - 48.40) \times 30\% +$
$(400 - 48.40) \times 70\%$
$1451.60 \times 30\% + 351.6 \times 70\%$
$435.48 + 246.12$
$681.60 = $ Outcome of Treatment One

Treatment Two $(1500 - 96.50) \times 50\% +$
$(400 - 96.50) \times 50\%$
$1403.5 \times 50\% + 303.50 \times 50\%$
$701.75 + 151.75$
$853.50 = $ Outcome of Treatment Two

Comparing the two values, it is clear that treatment two provides the greater economic benefit. The decision tree takes into account not only the cost of the treatment but also the differences in success.

In this example, clinical studies have been performed that provide the expected success of each treatment. The success on the given farm may not be identical to that of the controlled studies. In addition, changes may occur in the value of cattle and the cost of medications. Decision trees can be calculated on a spreadsheet computer program. Once programmed,

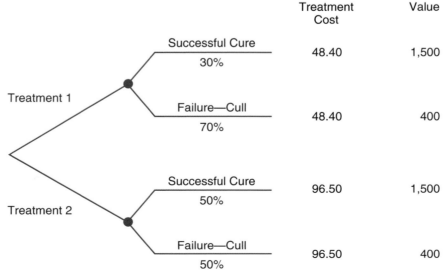

FIGURE A–1 Decision tree.

variables can be changed to determine if the same outcome is correct. For example, the success rate of treatment two can be lowered to determine at what level it is still the more economical decision. Decision trees can be made much more complex than this example. Computers make the complicated decision trees practical to design and evaluate.

Decision trees make an excellent scientific method to evaluate different treatment protocols. On the farm the veterinarian often lacks the detailed numbers to make the predictions of success and failure. Each farm situation must be evaluated to determine if a change of treatment is warranted. The results of controlled studies provide the foundation for the decision, but this information must be judged in a given setting.

For example, increasing the milking frequency of dairy cattle from twice to three times daily generally increases the milk output. This is often in the range of six to eight pounds of milk per cow per day. The change does not require a large investment, except for the cost of labor and running the milking equipment. In general the milk response is quite predictable, and straightforward calculations can be performed to determine if it is of economic value.

Once this change is implemented on a dairy farm, the success needs to be monitored. The author had the experience on one farm where the transition to three-times-a-day milking did not provide the expected results of an increased milk yield of 15%. Further evaluation of the individual farm determined that the milking parlor was very inefficient. By adding the extra milking, the cattle were forced to stand waiting for milking for long periods during the day. This kept them from resting, kept them from their ration, and increased lameness problems. While the vast majority of farms benefit from such a change, this farm did not. The underlying causes were determined and were not easily corrected. This farm returned to twice-a-day milking.

Production animals such as cattle, sheep, goats, and pigs have an inherent value based on the product they are producing. This value may be in the offspring, milk, or meat they can produce. This allows for specific numbers to be used in the calculations determining the economic return. Decision making in horses can also be based on economic return. The value of the horse is evaluated based on the specific use of each animal. The expected success of a racehorse can greatly influence the decision-making process in treating a significant illness or lameness. Likewise the value of a foal can greatly influence decisions on treating a brood mare or breeding stallion.

Many horses are used solely for pleasure purposes. The decision-making process in these animals is very similar to that in companion animals. Decisions in companion animals, such as dogs and cats, often become much less about economics and much more an emotional decision. Owners must place a value on the emotional attachment on their pets. This is not to say that economics does not play a role in the decision. Treatment may be too costly for an owner to afford. Health insurance for pets has yet to gain widespread acceptance, so owners are usually responsible for paying for any treatment.

Just as in production medicine, several treatment options may exist for a given condition. For example, a severe fracture may best be treated with a bone plate and screws. This method provides the most secure repair and the highest likelihood of success. Unfortunately, this means is quite costly and the owners may elect to try a splint or cast. The success is not as likely, but the owners are making an effort to give the animal a chance. In most situations, pets are strictly companions and the only returns on the owners' investments are more quality time together with their pets.

The veterinary profession uses euthanasia as a tool in providing relief to dying animals. Euthanasia, or putting an animal to sleep, is performed with an intravenous injection of an anesthetic. The medication is given in such high levels that it is fatal. The animal will feel the initial needle insertion into the vein, but the remainder of the procedure proves painless.

The decision for euthanasia is always a difficult one. Strong emotional attachment to pets exists in our society. Often owners refer to their pets as family. Whether the decision becomes necessary for an elderly pet with organ failure or a young animal with severe trauma, the natural reaction is to want more time with a loved one.

The decision does often come down to economics and the cost of treatment. But in addition, the likelihood that the animal can recover and the discomfort that the animal faces also play a large role in the decision of euthanasia. A term that is used to guide owners in this decision is *quality of life*. The owners are asked to evaluate the pet at home and judge whether the quality of life is up to their standard. This is not a clearly defined term. Owners must decide what this means to them and their pets.

Consider an elderly dog with such severe arthritis, secondary to hip dysplasia, that it can no longer rise on its own. The animal is developing sores from lying too long and is soiling its skin and coat with urine and feces. The animal appears bright otherwise and is still eating. This case is not clear cut, where euthanasia is essential. This animal's quality of life has definitely declined. Some owners will feel that the chronic pain and the dog's inability to rise warrant euthanasia. Other owners would see the dog's outward appearances, see it wag its tails and decide that the dog should not be put to sleep.

The veterinarian's role in the process is to educate the owners on their pets' conditions. Many aspects of a given disease can influence the final decision. These factors include the cost of treatments, the chance of successes, the commitment required by the owners, and the discomfort that the pets may feel. The veterinarian can provide medical answers about the animals' conditions, but ultimately the owner must make the final decision. Making the final decision to elect for euthanasia is often the most difficult decision that pet owners can make.

It is important to recognize that the loss of a pet, whether through euthanasia or natural death, is a loss of a loved one. Grieving is a natural process that occurs with the loss of a loved one, human or animal. This process is unique for every individual. At times, owners lack the support of other family members and friends. They may not understand the strong emotional attachment that builds between a pet and the owner. The individual facing the loss must understand that grief is a natural occurrence.

Emotionally almost all owners will feel sadness or sorrow over the loss of their pet. In addition, the owners may experience several other emotions. Some people will feel an entire range of emotions. Denial may be the first emotion faced when a life-threatening illness is diagnosed. The owners will not want to accept the fact that their pet may be dying. Denial can progress into guilt. When the owners face the fact that their pet has died, they may begin to question if they had done something wrong. The feeling of guilt makes the owners place the blame for all the pet's problems on themselves. With denial and guilt, it is important for veterinarians to support the owners and attempt to explain the medical facts surrounding the case.

Some owners will externalize the blame and become angry. This anger can easily be directed at the veterinarians, who failed to save the pet. In the heat of the moment, the owners might make comments that they later understand are not realistic. The extreme pain associated with losing a pet can make this possible.

The pain associated with this process can also lead to depression. The depression can lead to a feeling of despair, preventing the persons from handling normal events in their lives. Individuals can experience any or all of these emotions, to any degree of severity. It is important to understand that grief is a normal process that varies with every individual. Each person handles the crisis in different ways.

It is very helpful to find friends and family members who are supportive and willing to listen. It can be quite helpful to talk about the feelings that a person is experiencing. At some point the individual must recognize that grief is natural and that eventually the feelings will improve.

A major step in the process occurs when the individual obtains a new pet. It is important to recognize that a new pet is not designed to replace the last one. The new pet is not be identical and has different personality traits. The new pet does not replace a beloved pet; it becomes a new addition to the family. A strong, loving relationship can develop with the new pet without losing the memories of the recently deceased pet.

The amount of time required before desiring a new pet is also an individual preference. Some people will feel the need almost immediately; they desire to have a new loved one fill the loss in their lives. Others may require a much longer time or may never feel the need to have more pets. Each individual needs to pass through the grieving process at his or her own rate and decide when the time is right. It is improper to purchase a pet for others while they are grieving. People often do this in attempt to help folks forget about their last pets. Unfortunately, this can add to the guilt for moving on without properly grieving. This decision is best left to each individual.

Addendum B: Career Focus

The College Board Index of Majors and Graduate Degrees 2000 segments careers in veterinary science into three areas: veterinary medicine, veterinary assistant, and veterinary specialties. Statistics found in this section were also taken from the aforementioned guide. Dr. Baker, co-author and agricultural education instructor, often refers student to this helpful book, which can be found in most high school guidance offices or located at http://www.collegeboard.org.

Veterinary medical degrees, referred to as professional degrees (Veterinary Medical Doctor [VMD] and Doctor of Veterinary Medicine [DVM]), prepare individuals for the practice of veterinary medicine. Students selecting a school of veterinary medicine must first successfully complete a bachelor of science degree in a related field, such as animal science, biology, or premedicine. Many students also seek advanced degrees prior to admission to a veterinary medicine program. As of the 2000–2001 school year, 27 institutions granted veterinary medical degrees. The career outlook for veterinarians remains positive. As discretionary income rises, people are more willing to spend money on veterinary care for companion animals. Further, demand for veterinarians with specialties continues to increase, especially in metropolitan areas. On the other hand, rural practitioners willing to treat both large and small animals are in need, because most veterinarians prefer to work in suburban and urban locales.

CAREER FOCUS: PRIVATE PRACTITIONER

Dr. James Lawhead, a private practitioner in central Pennsylvania, graduated from the University of Pennsylvania School of Veterinary Medicine in 1987. Following graduation, he was employed as an associate veterinarian in a mixed animal practice. The practice provides service to 250 dairy farms. The primary emphasis of the practice is dairy cattle, but he also works on beef cattle, sheep, goats, pigs, and horses.

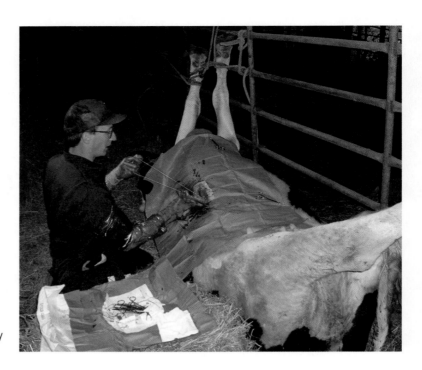

Dr. Lawhead performing a displaced abomasum surgery on a cow.

In addition, Dr. Lawhead spends approximately one third of his time in a small animal clinic consisting primarily of dog and cat patients.

When Dr. Lawhead joined the practice, two partners employed him. These two veterinarians were extremely supportive and patient as he gained experience in private practice. They helped to train Dr. Lawhead. Dr. Lawhead is now a partner as well, and the practice has grown to employ five veterinarians.

Dr. Lawhead experiences a tremendous variety of situations in his career. As a mixed practitioner, he finds value in each aspect of his job. The work with dairy farmers proves very rewarding. While working with the same clients for 15 years, a special relationship between Dr. Lawhead and his clients has developed, because both the veterinarian and farmer strive to maximize productivity and profitability. Dr. Lawhead finds that process challenging and stimulating. Dr. Lawhead has a special interest in nutrition and formulates the rations for several dairy clients.

The small animal portion of the profession also offers challenges and rewards. Although similar goals exist for maintaining the health of all animals, the companion animal portion of his job allows him to work on challenging medical and surgical cases. Further, the group practice environment is rewarding and allows for collaboration on cases.

Private practice does have its drawbacks. Dr. Lawhead is often on call and often is called at night. Even so, Dr. Lawhead must be at work the next day. Fortunately, the on-call responsibility is shared equally in his multiveterinarian practice. The work-day varies and is unpredictable. Dr. Lawhead's work-day often extends well beyond 12 hours.

CAREER FOCUS: ACADEMIA

Dr. Abby Maxson Sage is a 1987 graduate of the University of Pennsylvania School of Veterinary Medicine. Following her graduation, Dr. Sage elected to participate in advanced training in an internship and residency, also at the University of Pennsylvania. After the

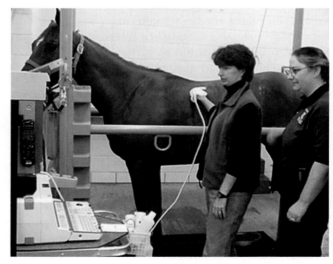

Dr. Sage, with the assistance of Dr. Ormond, is performing an ultrasound examination on a horse. *(Courtesy Dr. Abby Maxson Sage.)*

completion of her residency, Dr. Sage accepted teaching and research responsibilities at the University of Pennsylvania. Following extensive training, work, research, and testing, Dr. Sage gained certification as a specialist in the American College of Veterinary Internal Medicine. This honor is only given to the most highly trained specialists. In 1997 Dr. Sage joined the team at the University of Minnesota, School of Veterinary Medicine as an assistant clinical specialist.

Dr. Sage carries a variety of responsibilities in her career. In addition to admitting animals to the clinic, she uses these cases as teaching tools for training veterinary students. She also lectures to students in the classroom setting. At other times Dr. Sage performs research on equine-related topics.

CAREER FOCUS: PUBLIC HEALTH VETERINARIAN

Dr. Cathleen Hanlon is currently a veterinary medical officer in the Rabies Section at the Centers for Disease Control and Prevention (CDC) in Atlanta, Georgia. The CDC is the lead agency for disease outbreak investigations for such concerns as Ebola virus, Hantavirus, Lassa fever, foodborne and waterborne outbreaks, suspected bioterrorism attacks, and rabies. Dr. Hanlon re-

Dr. Hanlon working with a sedated raccoon that had been captured in a live trap. *(Courtesy Dr. Cathy Hanlon.)*

ceived a bachelor's degree in animal science at Rutgers University and a veterinary degree at the University of Pennsylvania in 1987. She subsequently received a doctorate in comparative medicine at the University of Pennsylvania. Her research studied a new type of rabies vaccine used for oral vaccination of raccoons. The vaccine took more than 10 years of work to develop. The vaccine is now being widely used to contain the spread of raccoon rabies in the eastern United States.

As a rabies researcher and public health professional, Dr. Hanlon engages in a wide range of activities. Her work includes field investigations involving species such as raccoons, skunks, foxes, dogs, and bats in the United States, as well as other countries. Dr. Hanlon also performs laboratory work including diagnosis and development of new methods for prevention of rabies and responds to questions about rabies from other public health professionals and the public.

CAREER FOCUS: VETERINARY SURGEON

Dr. David Sweet graduated in 1989 from the University of Pennsylvania School of Veterinary Medicine. Following his graduation, Dr. Sweet pursued further training as an intern at the University of Pennsylvania and a surgical residency at the North Carolina State University. Following that training, Dr. Sweet accepted an instructorship at Washington State University and returned to the University of Pennsylvania as an assistant professor. During his training, Dr. Sweet met the rigorous qualifications necessary to become a diplomate in the American College of Veterinary Surgeons. This honor earned by Dr. Sweet distinguishes him as a surgical specialist.

In 1996 Dr. Sweet joined the Veterinary Referral Center in Little Falls, New Jersey, as a surgical specialist. The center employs veterinary specialists in many fields, including surgery. The veterinary practice provides a service that allows private practitioners to refer difficult cases for more specialized treatment. Dr. Sweet performs both soft tissue and orthopedic surgery. He has performed many advanced surgeries, including hip replacements in dogs.

CAREER FOCUS: VETERINARY ASSISTANT

Veterinary assistants help the veterinary practice by performing a wide range of tasks. These individuals may greet patients, keep records, bill clients, and restrain animals, as well as feed, exercise, and provide basic health care for patients. Obviously, the responsibilities vary from employer to employer. Although most practices require the assistant to complete at least a certificate in postsecondary schooling, some may

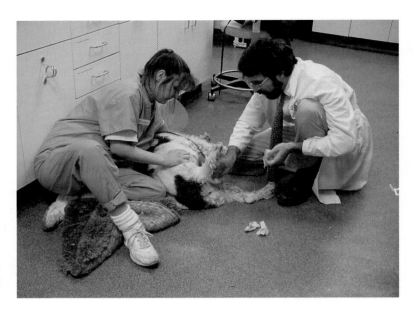

Dr. David Sweet, with the assistance of registered veterinary technician Michele Antoch, examines a surgical incision on a dog. *(Courtesy Dr. David Sweet.)*

hire assistants without additional education. More than 75 different schools offered veterinary assistant type programs in the 2000–2001 school year. Most of these institutions offer either associate degrees, which are typically completed in two years, or certificate programs, which vary in length according to the institution but usually average nine months. A few grant bachelor's degrees. Furthermore, some of these aforementioned programs grant degrees and certificates to veterinary technicians as opposed to veterinary assistants. Despite the differing titles, the required coursework remains somewhat similar. Numbers of available jobs for veterinary assistants and technicians will continue to grow with the demand for veterinarians.

CAREER FOCUS: VETERINARY EXTENSION SPECIALIST

Level of degree separates veterinary assistants from veterinary specialists. Almost 30 programs grant degrees in veterinary specialties. Most of these programs deliver master's and doctorate degrees, although a few award associate's and bachelor's

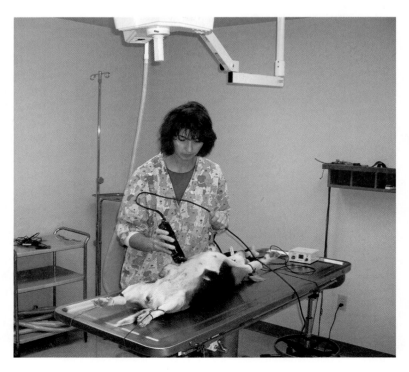

A veterinary assistant is shown preparing a dog for surgery.

Working on a computer is part of the job of an Extension veterinarian. *(Courtesy Dr. Lawrence Hutchinson.)*

degrees. Specialists may provide such supportive services as nutrition counseling, ration balancing, or radiology expertise to veterinary clinics. Conversely, other specialists may be employed in academia, where they perform research or Extension duties in veterinary-related areas. Further, specialists may seek employment as pharmacologists for drug companies or experts for agricultural businesses. Career opportunities remain positive for those seeking veterinary specialty degrees.

Students are encouraged to research employment in all veterinary related fields. Speaking with an agricultural education instructor, visiting the guidance or career placement department, interviewing local professionals employed in veterinary science–related fields, and searching the Web for information can provide an excellent start. From there, postsecondary school visitations may be in order. Best wishes in exploring the world of veterinary science.

Glossary

A

abscess   an accumulation of fluid (pus) in a dead space between tissues containing bacteria, white blood cells, and dead tissue

active immunity   immunity resulting from exposure of an animal to a pathogen either through disease contraction or vaccination

active transport   process that allows the pumping of a substance into an area of higher concentration

acute   sudden onset

Addison's disease   disease condition characterized by low levels of cortisol along with lethargy, weakness, weight loss, poor appetite, vomiting, and diarrhea (also known as hypoadrenocorticism)

adipose tissue   fat

alopecia   baldness

anabolism   cellular reactions that combine smaller molecules into larger ones

anaphylaxis   generalized life-threatening allergic reaction

anemia   low in red blood cells

anesthetize   to sedate animals so they lack sensitivity or awareness

anestrus   period when animal is not cycling through estrus

anthrax   a bacterial infection with skin, intestinal, and respiratory forms

antibiotics   drugs that fight bacterial infections

antibodies   infection-fighting proteins

antigen   any foreign material that is capable of stimulating an immune response

antimicrobial   agent that hinders the growth or kills microorganisms

antioxidant   oxidation-inhibiting substance

antiseptics   germicides that can be used on the skin of animals

appendicular skeleton   skeletal portion consisting of limb bones

arrhythmia   any change in rate, rhythm, or conduction within the heart

arteries   vessels that transport blood away from the heart

arthritis   disease condition in which the smooth lining of cartilage becomes rough and irregular

aseptic technique   general practices used to minimize the risk of infection that may occur following surgery

ataxic   uncoordinated muscle movement

atopy   sensitization to foreign antigens

atrophy   shrink

autoclave   piece of equipment used to sterilize instruments with pressurized steam

autoimmune disease   condition in which the body's immune systems destroys its own cells

autonomic system   involuntary portion of the peripheral nervous system

autopsy/postmortem   examination to determine cause of death; sometimes also referred to as necropsy

axial skeleton   skeletal portion consisting of skull, vertebrae, ribs, and sternum

azotemia   elevation of both urea nitrogen and creatinine in the blood

B

bacteriostatic   agent that slows the rate of growth of bacteria

band   means of nonsurgical castration in which a band is placed above the testicles, thus cutting off the blood supply, causing atrophy

benign   description of tumor that is localized and will not spread to other areas of the body

biosecurity   practices that protect herd health by preventing the spread of pathogens

bloat   accumulation of gas in the rumen

bolt   quick consumption of feed

borborygmi   the normal noise made within the intestinal tract as gas and fluid move through the tract (this is the noise that in humans is commonly described as stomach growling; singular is borborygmus)

botulism   food poisoning resulting from a toxin-producing bacteria

bronchodilators   medications that open the airways

bucks   male goat

C

calorie   unit of measure that defines the energy contained within a food

cancer   uncontrolled cell division

canine hip dysplasia   diseased ball and socket joint of a dog results in a shallow rather than normally deep socket

cardiac cycle   one complete contraction and relaxation of the heart

cardiopulmonary resuscitation (CPR)   procedure used to stimulate the heart to deliver oxygen to the lungs

carnivore   animals, such as dogs and cats, that require meat-based diets

castration   removal of testes

cat scratch fever   bacterial disease resulting from a cat scratch, which results in a soreness at the inflicted site, fever, and enlarged lymph nodes

catabolism   cellular reactions that break larger molecules into smaller ones

caudal   refers to the tail

central nervous system   brain and spinal cord

centrifuge   a machine that spins substances such as blood to separate differing densities within the spun matter

cervical disk disease   painful disease resulting from pressure of cervical disk on spinal cord

cesarean section   surgical removal of a newborn

chemistry profile   test that often evaluates blood sugar, electrolytes, protein, liver and pancreatic enzymes, bilirubin, and nitrogen-containing wastes

chronic   long term

colic   general term used to describe abdominal pain in horses

coliform   a related group of disease-causing bacteria commonly found in feces

colostrum   antibody rich milk that is first produced by the mother and secreted initially after parturition

coma   prolonged state of unconsciousness

comminuted fracture   bone break in which bone shatters in fragments

complete blood cell count (CBC)   test that evaluates red blood cell (RBC) count (often a packed cell volume is a part of this testing), size of RBCs, amount of hemoglobin, number of platelets, total white blood cell count, and a breakdown of the types of cells present

compound/open fracture   bone break in which the bone punctures the skin

concentrates   nonforage component of a diet consisting of grains, protein sources, vitamins, and minerals

congenital   present at birth

constipation   condition that occurs when the feces is too dry and moves too slowly

constricts   closes

contagious   disease condition that can be passed among organisms

cranial drawer sign   diagnostic test for a torn cruciate ligament

cranial   refers to the head

cribbing   wood chewing

cryptorchidism   condition occurs when one or both testes fail to enter the scrotum

Cushing's disease   elevated blood levels of cortisol with presentation of clinical signs such as excessive thirst, urination, and appetite, thin skin and hair coat, panting, enlarged abdomen, weakness, and lethargy (hyperadrenocorticism)

cutaneous larva migrans disease condition resulting from the infestation of hookworm larvae into the body (also called creeping eruption)

cyanosis blue color associated with low oxygen levels in the blood

D

débridement process in which damaged and contaminated tissue is removed from a wound prior to the suturing process

deciduous teeth initial set of teeth often referred to as the baby teeth

degenerative joint disease when a joint becomes so worn that the cartilage lining the joint thins and roughens

diabetes insipidus lack of antidiuretic hormone, which results in very diluted urine, uncontrolled thirst, and excessive urination

diabetes mellitus disease in which the animal is consistently hyperglycemic (elevated blood sugar)

diastole relaxation phase of the cardiac cycle

diffusion process that allows molecules to move across a membrane in an effort to equal concentration

dilates opens

disinfectant germicides that are too harsh to be used on skin but can be used on inanimate objects or surfaces

displaced abomasum condition commonly called a twisted stomach in which the fourth stomach of a cow fills with gas and is pulled upward; commonly occurs shortly after calving

doe female goat

dorsal toward, on, or in the back

dry matter percent of a feed that remains when all the water is removed

E

edema swelling due to an accumulation of fluid

electrocardiogram tracing made by the electrocardiograph

electrocardiograph electronic instrument that picks up the small electrical signal that runs through the body

ELISA test enzyme-linked immunosorbent assay

–emia suffix used to describe levels in the bloodstream

endocytosis process that allows the cell membrane to wrap around a particle and then section it into the cytoplasm as a vacuole

endotracheal tube tube, often used in administering anesthesia, that is passed through the trachea

enzymes protein molecules that speed chemical reactions in the body

epidural injection of local anesthetic into fluid around spinal cord

epilepsy seizure activity

epithelial tissues collection of cells that line the body's surface and openings

equine infectious anemia viral disease causing fever, red blood cell breakdown, depression, weight loss

equine protozoal myeloencephalitis disease in which a protozoa causes brain infection and clinical neurological signs

erythropoiesis production of red blood cells

eructate belch

estrous cycle series of events that occur in females in preparation for pregnancy

estrus state of sexual excitement

eukaryotic cells with membrane bound organelles such as a nucleus, mitochondria, and endoplasmic reticulum

exocytosis the process which takes a membrane bound sac, joins it to the cell membrane and then releases it to the extracellular fluid

expiration when air is forced from the lungs

F

fibrosarcoma cancer of the connective tissue

first intention healing wound closure that occurs when repaired during the first six to eight hours after trauma, when bacterial levels at site remain low, allowing wound to heal

flatulence an accumulation of gas in the intestinal tract

float filing of a horse's teeth to prevent discomfort from the edges of molars that sharpen with age

fomite inanimate object

foot and mouth disease highly infectious viral disease that selectively attacks epithelial tissue in cloven hoofed animals

forages high-fiber feed such as grasses, hay, or silages

free catch urine urine caught outside the body while the animal urinates

free choice diet having food available to animal at all times

free radicals atoms having a single free electron, which attracts another electron from neighboring atoms

G

gastric dilation–volvulus syndrome (GDV) condition typically occurring in large, deep-chested dogs that have consumed a large meal and or large volumes of water, which allows for the accumulation of gas and the rotation of the stomach; also called gastric torsion

gestation (gestating) pregnancy; carrying a pregnancy

glucose blood sugar

glycogen polysaccharide, which is used to store cellular energy

golden period time during the first six to eight hours after trauma when bacterial numbers remain at a moderate level, allowing the wound to be closed

gout painful disease resulting from high serum levels of uric acid

granulation tissue combination of capillaries and connective tissue in an open wound that takes on a fleshy red appearance

grazing consumption of forages such as grasses at a leisurely pace

H

hardware disease occurs when ruminants inadvertently consume metal, which migrates through their bodies, causing infection

heart failure condition in which the heart cannot meet the demands of the animal

heart murmur leakage within the heart, creating an abnormal heart sound

heaves noncontagious condition in horses that causes coughing, nasal discharge, labored breathing, and rapid fatigue; often caused by inhaling dust and molds

hematoma accumulation of blood in the dead space between tissues

hemolysis breakdown of red blood cells

hemophilia chromosomal defect resulting in a deficiency of one of the clotting factors

herbivore animals such as cows and horses that require plant-based diets

herd check reproductive exams and health maintenance work routinely done by veterinarians

high-rise syndrome occurs when cats fracture their lower jaws during falls from tall buildings

homeostasis maintenance of the extracellular fluid

Horner's syndrome nerve damage that causes several eye malfunctions, including pupil constriction, eyelid drooping, protrusion of the third eyelid, and sunken eyes

humoral immunity production of antibody in response to an antigen

hydrolysis process in which water is added to a molecule to cleave it into smaller parts

hyper– prefix that indicates above normal

hypocalcemia condition commonly referred to as milk fever; caused by low blood calcium occurring at parturition

hypoglycemia low blood sugar

hypo– prefix that indicates below normal

I

iatrogenic disease condition resulting from a treatment

idiopathic disease condition not explained by current medical knowledge

inspiration when air is taken into the lungs

integument skin

intervertebral disk disease occurs when the center of the disk becomes less pliable and pressure between the vertebrae causes the disk center to rupture through the fibrous outer layer

intestinal anastomosis procedure in which two regions of intestine are joined

intramedullary pin method of repairing broken bone in which a stainless steel pin is inserted into the medullary cavity

intranasally in the nose

intravenous in the vein

intussusception condition that occurs when a region of the intestine telescopes into itself

involution process by which the uterus returns to a normal state

isotonic the same concentration

J

joint ill occurs when bacteria enters a newborn's body through the umbilical opening and settles in the joints, with lameness resulting

K

kennel cough disease involving severe cough, which commonly occurs in dogs kept in close quarters

keratin specialized protein deposited in cells giving a typical hardness and durability

killed vaccines immunizations that are manufactured from dead versions of pathogens

Koch's postulates four principles that help define infectious disease, which were developed by Robert Koch

L

lidocaine anesthetic

ligaments connective tissue that attaches bones to bones

ligated tied

lipid fat

listeriosis (circling disease) brain infection commonly seen in cattle, sheep, and goats

Lyme disease bacterial infection that can result in human symptoms such as fatigue and joint pain

lymph transparent yellowish fluid that travels through the lymphatic system, which helps to remove bacteria and proteins, transport fat, and supply lymphocytes

M

maintenance energy requirement (MER) amount of energy required by an animal at rest plus any additional energy required for the normal activity

malignant description of a tumor that will invade other parts of the body

mastitis infection of the mammary gland

metabolism all reactions conducted in the cells

metastasis spread of cancer cells to other sites in the body

modified live vaccines immunizations that are manufactured from altered versions of pathogens

monogastric single stomached

myelinated nerves nerves with a myelin sheath

myelogram procedure in which dye is injected into the epidural space, followed with a radiograph to trace the path of the dye

myofiber muscle cell

N

necrotic dead

neoplasm tumor that develops when cells grow in an uncontrolled manner

neuron nerve cell

nystagmus condition in which the eyes jerk back and forth in rhythmic manner

O

obstetric having to do with pregnancy and delivery

ophthalmoscope instrument used to observe the structures in the interior of the eye, such as the optic nerve, retina, and retinal blood vessels

organs collections of tissue

orthopedic surgeon veterinarian or doctor who specializes in surgery of the bones

osmosis process that allows a solvent to move across a membrane in an effort to equal concentration but will not allow molecules to pass

ossification process in which bone is formed

P

pacemaker system maintains the regular rhythm of the heart

packed cell volume rapid test that provides the percent of the blood composed of red blood cells

palpated felt

parasympathetic systems part of the autonomic system that slows the body from the flight-or-fight mode, lowering heart rate and blood pressure

parturition giving birth

parvovirus viral disease that causes severe vomiting and diarrhea in dogs

passive immunity immunity resulting from transfer of antibodies from one animal to another.

pathologists scientists who interpret and diagnose changes in cells and tissues

peripheral nervous system all nerves outside the brain and spinal column

peristalsis organized set of muscle contractions in a hollow organ that propels its contents

peritonitis inflammation throughout the abdominal area

phagocytosis process in which a cell engulfs and ingests particles

phenobarbital drug used to control seizures

pheromones chemicals emitted by animals that serve as a means of sexual communication

pleural friction rub noise heard when listening to lungs with irritated pleura

plexus network of nerves

pneumonia inflammation of the lungs

pneumothorax condition in which air becomes trapped between the lungs and the chest wall

polarization condition in which one region of cell has different charge than adjacent region

polyestrus constant continuation of estrous cycle

porcine stress syndrome swine condition in which calcium leaks from the endoplasmic reticulum, causing pigs to shake involuntarily

primary response initial antibody production that occurs when an antigen is first introduced to the body

prodromal phase stage in disease when first signs of illness occur

prokaryotic cells that lack membrane-bound organelles

prolapsed uterus condition in which uterus turns inside out and is pushed through the vulva

proud flesh overgrowth of granulation tissue that prevents epithelial cells from growing across the wound

pruritus severe itchiness

puberty start of sexual maturation

pus accumulation of infection-fighting cells, destroyed pathogens, dying tissue cells, and tissue fluid that result at the site of infection

pyometra uterine infection that commonly occurs in older dogs and cats

Q

quarantine confinement of an animal separate from the herd in an effort to prevent spread of disease

R

rabies viral disease that infects the central nervous system

radiograph photograph taken when streams of x-rays pass through the body and expose film

radiology study of the images created on film by x-rays passing through the body

refractometer instrument that measures specific gravity

respiration exchange of gases between the animal and its environment

resting energy requirement (RER) amount of energy required by an animal at rest

retching strong, rapid abdominal contractions

retro– behind

rickets disease condition of deformed and weakened bones resulting from childhood deficiency of calcium

rigor mortis muscle stiffness occurring after death

ringworm fungal infection of the skin

roaring horse condition in which one of the vocal folds fails to open, thus causing an ensuing roaring noise when the horse breathes heavily during exertion

rodenticide poison used to control rodents such as rats and mice

rumination process in which the rumen and reticulum contract in a manner that forces some of the stomach contents back through the esophagus and into the mouth, where it is chewed

S

schistosomus reflexus congenital abnormality in which the affected fetus develops inside out, with internal organs exposed

seasonal polyestrus continuation of estrous cycle until pregnancy during only certain times of the year

second intention healing wound closure occurring after the first six to eight hours after trauma, when granulation tissue must first fill the gap between the skin edges and subsequently allow epithelial cells to grow

secondary response quick response mounted against a second exposure to an antigen, which typically prevents disease development

sensory somatic system operates the voluntary motor activity of the body

seroconversion change in titer reading by four times

serology measurement of the presence of antibodies against a specific organism

seroma accumulation of fluid in a dead space between tissues; similar to serum, with a small number of red blood cells

serum clear yellow substance obtained when separating blood components

shock condition in which not enough blood is pumped to vital tissues, associated with a drop in blood pressure

shunting moving

signalment basic description of an animal presented for evaluation

simple fracture clean bone break

skin turgor measure of hydration, which tests how quickly the skin returns to its normal position after being pinched

spay (ovariohysterectomy) removal of ovaries and uterus

spayed indicates a female animal with reproductive organs removed

specific gravity weight of a liquid as compared with distilled water

sterilization procedures such as application of pressurized steam and chemicals in which all microorganisms are destroyed

stocking up term used to describe an accumulation of fluid in the legs of horses that have been tied for excessive periods

subcutaneous under the skin

subluxate partially dislocate

sweeny nerve damage and resultant shoulder muscle shrinkage occurring in draft horses from pulling harnesses

symbiosis mutually beneficial relationship

sympathetic system the part of the autonomic system that stimulates organs for flight or fight

systemic affecting the entire body

systole contraction phase of the cardiac cycle

T

tachycardia elevated heart rate

tendons connective tissue that attaches muscles to bones

tetanus acute bacteria infection causing muscle stiffness and rigidity; often called lockjaw

tissue collection of cells organized for a particular function

titer measure of antibody levels in the bloodstream

total mixed ration (TMR) the mixing of all feeds stuffs in a diet

toxoplasmosis protozoal parasitic disease in which cats serve as the definitive host; of concern to humans, especially pregnant women and persons with compromised immune systems

tunnel ventilation system with air inlets at one end of a building and fans for outlet at the other

tying up or Monday morning disease cramping with potential muscle damage in working horses that occurs the Monday after a weekend of rest and full feed

U

uremia clinical signs associated with azotemia

urinalysis evaluation of urine

urinary incontinence leakage of urine at inappropriate times

V

vector organism that transmits disease

veins vessels that transport blood back to the heart

ventilation exchange of air from within a building and the outside

ventral below

vestibular system balance center

visceral larva migrans condition in which roundworm larvae migrate through the body, causing damage to internal organs

volt unit of electrical measurement

W

weaned removal from nursing

wet dewlap skin infection in the lower neck of rabbits

whelping birthing in dogs

X

x-rays electromagnetic radiation, which can pass through living tissue

Bibliography

Agar, Sandie. 2001. *Small Animal Nutrition.* Oxford. Butterworth-Heinemann.

Alexander, Joseph W., ed. May 1992. The Veterinary Clinics of North America: Small Animal Practice. *Canine Hip Dysplasia.* Vol. 22 (3). Philadelphia. W.B. Saunders.

August, John R., and Loar, Andrew S., eds. Jan. 1987. The Veterinary Clinics of North America: Small Animal Practice. *Zoonotic Diseases.* Vol. 17 (1). Philadelphia. W.B. Saunders.

Bacha, William J. Jr., and Bacha, Linda M. 2000. *Color Atlas of Veterinary Histology.* Philadelphia. Lippincott.

Bagley, Rodney S., ed. July 1996. The Veterinary Clinics of North America: Small Animal Practice. *Intracranial Disease.* Vol. 26 (4). Philadelphia. W.B. Saunders.

Baker, Edward, and Felsburg, Peter J., eds. July 1994. The Veterinary Clinics of North America: Small Animal Practice. *Immune-Associated Diseases and Nondermatologic Allergy.* Vol. 24 (4). Philadelphia. W.B. Saunders.

Baker, John C., ed. March 1987. The Veterinary Clinics of North America: Food Animal Practice. *Bovine Neurologic Diseases.* Vol. 3 (1). Philadelphia. W.B. Saunders.

Banks, William J. 1981. *Applied Veterinary Histology.* Baltimore. Lippincott.

Barrett, James T. 1978. *Textbook of Immunology: An Introduction to Immunohistochemistry and Immunobiology.* 3rd ed. St. Louis. Mosby.

Basmajian, John V, et al., eds. 1982. *Stedman's Medical Dictionary.* 24th ed. Baltimore. Lippincott.

Beasley, Val. R., ed. March 1990. The Veterinary Clinics of North America: Small Animal Practice. *Toxicology of Selected Pesticides, Drugs, and Chemicals.* Vol. 20 (2). Philadelphia. W.B. Saunders.

Behrand, Ellen N., and Kemppainen, Robert J., eds. Sept. 2001. The Veterinary Clinics of North America: Small Animal Practice. *Endocrinology.* Vol. 31 (5). Philadelphia. W.B. Saunders.

Benacerraf, Baruj, and Unanue, Emil R., 1979. *Textbook of Immunology.* Baltimore. Lippincott.

Benenson, Abram S. 1981. *Control of Communicable Diseases in Man.* 13th ed. Washington, D.C. American Public Health Association.

Berne, Robert M., and Levy, Matthew N., eds. 1998. *Physiology.* 4th ed. St. Louis. Mosby.

Blood, Douglas C., Henderson, James A., and Radostits, Otto M. 1979. *Veterinary Medicine: A Textbook of the Diseases of Cattle, Sheep, Pigs and Horses.* 5th ed. Philadelphia. Lea and Febiger.

Bonagura, John D., ed. 2000. *Kirk's Current Veterinary Therapy XIII: Small Animal Practice.* Philadelphia. W.B. Saunders.

Bonagura, John D., and Kirk, Robert W., eds. 1995. *Kirk's Current Veterinary Therapy XII: Small Animal Practice.* Philadelphia. W.B. Saunders.

Brock, Thomas D. 1979. *Biology of Microorganisms.* 3rd ed. Englewood Cliffs, N.J. Prentice-Hall.

Brockman, Daniel J., and Holt, David E. 2000. Management Protocol of Acute Canine Gastric Dilatation–Volvulus Syndrome in Dogs. *Compendium on Continuing Education for the Practicing Veterinarian.* Vol. 22 (11).

Brockman, Daniel J., Holt, David E., and Washabau, Robert J. 2000. Pathogenesis of Acute Canine Gastric Dilatation–Volvulus Syndrome: Is There a Unifying Hypothesis? *Compendium on Continuing Education for the Practicing Veterinarian.* Vol. 22 (12).

Burrows, George E., ed. July 1989. The Veterinary Clinics of North America: Food Animal Practice. *Clinical Toxicology.* Vol. 5 (2). Philadelphia. W.B. Saunders.

Callan, Robert J., 2001. Fundamental Considerations in Developing Vaccination Protocols. *34th Annual Convention Proceedings.* American Association of Bovine Practitioners.

Campbell, Karen L., ed. Nov. 1999. The Veterinary Clinics of North America: Small Animal Practice. *Dermatology.* Vol. 29 (6). Philadelphia. W.B. Saunders.

Campbell, Neil A. 1996. *Biology.* 4th ed. Menlo Park, Calif. Benjamin/Cummings Publishing.

Colville, Joann. 1991. *Diagnostic Parasitology for Veterinary Technicians.* Goleta, Calif. American Veterinary.

Compton, Robert W. 1998. *A Bovine Practitioner's Immunology Primer.* Larchwood, Iowa. Grand Laboratories.

Cullor, James S. 1995. Common Pathogens That Cause Foodborne Disease: Can They Be Controlled on the Dairy? *Veterinary Medicine.* Vol. 90 (2).

Dargatz, David A., ed. March 2002. The Veterinary Clinics of North America: Food Animal Practice. *Biosecurity of Cattle Operations.* Vol. 18 (1). Philadelphia. W.B. Saunders.

DeNayer, Sharon, and Downing, Robin. Sept. 2001. Ease Client's Pain. *Veterinary Economics.* Dennis, Stanley M., ed. March 1993. The Veterinary Clinics of North America: Food Animal Practice. *Congenital Abnormalities.* Vol. 9 (1). Philadelphia. W.B. Saunders.

Di Fiore, Mariano S.H. 1981. *Atlas of Human Histology.* 5th ed. Philadelphia. Lea and Febiger.

Ettinger, Stephen J., and Feldman, Edward C., eds. 2000. *Textbook of Veterinary Internal Medicine: Diseases of the Dog and Cat.* 5th ed. Vols. 1 and 2. Philadelphia. W.B. Saunders.

Evans, Howard E., and Christensen, George C. 1993. *Miller's Anatomy of the Dog.* 3rd ed. Philadelphia. W.B. Saunders.

Evans, Howard E., and De Lahunta, Alexander. 1980. *Miller's Guide to the Dissection of the Dog.* 2nd ed. Philadelphia. W.B. Saunders.

Ferguson, Duncan C., ed. May 1994. The Veterinary Clinics of North America: Small Animal Practice. *Thyroid Disorders.* Vol. 24(3). Philadelphia. W.B. Saunders.

Ford, Richard B., ed. May 2001. The Veterinary Clinics of North America: Small Animal Practice. *Vaccines and Vaccinations.* Vol. 31 (3). Philadelphia. W.B. Saunders.

Fossum, Theresa Welch, et al. 1997. *Small Animal Surgery.* St. Louis. Mosby.

Fraser, Clarence M., et al., eds. 1991. *The Merck Veterinary Manual.* 7th ed. Rathway, N.J. Merck and Co.

Getty, Robert. *Sisson and Grossman's The Anatomy of the Domestic Animals.* Vol. 1. 5th ed. Philadelphia. W.B. Saunders.

Gillespie, James H., and Timoney, John F. 1981. *Hagan and Bruner's Infectious Diseases of Domestic Animals.* 7th ed. Ithaca, N.Y. Cornell University.

Goodman, Louis S., and Gilman, Alfred. 1980. *Goodman and Gilman's The Pharmacological Basis of Therapeutics.* 6th ed. New York. Macmillan.

Greco, Deborah S., and Peterson, Mark E., eds. May 1995. The Veterinary Clinics of North America: Small Animal Practice. *Diabetes Mellitus.* Vol. 25 (3). Philadelphia. W.B. Saunders.

Guyton, Arthur C. 1976. *Textbook of Medical Physiology.* 5th ed. Philadelphia. W.B. Saunders.

Habel, Robert E. 1977. *Guide to the Dissection of Domestic Ruminants.* 3rd ed. Ithaca, N.Y. Robert Habel.

Habel, Robert E. 1981. *Applied Veterinary Anatomy.* Ithaca, N.Y. Robert Habel.

Habel, Robert E., and Sack, W.O. 1977. *Guide to the Dissection of the Horse.* Ithaca, N.Y. Veterinary Textbooks.

Hafs, Harold D., and Boyd, Louis J. 1976. *Dairy Cattle Fertility and Sterility.* Wisconsin. W.D. Hoard and Sons.

Harley, John P., and Prescott, Lansing M. 1999. *Laboratory Exercises in Microbiology.* 4th ed. Boston. WCB/McGraw-Hill.

Herdt, Thomas H., ed. July 1988. The Veterinary Clinics of North America: Food Animal Practice. *Metabolic Disorders of Ruminants.* Vol. 4 (2). Philadelphia. W.B. Saunders.

Herdt, Thomas H., ed. July 2000. The Veterinary Clinics of North America: Food Animal Practice. *Metabolic Disorders of Ruminants.* Vol. 16 (2). Philadelphia. W.B. Saunders.

Herren, Ray V., and Romich, Janet A. 2000. *Delmar's Veterinary Technician Dictionary.* New York. Delmar Thomson Learning.

Hildebrand, Milton. 1974. *Analysis of Vertebrate Structure.* New York. John Wiley and Sons.

Hill's Atlas of Veterinary Clinical Anatomy. 1989. Topeka, Kansas. Hill's Pet Products.

Holt, David E., ed. May 2000. The Veterinary Clinics of North America: Small Animal Practice. *Emergency Surgical Procedures.* Vol. 30 (3). Philadelphia. W.B. Saunders.

Hoskins, Johnny D., ed. July 1999. The Veterinary Clinics of North America: Small Animal Practice. *Pediatrics: Puppies and Kittens.* Vol. 29 (4). Philadelphia. W.B. Saunders.

Hoskins, Johnny D., ed. Jan. 1992. The Veterinary Clinics of North America: Small Animal Practice. *Feline Infectious Diseases.* Vol. 23 (1). Philadelphia. W.B. Saunders.

Howard, Jimmy L., ed. 1986. *Current Veterinary Therapy: Food Animal Practice 2.* Philadelphia. W.B. Saunders.

Howard, Jimmy L., ed. Nov. 1988. The Veterinary Clinics of North America: Food Animal Practice. *Stress and Disease in Cattle.* Vol. 4 (2). Philadelphia. W.B. Saunders.

Jones, Thomas C., and Hunt, Ronald D. 1983. *Veterinary Pathology.* Philadelphia. Lea and Febiger.

Junqueira, Luis C., Carneiro, Jose, and Contopoulos, Alexander N. 1977. *Basic Histology.* 2nd ed. Los Altos, Calif. Lange Medical.

Kandel, Eric. R., and Schwarts, James H., eds. 1981. *Principles of Neural Science.* New York. Elsevier North Holland.

Kilmer, Lee H., et al. 1985. *Dairy Profit Series: Reproduction Your key to Future Profits.* Ames, Iowa. Iowa State University Extension.

Kimball, John, W. 1978. *Biology.* 5th edition. Reading, Mass. Addison-Wesley.

King, A.S. 1978. *A Guide to the Physiological and Clinical Anatomy of the Central Nervous System.* 6th edition. Liverpool, England. University of Liverpool.

Kintzer, Peter P., ed. March 1997. The Veterinary Clinics of North America: Small Animal Practice. *Adrenal Disorders.* Vol. 27 (2). Philadelphia. W.B. Saunders.

Kirby, Rebecca, and Crowe, Dennis T. Nov. 1994. The Veterinary Clinics of North America: Small Animal Practice. *Emergency Medicine.* Vol. 24 (6). Philadelphia. W.B. Saunders.

Kirk, Robert W., and Bonagura, John D., eds. 1989. *Kirk's Current Veterinary Therapy XI: Small Animal Practice.* Philadelphia. W.B. Saunders.

Kirk, Robert W., and Bonagura, John D., eds. 1992. *Kirk's Current Veterinary Therapy X: Small Animal Practice.* Philadelphia. W.B. Saunders.

Lagoni, Laurel. Oct. 2001. Preparing for Euthanasia of the Older Pet. *Practice Builder, A Supplement to DVM Magazine.*

Lawlor, Dennis F, and Colby, Emerson D., eds. May 1987. The Veterinary Clinics of North America: Small Animal Practice. *Pediatrics.* Vol. 17 (3). Philadelphia. W.B. Saunders.

Leib, Michael S., ed. May 1993. The Veterinary Clinics of North America: Small Animal Practice. *Gastroenterology: The 1990s.* Vol. 23 (3). Philadelphia. W.B. Saunders.

Lessard, Pierre R., and Perry, Brian D., eds. March 1988. The Veterinary Clinics of North America: Food Animal Practice. *Investigation of Disease Outbreaks and Impaired Productivity.* Vol. 4 (1). Philadelphia. W.B. Saunders.

Lewis, Lon D., and Morris, Mark L. 1984. *Small Animal Clinical Nutrition.* 2nd ed. Topeka, Kansas. Mark Morris Associates.

Liebman, Michael, 1986. *Neuroanatomy Made Easy and Understandable.* 3rd ed. Rockville, Md. Aspen Publishers.

Linn, James G., et al. 1988. *Feeding the Dairy Herd.* Cooperative Extension Services of the Universities of Illinois, Iowa State, Minnesota, and Wisconsin.

Linn, James G., et al. 1994. *Feeding the Dairy Herd.* Minneapolis. Minnesota Extension Service.

Lodish, Harvey, et al. 2000. *Molecular Cell Biology.* 4th ed. New York. W.H. Freeman and Co.

Luttgen, Patricia J., ed. May 1988. The Veterinary Clinics of North America: Small Animal Practice. *Common Neurologic Problems.* Vol. 18 (3). Philadelphia. W.B. Saunders.

Maloney, Beth. 1997. *Nutrition and the Feeding of Horses.* Shrewsbury, England. Swan Hill.

Martinez, Steven A., ed. Sept. 1999. The Veterinary Clinics of North America: Small Animal Practice. *Fracture Management and Bone Healing.* Vol. 29 (5). Philadelphia. W.B. Saunders.

Mayers, Michelle. 2002. Feline Urinary Tract Obstruction: Relief is a Stone Thrown Away! *Veterinary Technician.* Vol. 23 (5).

Maynard, Leonard A., et al. 1979. *Animal Nutrition.* 7th edition. New York. McGraw-Hill.

McKiernan, Brendan C., ed. Sept. 1992. The Veterinary Clinics of North America: Small Animal Practice. *Update on Respiratory Diseases.* Vol. 22 (5). Philadelphia. W.B. Saunders.

Moore, Michael P., ed. July 1992. The Veterinary Clinics of North America: Small Animal Practice. *Diseases of the Spine.* Vol. 22 (4). Philadelphia. W.B. Saunders.

Morrow, David A., ed. 1986. *Current Therapy in Theriogenology.* Vol. 2. Philadelphia. W.B. Saunders.

Murphy, Michael. 1996. *A Field Guide to Common Animal Poisons.* Ames, Iowa. Iowa State University.

Nester, Eugene W., Gilstrap, Marie, and Kleyn, John G. 1978. *Experiments in Microbiology.* New York. Holt, Rinehart and Winston.

New England Committee on Dairy Nutrition. 1996. *Dairy Nutrition Manual Bulletin #2107.* University of Maine Cooperative Extension.

Nutrient Requirements of Dairy Cattle. 2001. 7th rev. ed. Washington, D.C. National Academy.

Osborne, Carl A., and Stevens, Jerry B. 1999. *Urinalysis: A Clinical Guide to Compassionate Patient Care.* Kansas City, Mo. Bayer Corporation.

Osborne, Carl A., Kruger, John M., and Lulich, Jody P., eds. March 1996. The Veterinary Clinics of North America: Small Animal Practice. Disorders of the Feline Lower Urinary Tract I. *Etiology and Pathophysiology.* Vol. 26 (2). Philadelphia. W.B. Saunders.

Osborne, Carl A., Kruger, John M., and Lulich, Jody P., eds. May 1996. The Veterinary Clinics of North America: Small Animal Practice. Disorders of the Feline Lower Urinary Tract II. *Diagnosis and Therapy.* Vol. 26 (3). Philadelphia. W.B. Saunders.

Osweiler, Gary D., and Galey, Francis D., eds. Nov. 2000. The Veterinary Clinics of North America: Food Animal Practice. *Toxicology.* Vol. 16 (3). Philadelphia. W.B. Saunders.

Parry, Bruce W., ed. July 1989. The Veterinary Clinics of North America: Small Animal Practice. *Clinical Pathology: Part I.* Vol. 19 (4). Philadelphia. W.B. Saunders.

Pavletic, Michael M., ed. Jan. 1990. The Veterinary Clinics of North America: Small Animal Practice. *Plastic and Reconstructive Surgery.* Vol. 20 (1). Philadelphia. W.B. Saunders.

Pennsylvania Department of Agriculture. May/June 2001. Foot and Mouth Disease Prevention, Recommendations for Travelers. *PVMA News.*

Peterson, Michael E., and Talcott, Patricia A. 2001. *Small Animal Toxicology.* Philadelphia. W.B. Saunders.

Polzin, David J., ed. Nov. 1996. The Veterinary Clinics of North America: Small Animal Practice. *Renal Dysfunction.* Vol. 26 (6). Philadelphia. W.B. Saunders.

Poppenga, Robert H., and Volmer, Petra A., eds. March 2002. The Veterinary Clinics of North America: Small Animal Practice. *Toxicology.* Vol. 32 (2). Philadelphia. W.B. Saunders.

Radostits, Otto M., and Blood, Douglas C., eds. 1985. *Herd Health: A Textbook of Health and Production Management of Agricultural Animals.* Philadelphia. W.B. Saunders.

Reinhart, Gregory A., and Carey, Daniel P., eds. 1998. *Recent Advances in Canine and Feline Nutrition.* Vol. 2. Wilmington, Ohio. Orange Frazer.

Reinhart, Gregory A., and Carey, Daniel P., eds. 2000. *Recent Advances in Canine and Feline Nutrition.* Vol. 3. Wilmington, Ohio. Orange Frazer.

Richardson, Jill A. 2002. Poison Prevention and Management Primer. *Veterinary Technician.* Vol. 23 (3).

Robbins, Stanley L, Cotran, Ramzi S., and Kumar, Vinay. 1984. *Pathologic Basis of Disease.* 3rd ed. Philadelphia. W.B. Saunders.

Romich, Janet A. 2000. *An Illustrated Guide to Veterinary Medical Terminology.* New York. Delmar Thomson Learning.

Ross, Michael H., and Reith, Edward J. 1977. *Atlas of Descriptive Histology.* 3rd ed. New York. Harper and Row.

Rosychuk, Rod A. W., and Merchant, Sandra R., eds. Sept. 1994. The Veterinary Clinics of North America: Small Animal Practice. *Ear, Nose and Throat.* Vol. 24 (5). Philadelphia. W.B. Saunders.

Roth, James A., ed. Nov. 2001. The Veterinary Clinics of North America: Food Animal Practice. *Immunology.* Vol. 17 (3). Philadelphia. W.B. Saunders.

Roudebush, Philip. 2001. Flatulence: Causes and Management Options. *Compendium on Continuing Education for the Practicing Veterinarian.* Vol. 23(12).

Roussel, Allen J., and Constable, Peter D., eds. Nov. 1999. The Veterinary Clinics of North America: Food Animal Practice. *Fluid and Electrolyte Therapy.* Vol. 15 (3). Philadelphia. W.B. Saunders.

Roussel, Allen J., and Hjerpe, Charles A., eds. March 1990. The Veterinary Clinics of North America: Food Animal Practice. *Fluid and Electrolyte Therapy, Bovine Herd Vaccination Programs.* Vol. 6 (1). Philadelphia. W.B. Saunders.

Schaer, Michael, ed. May 1988. The Veterinary Clinics of North America: Food Animal Practice. *Advances in Fluid and Electrolyte Disorders.* Vol. 28 (3). Philadelphia. W.B. Saunders.

Shores, Andy, ed. March 1993. The Veterinary Clinics of North America: Small Animal Practice. *Diagnostic Imaging.* Vol. 23 (2). Philadelphia. W.B. Saunders.

Slatter, Douglas H., ed. 1985. *Textbook of Small Animal Surgery.* Vols. 1 and 2. Philadelphia. W.B. Saunders.

Smith, Bradford P., ed. 2002. *Large Animal Internal Medicine.* 3rd ed. St. Louis. Mosby.

Solomon, Eldra P., Berg, Linda R., and Martin, Diana W. 1999. *Biology.* 5th ed. Fort Worth, Tex. Saunders.

Soulsby, E. J. L. 1982. *Helminths, Arthropods and Protozoa of Domesticated Animals.* 7th ed. Philadelphia. Lea and Febiger.

Stokka, Gerald L., ed. July 1998. The Veterinary Clinics of North America: Food Animal Practice. *Feedlot Medicine and Management.* Vol. 14 (2). Philadelphia. W.B. Saunders.

Swaim, Steven F., Hinkle, Sherri H., and Bradley, Dino M. 2001. Wound Contraction: Basic and Clinical Factors. *Compendium on Continuing Education for the Practicing Veterinarian.* Vol. 23 (1).

Swift, Nigel C., and Johnston, Karen L. 2000. *Gastrointestinal Disease Management.* Trenton, N. J. Veterinary Learning Systems.

Thiel, C.C., and Dodd, F.H., eds. 1979. *Machine Milking.* Reading, England. The National Institute for Research in Dairying.

Thomas, William B., ed. Jan. 2000. The Veterinary Clinics of North America: Small Animal Practice. *Common Neurologic Problems.* Vol. 30 (1). Philadelphia. W.B. Saunders.

Tobey, Marilyn. 2002. Euthanasia: Easing Your Client's Grief. *Veterinary Technician.* Vol. 23 (4).

Van Gelder, Gary A., ed. 1973. *Clinical and Diagnostic Veterinary Toxicology.* 2nd ed. Dubuque, Iowa. Kendall/Hunt Publishing.

Van Meter, Margaret, and Lavine, Peter G. 1981. *Reading EKGs Correctly.* 8th ed. Horsham, Pa. Intermed Communications, Inc.

Van Soest, Peter J. 1982. *Nutritional Ecology of the Ruminant.* 2nd ed. Ithaca, N.Y. Cornell University.

Weaver, A. David. 1986. *Bovine Surgery and Lameness.* London. Blackwell Scientific.

Weikel, Bill, ed. 1979. *Know Practical Horse Feeding.* Omaha. Farnam Horse Library.

White, Stephen D., ed. Sep. 1988. The Veterinary Clinics of North America: Small Animal Practice. *Pruritus.* Vol. 18 (5). Philadelphia. W.B. Saunders.

Wilson, James A. 1979. *Principles of Animal Physiology.* 2nd ed. New York. Macmillan.

Wilson, Julia H., ed. July 1992. The Veterinary Clinics of North America: Food Animal Practice. *Physical Examination.* Vol. 8 (2). Philadelphia. W.B. Saunders.

Wren, Geni. March–April 2002. Getting the Most out of Serology. *Bovine Veterinarian.*

Index

t = table, f = figure